Y0-CQG-070
SEP 2 1981
DALLAS, PENNSYLVANIA

free radical mechanisms in tissue injury

T. F. Slater

Series editor J.R.Lagnado

1 **Free radical mechanisms in tissue injury** T.F.Slater

2 **Biochemical reactors** B.Atkinson

3 **Brain function and macromolecular synthesis** B.Jakoubek

4 **Ruminant metabolism** D.B.Lindsay

5 **Transmissible drug resistance in bacteria** S.Falkow

6 **Optical properties of proteins** Catherine Rice-Evans

free radical mechanisms in tissue injury

T. F. Slater

p Pion ... rk, London NW2 5JN

Published in the USA by
Methuen, Inc.
733 Third Avenue,
New York, New York 10017

Library edition SBN 85086 031 8
Student edition SBN 85086 032 6

Set on IBM 72 Composers by Pion Limited, London
Printed in Great Britain by J.W.Arrowsmith Limited, Bristol

Preface

The diversity of disease processes that occur in man and other living organisms include without doubt disturbances of biochemical control mechanisms of every possible kind. To some extent it is already possible to classify such diversity into more easily considered sections: disorders of amino-acid metabolism; storage diseases; hormonal imbalance, and so on. One other such section concerns the action of free radicals, produced *in situ*, on neighbouring structures and molecules. A considerable number of tissue lesions have been suggested to involve free-radical mechanisms at the primary stage of development; examples are spread throughout the biological and chemical literature. It is this particular area of biochemical pathology that continues to be of great interest to the author and it is felt that a general review of the subject is now very timely, not only to collate information on some of the major examples but in addition to publicise the importance of this developing field in an attempt to interest new minds from the biological and chemical sciences. These, then, are the justifications for the book: timeliness for review, self-interest in the field, and the hope of stimulating research in the general area of biochemical pathology.

The work summarised in this book crosses the boundaries of many classical disciplines previously held inviolate to the scientific traveller. Only a decade or so ago such movement in a scientific sense would have rendered one suspect as an infiltrator or, at worst, a dilettante. Fortunately, attitudes change; to explore thoroughly such a subject as pathology involves biology in the widest sense, and in such a way that artificial barriers between subjects are a hindrance. Nevertheless, to cross too many barriers too frequently raises considerable difficulties with regard to language: each discipline has its own vocabulary that it guards somewhat jealously and translation must be made with care.

Many aspects of free radical chemistry that have developed into large technologies of biological interest (for example, antioxidants and food additives) are considered here in only a brief manner. Other authors have devoted extensive works to such subjects and reference to them is given at an appropriate position in the text. Obviously to have attempted to cover such areas in any detailed manner in a work of this kind would have been impossible; on the other hand, it is only too easy to present an abbreviated treatment that lacks any scientific depth at all. A major aim of this work can be stated thus: to avoid excessive length and detail and yet to present an essential picture that is neither trivial nor shallow. Each reader must judge for himself in view of his own special interests and experience whether the author has been successful in achieving this aim. No doubt chemists who read this book in search of details of the organic chemistry of free radicals will be disappointed, although they should find some aspects of cell pathology that are both new to them and of interest. Likewise, pathologists or cell biologists may well regret the lack of depth with which some histological aspects are considered, but may profit by gaining some general knowledge of the growing importance of free radical reactions to their chosen field of study. In part I of this book I have drawn on several excellent textbooks for background data. These works are: *Free Radicals* by W. A. Pryor; *Free Radicals in Solution* by Ch. Walling; *The Strength of Chemical Bonds* by T. L. Cottrell; and *Biological and Biochemical Applications of Electron Spin Resonance* by D. J. E. Ingram. Full acknowledgments to these authors are made at the relevant positions in the text.

The major part of this monograph covers biological effects produced by the formation of free radicals within living cells. Of the many examples quoted, one particular problem has been selected for special reference; this concerns the hepatotoxicity of carbon tetrachloride. The reasons for choosing this example as the major illustration of free-radical motivated tissue damage are twofold. Firstly, it is a problem that has interested the author deeply for many years. Secondly, it is the outstanding example of a cellular lesion that has been studied in depth by numerous investigators, so that many of the earliest actions of carbon tetrachloride on the liver are now well understood in biochemical terms.

Although I bear the full responsibility for any mistakes that remain, I acknowledge with gratitude the assistance and enthusiastic support of many research students and colleagues, particularly Professor Kenneth Rees who, by his outstanding contributions to biochemical pathology sparked off my interest in the subject. I have a special debt of gratitude to Mrs. Barbara Sawyer, who has been my research assistant for ten years and who has carried out with unfailing precision the experiments yielding most of the data recorded under our joint names in this book.
Dr. P. A. Riley has read parts of the manuscript and has made helpful criticisms; the speculations put forward in Chapter 13 concerning the carcinogenic action of polycyclic hydrocarbons were developed conjointly with him. I am indebted to Dr. Gillian Bullock for the electron micrographs presented here and for her enthusiastic co-operation over the last five years. I am grateful to the Medical Research Council, Science Research Council, British Cancer Campaign, Agricultural Research Council, New Zealand Department of Agriculture, and the Central Research Funds Committee University of London for the financial support of many of the investigations reported here. Finally, I thank Mrs. Rowena Horsey for typing the many editions of this manuscript with undiminishing good humour and expertise.

T.F.Slater
Professor of Biochemistry, Brunel University

To my wife and parents
without whose encouragement
this book would not have been written

Contents

Part 1

What are free radicals?

1

Introductory remarks

1.1 Introduction

Part 1 (Chapters 1–5) of the Monograph presents a very brief summary of free-radical chemistry. It is intended to serve as an introduction for the reader who, whilst more interested in the biological aspects presented in Part 3, requires some background knowledge concerning free-radical behaviour.

A covalent single bond between two molecules or atoms A and B can be considered to entail the sharing of a pair of electrons by the covalently linked groups. The two electrons of the bond may both be provided by one molecule or each molecule may contribute an electron. The first process may be called coordination and the latter process colligation:

$$\mathrm{A}{:}^{-} + \mathrm{B}^{+} \longrightarrow \mathrm{A}{:}\mathrm{B} \tag{1.1}$$

$$\mathrm{A}^{\cdot} + \mathrm{B}{\cdot} \longrightarrow \mathrm{A}{:}\mathrm{B} \tag{1.2}$$

The reverse process involves the dissociation of a covalent bond; an input of free energy is required for this process to occur. The reverse of Equation (1.1) is called heterolysis and that of Equation (1.2) is homolysis.

Homolytic dissociation produces molecules or atoms containing a single unpaired electron. Such molecules often have high chemical reactivity and consequently only transient lifetimes; they are called free radicals. The free-radical character of a molecular structure is conventionally illustrated in formulae by a bold point ($^{\bullet}$) to indicate the presence of the unpaired electron. Some examples of reactive free radicals are trichloromethyl ($CCl_3^{\bullet}$), chlorine atoms ($Cl^{\bullet}$), hydroxyl radical ($OH^{\bullet}$) and flavin semiquinone ($FH_2^{\bullet}$). Many free radicals are stable, however, even in solution at room temperature; examples of stable free radicals are: oxygen (a stable diradical, see Figure 3.1), and diphenylpicryl hydrazyl (see Figure 3.2).

In this Monograph we will be concerned mainly with those processes that result in the homolytic dissociation of covalent bonds in biological components to produce free-radical products of a potentially dangerous nature. The energy required to initiate such dissociations is usually provided by irradiation[(1)] (e.g. ultraviolet radiation) or by localised high temperature (as occurs for example when a laser beam impinges on solid tissue). In some instances, however, the energy source may involve chemical reactions such as the iron-catalysed dissociation of hydrogen peroxide (see Section 2.4).

[(1)] The principles of molecular excitation, including photochemical reaction mechanisms, cannot be dealt with in any detail in this Monograph, where only a simplified treatment will be presented. The reader interested in thorough treatment of such excitation phenomena is referred to standard works. General introductions are by Cottrell (1965) and Reid (1957).

1.2 Membranes

The *biological* significance of free radicals was recognised some 40 years ago by Michaelis but his ideas were slow to find general acceptance for two main reasons. First, for many people it was hard to visualise how free radicals with high chemical reactivities (and short lifetimes) could play any significant *specific* functional role in living processes as opposed to being concerned simply with unspecific degradative reactions. Secondly, the lack of a suitably sensitive method for detecting free radicals ensured that the evidence that was available was indirect and contentious. The development of electron spin resonance (Section 4.2), however, has led to the widespread recognition of free-radical components as intermediates of normal metabolism. In many cases it has been found that the steady-state concentration of these free-radical intermediates is very low; nevertheless the turnover of such components is often very rapid and they play a vital role in many metabolic events.

The biological importance of **membranes** in allowing ordered sequences of enzymes to be built up and in separating susceptible components from degradative influences can hardly be overstated. Since biological membranes are often rich in unsaturated lipids and are bathed in an oxygen-rich, metal-containing fluid they are susceptible to a peroxidative attack that involves free-radical reactions. Such reactions are of relatively common occurrence in pathological events and membrane disorders will be considered in some detail in Part 3. The study of such disturbances is of value not only to our understanding of pathological events, but also affords considerable insight into the *normal* structure of membranes and of the mechanisms that control their functional integrity. This result of pathological studies was clearly seen by Claude Bernard: "Les poisons peuvent être employés comme agents de destruction de la vie ou comme moyens de guérison des maladies; mais, outre ces deux usages bien connus de tout le monde, il en est un troisième qui intéresse particulièrement le physiologiste. Pour lui, le poison devient un instrument qui dissocie et analyse les phénomènes les plus délicats de la machine vivante, et, en étudiant attentivement le méchanisme de la mort dans les divers empoisonnements, il s'enstruit par voie indirecte sur le mécanisme physiologique de la vie" (1878).

Because of the interest in membrane disturbances in Part 3 of this Monograph it will be of value to mention very briefly the different types of membrane-bounded structures that occur in mammalian cells.

Solid tissues like liver contain a number of types of inclusions that are separated from surrounding components or fluid by a limiting membrane. The most important of these, from our viewpoint of tissue damage, are the mitochondria, the lysosomes, the endoplasmic reticulum and the nucleus. Although the cell membrane itself can also be damaged under certain conditions this aspect has been particularly studied only for the

Table 1.1. Localisation of important enzymes and coenzymes in different intracellular fractions of liver. A plus-symbol (+) indicates that the component is mainly restricted to the fraction indicated.
Abbreviations: ER, endoplasmic reticulum. BC, bile canaliculus. TCA, tricarboxylic acid cycle.

Component	Nucleus	Mitochondria	Lysosomes	Peroxisomes	ER	Cell Sap	BC
1 DNA	+						
2 NAD-synthetase	+						
3 DNA polymerase	+						
4 Respiratory chain		+					
5 TCA cycle		+					
6 Oxidative phosphorylation		+					
7 $NADP^+$ + NADPH		+					
8 Acid hydrolases			+				
9 Catalase				+			
10 Uricase				+			
11 Glucose-6-phosphatase					+		
12 Protein synthesis					+		
13 RNA					+		
14 Drug detoxication					+		
15 Glucuronyl transferase					+		
16 Glycolysis						+	
17 Pentose shunt						+	
18 Fatty acid synthesis						+	
19 NAD^+ + NADH						+	
20 Adenosine triphosphatase							+

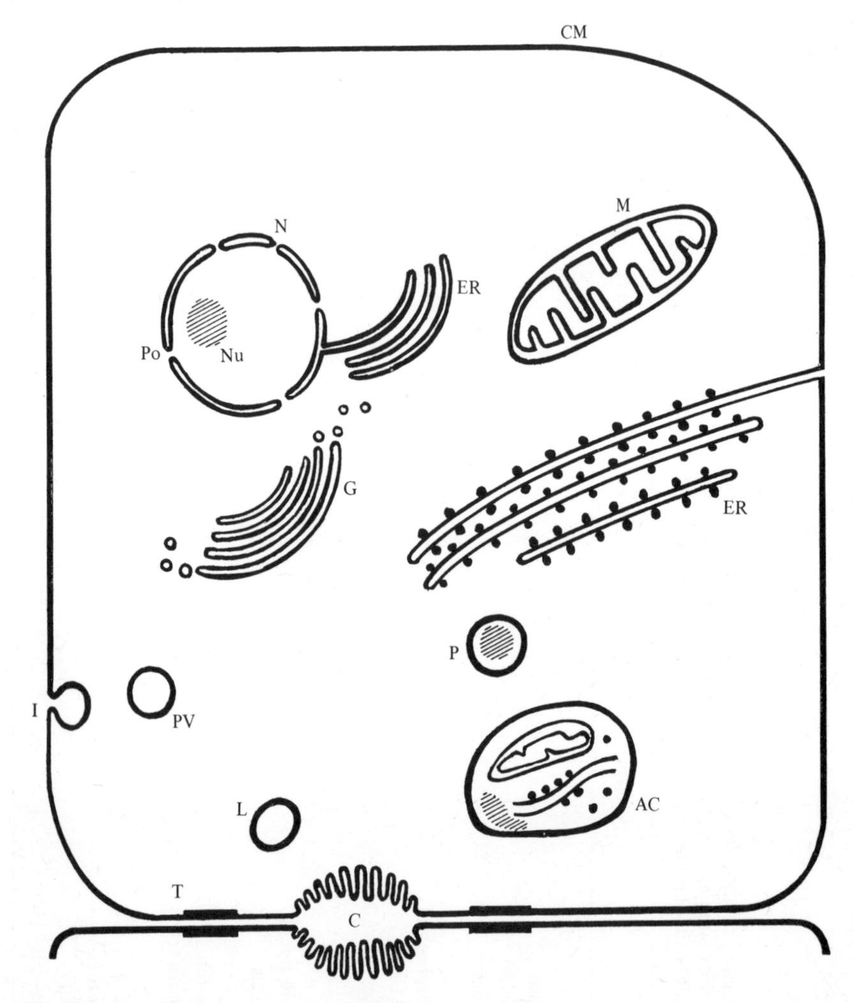

Figure 1.1. Diagrammatic representation of a liver parenchymal cell. The cell membrane (CM) is shown invaginating at I to form pinocytotic vesicles (PV). At the lower part of the cell and adjacent to the neighbouring parenchymal cell the celi membrane forms the specialised biliary canaliculus (C) with numerous microvilli. The tight junctions between the neighbouring cells are indicated by T. A mitochondrion (M), peroxisome with crystalline inclusion (P) and lysosome (L) are indicated with no attempt to reproduce relative sizes. An autophagic cytosegresome (secondary lysosome) containing a degraded mitochondrion and some fragments of endoplasmic reticulum with lipid is shown at AC. The Golgi structure with associated microvesicles is shown at G. The endoplasmic reticulum (ER) with attached ribosomes is shown either terminating in a 'blind end' or disappearing out of the plane of the section. The nucleus is indicated at N with a nucleolus (Nu). The nuclear membrane is shown to contain pores (Po) and to be in contact with membranes of the ER, some of which may be continuous with the cell membrane as indicated.

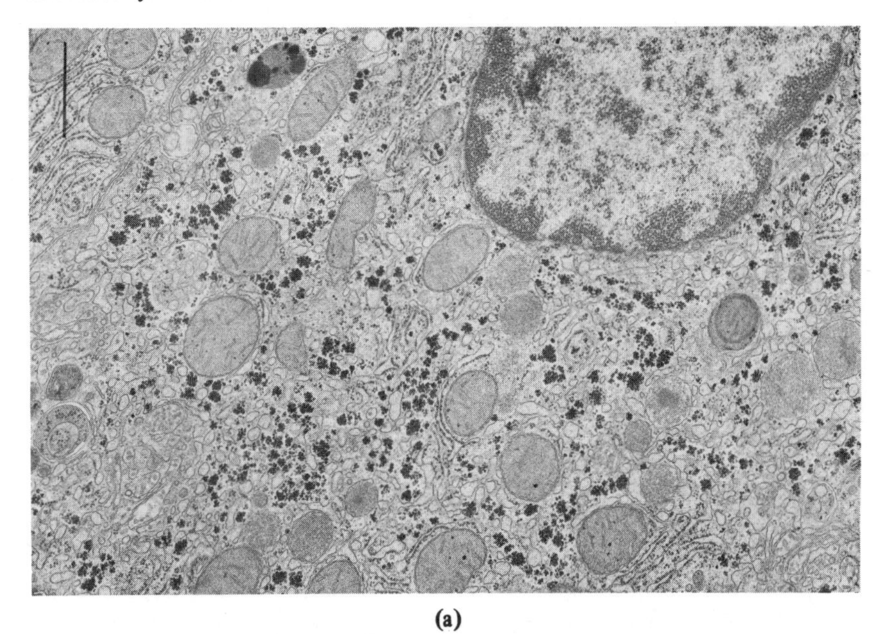

(a)

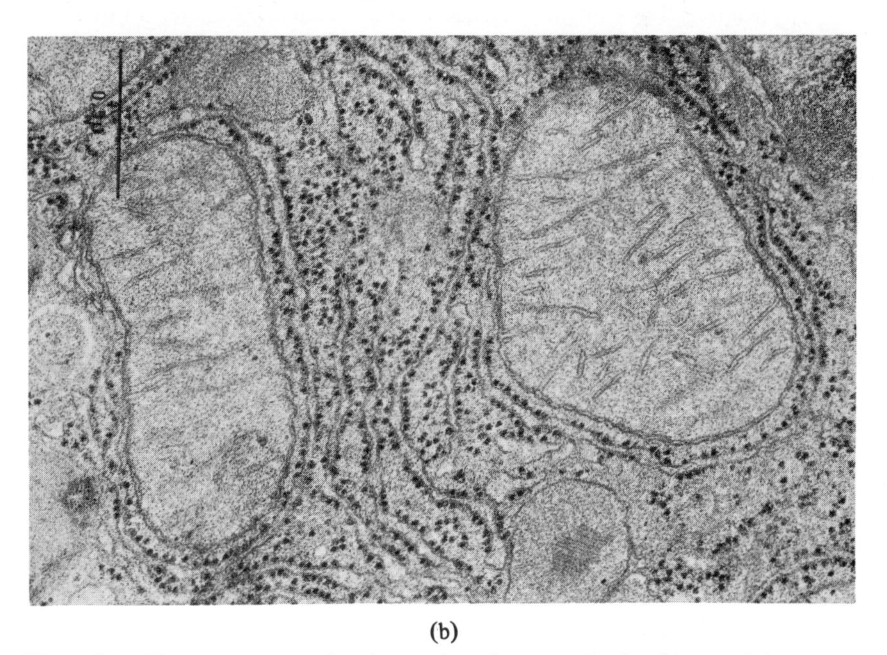

(b)

Figure 1.2. Electron micrographs of normal rat liver samples fixed in cacodylate-glutaraldehyde and stained with osmic acid. In (a) a low magnification picture illustrates the diverse types of membrane-bound structures occurring in the liver cell cytoplasm. Numerous mitochondria and areas of endoplasmic reticulum can be seen. The nucleus is lying in the top right hand corner of the picture. In (b) a high magnification picture of normal liver cytoplasm is shown. Two mitochondria with internal cristae can be seen with numerous profiles of endoplasmic reticulum. Above the mitochondria on the left of the picture is a peroxisome (unpublished data of T. F. Slater and G. Bullock).

rather special case of the non-nucleated erythrocyte. The diagrammatic appearance of the intracellular structures within a parenchymal liver cell is indicated in Figure 1.1: an electron micrograph of liver is shown in Figure 1.2.

It has been found as a result of numerous studies that many enzymes and coenzymes are often localised in a particular intracellular organelle or compartment. This distribution may have obvious advantages such as the segregation of destructive acid hydrolytic enzymes within the confines of the lysosome, or may reflect the need to supply a particular substrate required to activate a membrane-linked metabolic pathway. A simplified version of the intracellular distribution pattern for rat liver is given in Table 1.1. It may be seen from a consideration of Table 1.1 and Figure 1.2 that not only do the various intracellular organelles have very different morphological appearances but that they also differ markedly in their biochemical composition and function. As a consequence, an injury restricted to a particular intracellular organelle may result in remarkably specific effects on the behaviour and biochemical activities of the cell (or tissue) as a whole.

In Part 3 we shall consider examples of damage to each of the intracellular structures mentioned above through mechanisms that involve free-radical intermediates. The importance of such free-radical intermediates to human biology is indicated by the following abbreviated list of disturbances or essential biological phenomena that have been suggested to involve free-radical initiation stages: chemical carcinogenesis; ageing; central nervous system function; atheroma; photosynthesis; photosensitisation; oxidative phosphorylation and electron transport; radiation-induced genetic changes; the action of some hormones. First, however, in Part 1 the routes of free-radical formation, their properties and measurement will be described.

References

Bernard, C., 1878, *La Science Expérimentale* (J. B. Baillière et fils, Paris).
Cottrell, T. L., 1965, *Dynamic Aspects of Molecular Energy States* (Oliver and Boyd, Edinburgh).
Reid, C., 1957, *Excited States in Chemistry and Biology* (Butterworths, London).

Production of free radicals

2.1 Introduction

The net formation of free radicals generally involves the homolytic fission of covalent bonds (Figure 2.1a); free radicals may also be produced by electron capture (Figure 2.1b).

The homolysis of a covalent bond (as in Figure 2.1a) requires an input of energy and this is called the bond-dissociation energy $D(CH_3{-}H)$. This is different from the average bond strength $\Delta H/4$ (Figure 2.1c), where ΔH is the change in enthalpy, since the bond-dissociation energy is dependent on the nature of the groups surrounding the bond that is broken. This is shown clearly in Table 2.1 for the successive removal of hydrogen atoms from methane.

A new species of free radical may arise by the interaction of a free radical $A^{\bullet}$ with the molecule $B{-}C$ to yield the products $A{-}B + C^{\bullet}$. If the reaction passes through a transition state $A \cdots B \cdots C$ then the energy changes involved can be represented as shown in Figure 2.2, where ΔH $(= H_1 - H_2)$ represents the enthalpy change of the reaction, H_1 the activation energy of the forward reaction and H_2 the activation energy of the back reaction. The reaction involves: (a) the breaking of the bond $B{-}C$ and (b) the

(a) $CH_3{:}H \xrightarrow{\text{homolysis}} CH_3^{\bullet} + H^{\bullet}$

(b) $CCl_4 + e^- \xrightarrow{\text{electron capture}} CCl_3^{\bullet} + Cl^-$

(c) $CH_4 \longrightarrow C_{(solid)} + 4H^{\bullet} \qquad \frac{\Delta H}{4} = 72 \text{ kcal mol}^{-1}$

Figure 2.1. Examples of (a) homolytic dissociation, (b) electron capture and (c) average bond strength.

Table 2.1. Bond-dissociation energies (after Cottrell, 1954).

Bond	D (kcal mol^{-1})	Bond	D (kcal mol^{-1})
$CH_3{-}H$	101	$CCl_3{-}F$	102
$-CH_2{-}H$	92	$OH{-}H$	117
$=CH{-}H$	120	$OH{-}OH$	52
$\equiv C{-}H$	80	$HOO{-}H$	90
$Cl{-}Cl$	58	$CH_3{-}OH$	90
$CH_3{-}Cl$	81	$CH{\equiv}CH$	230
$CCl_3{-}Cl$	68	$CH_3{-}CH_3$	83
$CCl_3{-}Br$	49	$CH_3{-}NH_2$	80
$CCl_3{-}H$	90	$NH_2{-}NH_2$	60

formation of the bond A—B so that

$$D(\mathrm{B-C}) - D(\mathrm{A-B}) = \Delta H\,. \tag{2.1}$$

In the homolytic dissociation of A—B to yield the free-radical products $A^{\bullet}$ and $B^{\bullet}$ the activation energy for the recombination reaction (H_2) of $A^{\bullet}$ and $B^{\bullet}$ may be very low indeed. In such a case the bond-dissociation energy $D(\mathrm{A-B})$ would be the activation energy H_1.

A discussion of bond-dissociation energies with a comprehensive list of measurements is given in Cottrell (1958), from which are taken the values shown in Table 2.1.

The energy required to cause the dissociation of a covalent bond may be provided by (a) thermal sources, (b) electromagnetic radiation or (c) by oxidation–reduction coupling with a source of free radicals. Examples of free-radical production by these three mechanisms will now be given.

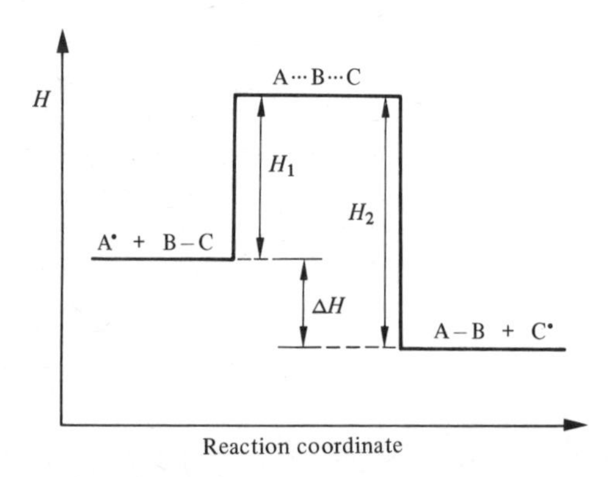

Figure 2.2. Energy diagram for reaction of $A^{\bullet}$ + B—C to yield A—B + $C^{\bullet}$ through an intermediate transition state A—B—C. Changes in enthalpy (H) are shown.

2.2 Thermal production of free radicals

The bond-dissociation energy for a C—C, a C—H or C—O bond is generally about 90 kcal mol^{-1} (see Table 2.1). To provide sufficient energy for the homolytic dissociation of such bonds by thermal means requires temperatures of about 450–650°C. This is the temperature range at which the cracking of petroleum is carried out to produce a wide range of economically useful products. The increased homolytic dissociation of organic material with increasing temperature is shown very clearly in Figure 2.3: as the temperature of pyrolysis is increased the number of free radicals produced reaches a maximum and then falls. The *increase* in free radicals is ascribed by Ingram (1958) to the formation of resonating carbon ring structures, and the *decrease* at high temperatures to the cross-linking of such rings to form graphitic sheets and lattices with destruction of the free radicals concerned.

Most covalent bonds require high temperatures for dissociation but under certain conditions dissociation may occur at quite low temperatures. The free radical(s) formed may be stabilised by resonance, or the reaction may involve the formation of a very stable molecule, such as nitrogen, which has a high heat of formation (ΔH_f). As a consequence of the comparatively easy production of free-radical products $R^{\bullet}$ from such compounds, they may be used to initiate the homolytic dissociation of other suitable compounds at relatively low temperature. This follows since Reaction (2.2)

$$R^{\bullet} + A{-}B \longrightarrow R{-}A + B^{\bullet} \qquad (2.2)$$

may be more easily achieved than the direct homolysis

$$A{-}B \longrightarrow A^{\bullet} + B^{\bullet} \qquad (2.3)$$

for in Reaction (2.2) the energy required to dissociate A—B is reduced by the energy 'gained' in the formation of the new covalent bond R—A. This explains the common use of thermally unstable materials as initiators of free-radical reactions, such as radical polymerisations, where the initial dissociation of A—B may result in a chain reaction of considerable length (see 3.1.2). The following discussion illustrates these points.

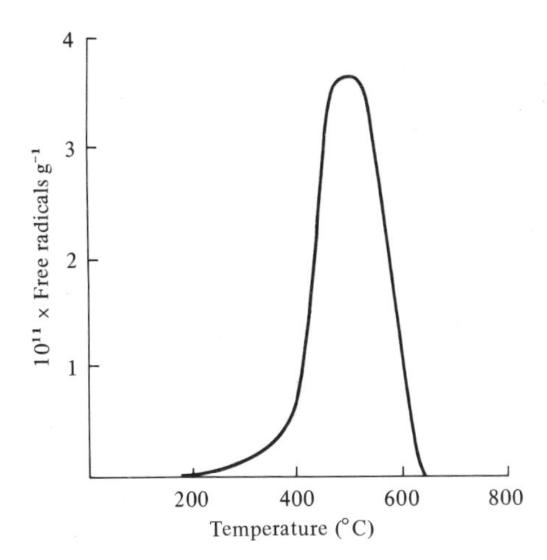

Figure 2.3. Increase in free-radical concentration on increasing the temperature of pyrolysis of oil-distillation residues from 200°C to approximately 500°C (after Ingram, 1969).

2.2.1 Peroxides

The O—O bond in peroxides is weaker than many C—C or C—O bonds (Table 2.2). The data given in Table 2.2 for two commonly used peroxide initiators are given by Pryor (1966). There are at least three different mechanisms of peroxide decomposition:

1. The straightforward unimolecular homolysis described in Table 2.2:

$$R_1{-}O{-}O{-}R_2 \longrightarrow R_1{-}O^{\bullet} + R_2{-}O^{\bullet} . \qquad (2.4)$$

For the relatively stable t-butyl peroxide the rate constant for decomposition is given (Pryor, 1966) by:

$$K = 10^{16}\,e^{-37000/RT}\,s^{-1} .$$

This leads to half-life times at 20°C, 60°C, 100°C and 180°C of approximately 9500 years, 5 years, 4 days and 40 s respectively.

2. An induced decomposition by reaction of the peroxide with an exogenous radical (e.g. hydrogen abstraction).

$$R_1^{\bullet} + R_2{-}CH_2{-}O{-}O{-}CH_2{-}R_3 \longrightarrow R_1H + R_2{-}CH^{\bullet}{-}O{-}O{-}CH_2{-}R_3$$

$$R_2{-}CH^{\bullet}{-}O{-}O{-}CH_2{-}R_3 \longrightarrow R_2CHO + R_3CH_2O^{\bullet} . \qquad (2.5)$$

3. By the poorly understood detonation pathway—a common hazard when using peroxides.

In this section on peroxides the phenomenon of **molecule-induced homolysis** (Pryor, 1966) is relevant, where the homolytic decomposition of a peroxide may be greatly facilitated by the admixture with olefins or amines. The homolysis of benzoyl peroxide is greatly accelerated by the presence of amines, and tertiary amines in particular are often used as catalysts in polymerisation processes that are initiated by the homolysis of a peroxide bond. Benzoyl peroxide (0·01 M) has a half-life at 20°C of 60000 h but in the presence of 0·01 M *NN'*-dimethylaniline the half-life is reduced to 13 min. The mechanism is believed to be through the intermediate formation of an ionic complex as shown in Figure 2.4; the reaction mechanism is discussed in considerable detail by Pryor (1966) and Walling (1957).

Table 2.2. Data for the thermal homolysis of benzoyl peroxide and t-butyl peroxide.

	Activation energy for homolytic dissociation (kcal mol^{-1})	Temperature for a half-life of 1 h in inert solvent (°C)
Benzoyl peroxide	30	95
t-Butyl peroxide	37	150

$$
\begin{array}{l}
\quad\; CH_3 \\
\quad\;\; | \\
C_6H_5{-}N \quad + \; C_6H_5{-}CO{-}O{-}O{-}CO{-}C_6H_5 \\
\quad\;\; | \\
\quad\; CH_3 \\
\qquad\qquad\qquad \downarrow \\
\quad\; CH_3 \\
\quad\;\; | \\
C_6H_5{-}N^{+}{-}O{-}CO{-}C_6H_5 \; + \; C_6H_5COO^{-} \\
\quad\;\; | \\
\quad\; CH_3 \\
\qquad\qquad\qquad \downarrow \\
\quad\; CH_3 \\
\quad\;\; | \\
C_6H_5{-}N^{\bullet +} \; + \; C_6H_5COO^{\bullet} \\
\quad\;\; | \\
\quad\; CH_3 \\
\qquad\qquad\qquad \downarrow \\
\quad\; CH_3 \\
\quad\;\; | \\
C_6H_5{-}N \qquad\qquad\qquad\qquad + \; C_6H_5COOH \; + \; CH_2O \\
\quad\;\; | \\
\quad\; CH_2{-}O{-}CO{-}C_6H_5 \; + \; C_6H_5{-}NH{-}CH_3 \\
\qquad\qquad\qquad\qquad\qquad\quad + \; C_6H_5{-}CO{-}O{-}C_6H_4{-}N(CH_3)_2
\end{array}
$$

Figure 2.4. Molecule-induced homolytic dissociation of benzoyl peroxide by dimethylaniline showing the major products formed (see Horner and Betzel, 1953).

2.2.2 Hydroperoxides

It is well known that hydrogen peroxide decomposes on heating to provide a source of hydroxyl radicals as well as oxygen and water.

$$H_2O_2 \longrightarrow 2OH^{\bullet} \qquad D(O{-}O) = 52\ \text{kcal mol}^{-1}\,. \tag{2.6}$$

Since the bond-dissociation energy is 52 kcal mol^{-1} the homolysis proceeds at relatively low temperatures and is accelerated by the presence of metal ions. Organic materials can be hydroxylated at room temperature by a mixture of hydrogen peroxide and ferrous ions, as will be discussed in Section 2.4.

Organic hydroperoxides (R—OOH) also decompose on heating to yield $RO^{\bullet}$ and $OH^{\bullet}$. The decomposition may follow a chain mechanism:

$$\text{t-butyl}{-}OOH \longrightarrow \text{t-butyl}{-}O^{\bullet} + OH^{\bullet} \tag{2.7}$$

$$\text{t-butyl}{-}O^{\bullet} + \text{t-butyl}{-}OOH \longrightarrow \text{t-butyl}{-}OH + \text{t-butyl}{-}OO^{\bullet} \tag{2.8}$$

$$2(\text{t-butyl}{-}OO^{\bullet}) \longrightarrow 2(\text{t-butyl}{-}O^{\bullet}) + O_2 \tag{2.9}$$

Solvents such as alcohols can stimulate the reaction by providing a source of hydrogen atoms.

$$\text{t-butyl}{-}O^{\bullet} + ROH \longrightarrow \text{t-butyl}{-}OH + RO^{\bullet}\,. \tag{2.10}$$

2.2.3 Azo-compounds

Although the dissociation energies of the C—N bond (80 kcal mol^{-1}) and the N—N bond (100 kcal mol^{-1}) may have normal values in azo-compounds, the dissociation of a C—N bond in such a compound is favoured by the strong heat of formation of the product nitrogen (ΔH_f = 225 kcal mol^{-1}).

With azomethane, the nitrogen effect leads to a lowering of the activation energy to 51 kcal mol^{-1} but, even so, thermal homolysis would only be appreciable above 400°C.

With more complex azo-compounds the radicals formed are stabilised by resonance and this further lowers the activation energy required for homolysis. For example, **azoisobutyronitrile**, a commonly used radical initiator, dissociates with an activation energy of 31 kcal mol^{-1}; the half-life at 80°C is 1·3 h (Figure 2.5). The ease of free-radical formation from such azo-compounds may be of relevance to the biological toxicity of some azo-dyestuffs, for example dimethylaminoazobenzene (butter yellow), as discussed in Section 13.2.

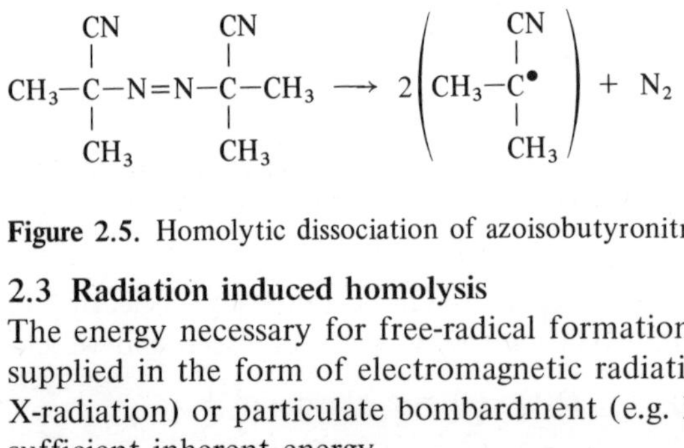

$$CH_3-\underset{CH_3}{\overset{CN}{C}}-N=N-\underset{CH_3}{\overset{CN}{C}}-CH_3 \longrightarrow 2\left(CH_3-\underset{CH_3}{\overset{CN}{C}}{}^\bullet\right) + N_2$$

Figure 2.5. Homolytic dissociation of azoisobutyronitrile.

2.3 Radiation induced homolysis

The energy necessary for free-radical formation by bond scission may be supplied in the form of electromagnetic radiation (visible, ultraviolet, X-radiation) or particulate bombardment (e.g. high energy electrons) of sufficient inherent energy.

A quantum of electromagnetic radiation of frequency ν has energy $E = h\nu$ and a mol of quanta $Nh\nu$ = 1 einstein, where N is Avogadro's number. The energy associated with electromagnetic radiation of wavelength λ is given by the relationship:

$$E = Nh\nu = \frac{2\cdot 86 \times 10^5}{\lambda}\ \text{kcal einstein}^{-1}\ .$$

This relationship is shown graphically in Figure 2.6.

It can be seen from Figure 2.6 that incident radiation must have a wavelength less than 3000 Å to break covalent bonds with bond-dissociation energy of say 100 kcal mol^{-1}.

The Grotthus–Draper law states: "only radiation that is absorbed by a molecule can produce a chemical reaction"; this does not apply, however, to high energy particulate material where impact phenomena may produce the energy necessary for bond fission. The energy absorbed by a compound from radiation raises it to an excited state which may lose energy by internal decay or by emission of radiation usually over a period of $10^{-6}-10^{-9}$ s and with a longer wavelength than the incident radiation; this process is called **fluorescence**. The excited state may, however, pass over via a so-called 'forbidden' transition to a **triplet state**, which, although having a higher energy state than the ground state of the molecule may have a quite appreciable lifetime since the decay to the ground state by light-emission **(phosphorescence)** again involves a 'forbidden' transition. The triplet state arises when the normally opposing spins of two electrons in the same orbital are changed so that the spins have the same directional sense and this produces, in effect, a diradical. A discussion of the relationship between the triplet state and diradicals is given by Reid (1957). In this sense the absorption of radiation can produce free-radical intermediates without first producing bond fission.

Absorption of radiation of sufficient inherent energy can, however, lead to homolysis directly: for example, the long known photochemical halogenation reaction shown in Figure 2.7. It is also possible to cause homolysis of a compound A—B with radiation having a monochromatic wavelength that by itself is not directly absorbed by A—B. This apparent contradiction of the Grotthus–Draper law is achieved by including in the reaction mixture another molecular species C that is excited by the incident radiation and can pass on the excitation energy to cause a

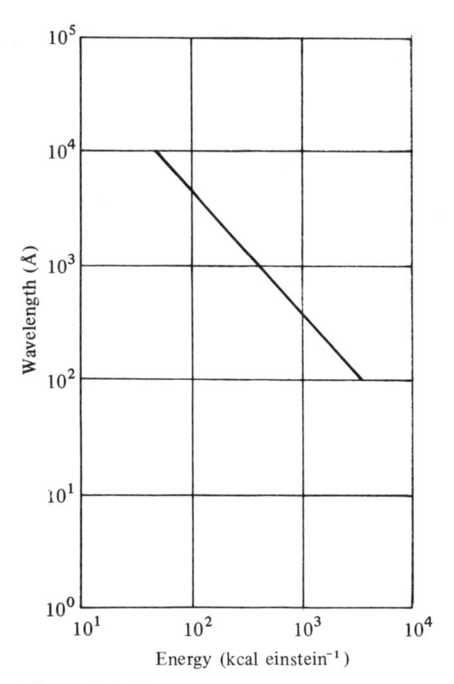

Figure 2.6. Relationship between wavelength of electromagnetic radiation and the associated energy.

homolysis of A—B. This process is called **photosensitisation**. A well-known example is the mercury-catalysed decomposition of hydrogen with light of wavelength 2537 Å.

$$\text{Hg} + 2537\ \text{Å} \longrightarrow \text{Hg}^* \tag{2.11}$$

$$\text{Hg}^* + \text{H}_2 \longrightarrow \text{Hg} + 2\text{H}^\bullet \tag{2.12}$$

The energy liberated in the transition of the mercury atom from the excited state ($^3p_1$) to the ground state ($^1s_0$) is 112 kcal mol^{-1}; this is sufficient to cause homolysis of hydrogen (bond-dissociation energy 103 kcal mol^{-1}).

$$\text{Cl}_2 + 4875\ \text{Å radiation} \longrightarrow \text{Cl}_2^* \longrightarrow 2\text{Cl}^\bullet$$

$$\text{Cl}^\bullet + \text{RH} \longrightarrow \text{R}^\bullet + \text{HCl}$$

$$\text{Cl}_2 + \text{R}^\bullet \longrightarrow \text{RCl} + \text{Cl}^\bullet$$

Figure 2.7. Homolysis of chlorine by radiation at 4875 Å; the chlorine atoms so formed then produce a chain reaction involving the homolysis of R—H. The bond-dissociation energy of Cl—Cl is 57 kcal mol^{-1} and the energy of a quantum of light of wavelength 4875 Å is 59 kcal mol^{-1}.

A further interesting example is the decomposition of oxalic acid catalysed by the uranyl ion in the presence of light. This reaction has been used as a calibration technique for measuring radiation intensity (Volman and Seed, 1964).

$$\text{UO}_2^{2+} + \text{radiation} \longrightarrow (\text{UO}_2^{2+})^*\ \text{excited} \tag{2.13}$$

$$(\text{UO}_2^{2+})^*\ \text{excited} + \text{oxalic acid} \longrightarrow \text{UO}_2^{2+} + \text{CO}_2 + \text{CO} + \text{H}_2\text{O} \tag{2.14}$$

Many photochemical reactions have a requirement for oxygen. For example, the production of the transannular peroxide ascaridole from α-terpinene (Figure 2.8), where the photosensitiser (PS) can be methylene

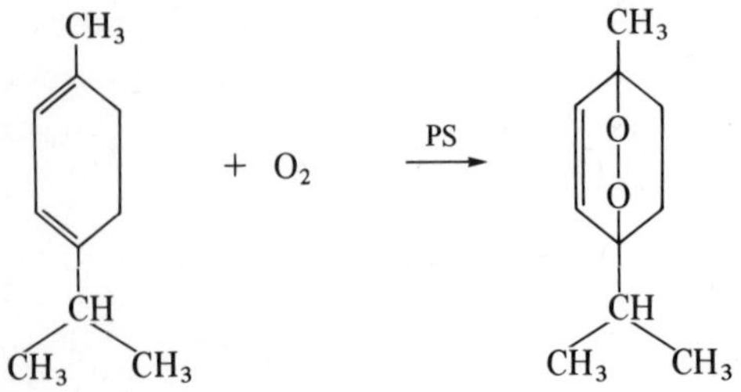

CH$_3$ CH$_3$
+ O$_2$ $\xrightarrow{\text{PS}}$
O O
CH CH
CH$_3$ CH$_3$ CH$_3$ CH$_3$

Figure 2.8. Formation of the transannular peroxide ascaridole by the radiation of α-terpenene in the presence of a photosensitiser (PS).

blue or chlorophyll. α-Terpinene occurs naturally (e.g. in oil of lemons) and ascaridole is found in the oil of *Chenopodium ambrosioides*. Ascaridole has been used therapeutically as an anthelminthic and for amoebic dysentery but has very toxic side-effects. It has been reported to be a weak carcinogen (Van Duuren *et al.*, 1965).

In biological systems photosensitised reactions involving oxygen are usually called **photodynamic effects** (see Blum, 1964, for a comprehensive discussion of these phenomena). For example red blood cells exposed to visible light greater than 3300 Å in the presence of oxygen do not lyse at an appreciable rate; in the presence of the photosensitiser eosin, however, rapid haemolysis occurs. These effects are discussed in some detail in Chapter 12.

2.4 Homolysis by redox coupling with metal ions

Free radicals are produced in these reactions by the interaction of a substrate (for example hydrogen peroxide or vitamin C) with a metal ion that can undergo univalent redox change. In 1894 Fenton discovered the reaction that now bears his name whilst investigating the oxidation of tartaric acid. He found that a mixture of ferrous ions and hydrogen peroxide can **hydroxylate** organic compounds at room temperature; the mechanism of Fenton's reaction is believed (Green *et al.*, 1963; Smith and Norman, 1963) to be as shown in Figure 2.9. The *decomposition* of hydrogen peroxide by ferrous ions probably follows the course illustrated in Figure 2.10 (Uri, 1952). The scheme is consistent with the production of oxygen during the decomposition reaction. The reaction is not specific for ferrous ions and hydrogen peroxide, and the scheme may be generalised to include other transitional metals that undergo univalent redox change (e.g. copper) as well as hydroperoxides, peroxides and peresters ($R_1{-}O{-}O{-}R_2$) [Reaction (2.15)]

$$\text{Metal}^{+} + R_1{-}O{-}O{-}R_2 \longrightarrow \text{metal}^{2+} + R_1{-}O^{\bullet} + R_2O^{-} \qquad (2.15)$$

where R_1 = alkyl or aryl; R_2 = hydrogen, alkyl, aryl, —CO—alkyl or —CO—aryl.

Another example of redox induced homolysis is the system described by Udenfriend *et al.* (1954) consisting of buffer, oxygen, a metal complex

$$Fe^{2+} + H_2O_2 \longrightarrow Fe^{3+} + OH^{-} + OH^{\bullet}$$

$$C_6H_6 + OH^{\bullet} \longrightarrow C_6H_6OH^{\bullet}$$

$$C_6H_6OH^{\bullet} + Fe^{3+} \longrightarrow C_6H_5OH + Fe^{2+} + H^{+}$$

$$C_6H_6OH^{\bullet} + C_6H_6 \longrightarrow C_{12}H_{12}OH^{\bullet}$$

$$C_{12}H_{12}OH^{\bullet} + A \longrightarrow C_{12}H_{10} + AH + H_2O$$

Figure 2.9. Hydroxylation of benzene by Fenton's reagent. The formation of biphenyl by reaction with an acceptor A is also shown (Smith and Norman, 1963).

and vitamin C, which can hydroxylate aromatic compounds to give similar products to those produced *in vivo* by the interaction of the substrate with the cytochrome P_{450} chain in liver endoplasmic reticulum. The Udenfriend system is believed to act by the production of the perhydroxyl radical $O_2H^{\bullet}$ or its anion $O_2^{\bullet}$. The pK_a of the dissociation

$$O_2H^{\bullet} \rightleftarrows O_2^{\bullet -} + H^+ \qquad (2.16)$$

is approximately 4·5. For a review of hydroxylation mechanisms with model systems see Staudinger *et al.* (1965). The subject of aromatic hydroxylation by enzyme systems has been reviewed by Hayaishi (1969).

Finally in this section we can mention the well-known accelerating role of metal ions in many autoxidative processes, particularly in deteriorative reactions of foods. A full discussion of such catalytic behaviour is given by Walling (1957); specific examples relating to intracellular components are given in later Sections (12.4; 15.2).

$$Fe^{2+} + H_2O_2 \longrightarrow Fe^{3+} + OH^- + OH^{\bullet}$$
$$Fe^{3+} + H_2O_2 \longrightarrow Fe^{2+} + HO_2^{\bullet} + H^+$$
$$Fe^{2+} + OH^{\bullet} \longrightarrow Fe^{3+} + OH^-$$
$$Fe^{2+} + HO_2^{\bullet} \longrightarrow Fe^{3+} + HO_2^-$$
$$Fe^{3+} + HO_2^- \longrightarrow Fe^{2+} + H^+ + O_2^{\bullet -}$$
$$OH^{\bullet} + H_2O_2 \longrightarrow HO_2^{\bullet} + H_2O$$
$$O_2^{\bullet -} + H_2O_2 \longrightarrow O_2 + OH^- + OH^{\bullet}$$

Figure 2.10. Major reactions occurring in the iron-catalysed decomposition of hydrogen peroxide (Fenton's reagent). For details see Uri (1952) and Pryor (1966).

2.5 Production of free radicals by hydrogen addition or abstraction

As we have already seen free radicals may be produced by simple addition or abstraction of hydrogen, and although this does not result (directly) in the *net* formation of free radicals it is a process of considerable biological importance.

One example of such behaviour concerns the flavin semiquinone radicals formed during cyclic oxidation and reduction of the flavin coenzymes of electron-transport chains in intracellular organelles (for references see Massey and Gibson, 1964). If the oxidised and fully reduced forms of the flavins are written FH and FH_3 respectively then the half-reduced semiquinone radical is $FH_2^{\bullet}$.

$$FH \xrightarrow{H^+ + e^-} FH_2^{\bullet} \xrightarrow{H^+ + e^-} FH_3 \qquad (2.17)$$

Here the proton and electron come from enzyme-catalysed oxidation of a reduced coenzyme (NADH or NADPH). The flavin acts as an intermediary in the transfer of reducing potential to an acceptor such as a cytochrome.

Possible structures of the flavins $FH_2^{\bullet}$, FH and FH_3 in the un-ionised forms are shown in Figure 2.11 but several resonance hybrids exist for each level of oxidation, and ionisation to the anionic forms also complicates the picture. This is discussed by Mahler and Cordes (1966) and by Land and Swallow (1969).

It is interesting in view of the quite widespread occurrence of metal ions in flavoproteins (Malkin and Rabinowitz, 1967) that Hemmerich *et al.* (1963) found the equilibrium between the semiquinone radical and the quinone and hydroquinone forms to be strongly stabilised by metal chelation in favour of the radical.

$$FH_3 + FH \underset{}{\overset{\text{metal ion}}{\rightleftharpoons}} 2FH_2^{\bullet} . \qquad (2.18)$$

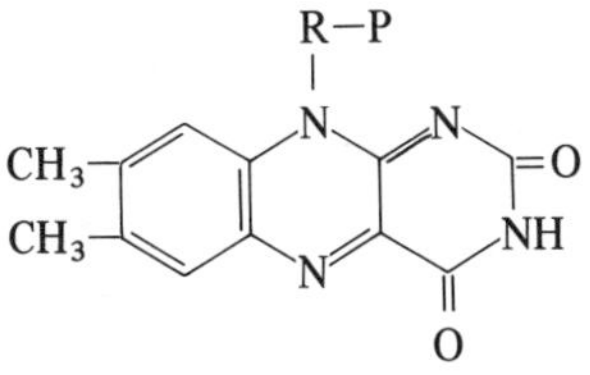

(FH)

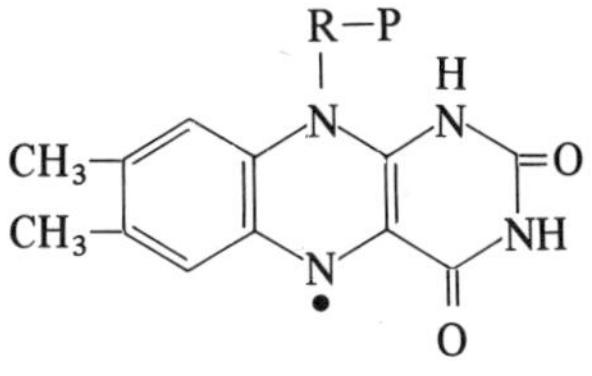

(FH_2)

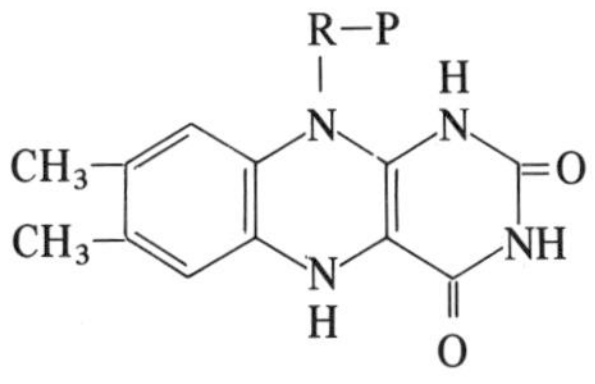

(FH_3)

Figure 2.11. Structures of the fully reduced (FH_3), half-reduced ($FH_2^{\bullet}$) and fully oxidised (FH) forms of flavin mononucleotide; R and P stand for *D*-ribitol and inorganic phosphate respectively.

References

Blum, H. F., 1964, *Photodynamic Action and Diseases Caused by Light* (Hafner, New York).
Cottrell, T. L., 1958, *The Strength of Chemical Bonds* (Butterworths, London).
Green, J. H., Ralph, B. J., Schofield, P. J., 1963, *Nature, Lond.*, **198**, 754.
Hayaishi, O., 1969, *A. Rev. Biochem.*, **38**, 21.
Hemmerich, P., Devartanian, D. V., Veeger, C., Vorst, J. D. W. van, 1963, *Biochim. biophys. Acta*, **77**, 504.
Horner, L., Betzel, C., 1953, *Annln Chemie*, **579**, 175.
Ingram, D. J. E., 1958, *Free Radicals as Studied by Electron Spin Resonance* (Butterworths, London).

Ingram, D. J. E., 1969, *Biological and Biochemical Applications of Electron Spin Resonance* (Adam Hilger, London), p.158.
Land, E. J., Swallow, A. J., 1969, *Biochemistry, Easton,* **8**, 2117.
Mahler, H. R., Cordes, E. H., 1966, *Biological Chemistry* (Harper International Edition), p.576.
Malkin, R., Rabinowitz, J. C., 1967, *A. Rev. Biochem.,* **36**, 113.
Massey, V., Gibson, Q. H., 1964, *Fedn. Proc. Fedn. Am. Socs. exp. Biol.,* **23**, 18.
Pryor, W. A., 1966, *Free Radicals* (McGraw-Hill, New York).
Reid, C., 1957, *Excited States in Chemistry and Biology* (Butterworths, London).
Smith, J. R. L., Norman, R. O. C., 1963, *J. Chem. Soc.,* 2897.
Staudinger, H., Kerekjártó, B., Ullrich, V., Zubrzycki, Z., 1965, in *Oxidases and Related Systems,* **2**, Eds. T. E. King, H. S. Mason, M. Morrison (John Wiley, New York), p.815.
Udenfriend, S., Clark, C. T., Axelrod, U., Brodie, B. B., 1954, *J. biol. Chem.,* **208**, 731.
Uri, N., 1952, *Chem. Rev.,* **50**, 375.
Van Duuren, B. L., Orris, L., Nelson, N., 1965, *J. natn. Cancer Inst.,* **35**, 707.
Volman, D. H., Seed, J. R., 1964, *J. Am. chem. Soc.,* **86**, 5095.
Walling, C., 1957, *Free Radicals in Solution* (John Wiley, New York).

3

Chemical reactions of free radicals

The reactions characteristic of free radicals will be considered in this Chapter. The treatment will necessarily be brief and is intended to provide only a very general background to facilitate the biological discussions that are dealt with in Part 3. The reader interested in obtaining more comprehensive discussion of the reactions of free radicals is referred to texts by Walling (1957), Pryor (1966) and Ingram (1969).

3.1 Chemical reactivity

The unpaired electron that characterises a free radical in general confers on the molecule an extremely high chemical reactivity. Free radicals, particularly when produced in solution, often react very rapidly with neighbouring substances so that the half-life of the radical may be very short indeed. The extreme reactivity of free radicals with many types of neighbouring molecule often results in a very low steady-state concentration of the free radical even though it is being continuously produced (e.g. by a metabolic reaction *in vivo*). This means that sensitive methods for free-radical detection are necessary and this point is taken up in Chapter 4. Some important reactions that free radicals undergo are outlined in Sections 3.1.1–3.1.6.

3.1.1 Combination reactions

Here two free radicals undergo a self-annihilation reaction:

$$R^{\bullet} + R^{\bullet} \longrightarrow R{-}R \tag{3.1}$$

This type of reaction is not observed with carbonium ions or with carbanions owing to the energy barrier created by the electrostatic repulsion between approaching like charges. A typical example of self-annihilation is the original observation of Gomberg that hexaphenylethane was a product of the formation of triphenylmethyl.

$$C(C_6H_5)_3^{\bullet} + C(C_6H_5)_3^{\bullet} \longrightarrow (C_6H_5)_3C{-}C(C_6H_5)_3 \tag{3.2}$$

This type of reaction often has a very low energy of activation; the combination of two methyl radicals, for example, has been found to have *zero* energy of activation.

3.1.2 Transfer reactions including hydrogen abstraction

This is a very common reaction of free radicals, the reactive free radical $R^{\bullet}$ withdrawing a hydrogen or halogen atom as shown:

$$R^{\bullet} + A{-}H \longrightarrow R{-}H + A^{\bullet} \tag{3.3}$$

$$CH_3^{\bullet} + Cl{-}Cl \longrightarrow CH_3Cl + Cl^{\bullet} \tag{3.4}$$

Reaction (3.4) occurs in chain reactions characteristic of halogenation; this process has been extensively studied and is of considerable synthetic value.

$$Cl_2 \xrightarrow[\text{or radiation}]{\text{high temperature}} 2Cl^{\bullet} \tag{3.5}$$

$$Cl^{\bullet} + CH_4 \longrightarrow HCl + CH_3^{\bullet} \tag{3.6}$$

$$CH_3^{\bullet} + Cl_2 \longrightarrow CH_3Cl + Cl^{\bullet} \tag{3.7}$$

In Reaction (3.3) the strength of the bond A—H undergoing dissociation may be relatively high and in this situation the reaction is favoured by an increasing bond strength of the product R—H. Pryor (1966) quotes an interesting example of this behaviour between halogen atoms and ethane.

$$Hal^{\bullet} + C_2H_6 \longrightarrow H{-}Hal + C_2H_5^{\bullet} \tag{3.8}$$

Here the bond-dissociation energy of H—Hal is 136 kcal mol^{-1} for H—F and 87 kcal mol^{-1} for H—Br; the corresponding activation energies for the hydrogen-abstraction reaction are 0·2 and 14 kcal mol^{-1} respectively. This situation is of course analogous to the dissociation of the C—N bond in azo-compounds which is favoured by the high heat of formation of nitrogen (see Section 2.2.3).

A hydrogen-abstraction reaction that is of considerable biological importance is the one between a free radical in solution (produced for example by high energy radiation) and a free-radical scavenger. Thiols are particularly effective as free-radical scavengers in biological systems (see Chapter 5). The thiyl radical so produced may undergo a number of reactions as shown in Reactions (3.9)–(3.11).

$$R^{\bullet} + X{-}SH \longrightarrow R{-}H + X{-}S^{\bullet} \tag{3.9}$$

$$2X{-}S^{\bullet} \longrightarrow X{-}S{-}S{-}X \tag{3.10}$$

$$X{-}S^{\bullet} + O_2 \longrightarrow X{-}SO_2^{\bullet} \tag{3.11}$$

3.1.3 Aromatic substitution

The rules that apply to ionic mechanisms of aromatic substitution (see Ingold, 1970) do not apply generally to homolytic mechanisms. Not only does a previous substitution in the aromatic ring differently affect the rate and position of a subsequent substitution in ionic compared with homolytic mechanisms, but the homolytic reaction may entail the formation of a considerable variety of additional products. Phenylation offers a good example of these effects. Chlorobenzene is attacked rather more rapidly than the parent benzene by phenyl radicals and the product is mainly *o*-phenyl chlorobenzene. In an ionic mechanism, however, chlorobenzene is attacked much less readily than is benzene by the nitronium ion NO_2^+ and the product is largely *p*-nitrochlorobenzene.

Further, the production of phenyl radicals in the presence of an aromatic species Ar—H leads to the complexities in product formation shown in Reactions (3.12)-(3.15).

$$\text{Ar—H} + C_6H_5^{\bullet} \longrightarrow \text{Ar—}C_6H_5 + H^{\bullet} \tag{3.12}$$

$$2C_6H_5^{\bullet} \longrightarrow C_{12}H_{10} \tag{3.13}$$

$$\text{Ar—H} + C_6H_5^{\bullet} \longrightarrow \text{Ar}^{\bullet} + C_6H_6 \tag{3.14}$$

$$2\text{Ar}^{\bullet} \longrightarrow \text{Ar—Ar} \tag{3.15}$$

A detailed discussion of the mechanisms of phenylation and other aromatic substitution reactions by free radicals can be found in Pryor (1966) and Williams (1960).

3.1.4 Oxygen uptake

Oxygen has the electronic configuration $(1s)^2(2s)^2(2p)^4$ and is a biradical as demonstrated by its paramagnetic properties. The presence of two unpaired electrons of the same spin category can be illustrated by using Linnett's novel visualisation of electronic dispositions (Figure 3.1). Oxygen may react in a number of ways with excited molecules. For example it can quench fluorescence (see Caldin, 1964); it can act as a radical scavenger and it is a prerequisite of many photosensitised oxidations. The scavenging action of oxygen is illustrated by its ready reaction with many radical species to yield the corresponding peroxy derivatives:

$$R^{\bullet} + O_2 \longrightarrow RO_2^{\bullet}\,. \tag{3.16}$$

Such peroxy radicals may abstract hydrogen from a neighbouring molecule to form the hydroperoxide.

$$RO_2^{\bullet} + \text{X—H} \longrightarrow RO_2H + X^{\bullet}\,. \tag{3.17}$$

This mechanism will be discussed in more detail in Section 3.3.2 in connection with lipid peroxidation. The ready reaction of oxygen with

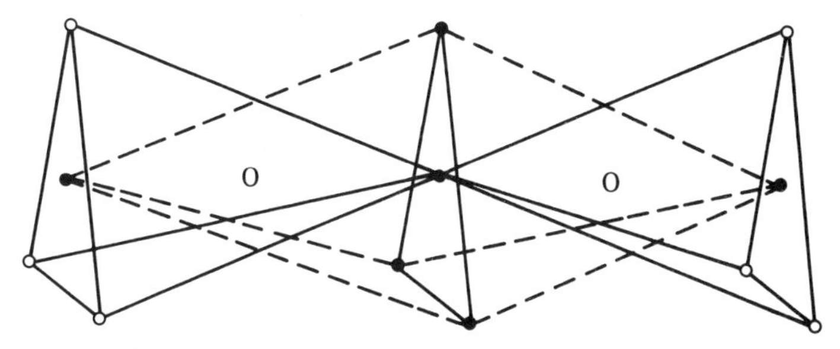

Figure 3.1. Electronic distribution in the oxygen molecule as suggested by Linnett (1964). The electrons are arranged in four tetrahedra, one pair sharing a common apex, the other pair sharing a common base.

many types of free radical may be used to indicate that a particular mechanism proceeds through the formation of free-radical intermediates, for in such cases the reaction is inhibited by the addition of the scavenger molecule; other scavengers used in this respect are iodine, sulphur and quinones.

3.1.5 Disproportionation reactions

This type of free-radical reaction is illustrated by Reaction (3.18).

$$2CH_3{-}CH_2^{\bullet} \longrightarrow CH_2{=}CH_2 + CH_3{-}CH_3 . \qquad (3.18)$$

Here it can be seen that hydrogen abstraction has occurred from the β-carbon atom of one radical resulting in the formation of an unsaturated linkage.

3.1.6 Addition reactions

A large variety of addition reactions involving free-radical intermediates and unsaturated compounds is known. The general form of the reaction is as shown in Reactions (3.19)–(3.21).

$$R{-}Q \longrightarrow R^{\bullet} + Q^{\bullet} \qquad (3.19)$$

$$A{-}C{=}C{-}B + R^{\bullet} \longrightarrow A{-}\underset{\substack{|\\R}}{C}{-}\dot{C}{-}B \qquad (3.20)$$

$$A{-}\underset{\substack{|\\R}}{C}{-}\dot{C}{-}B + \underset{(\text{or } R{-}Q)}{Q^{\bullet}} \longrightarrow A{-}\underset{\substack{|\\R}}{C}{-}\underset{\substack{|\\Q}}{C}{-}B \ (\text{or} + R^{\bullet}) \qquad (3.21)$$

A well-known example of radical addition occurs in the reaction of hydrogen bromide with an alkene. Hydrogen bromide may also add by an ionic mechanism and in such a case the product is as specified by the **Markovnikov** rule where hydrogen adds to the carbon atom of the double bond that is already the richer in hydrogen attachments:

$$H{-}Br + CH_2{=}CH{-}CH_3 \longrightarrow CH_3{-}\underset{\substack{|\\Br}}{CH}{-}CH_3 \qquad (3.22)$$

However, the radical mechanism proceeds through an anti-Markovnikoff pathway:

$$H{-}Br + CH_2{=}CH{-}CH_3 \longrightarrow \underset{\substack{|\\Br}}{CH_2}{-}CH_2{-}CH_3 \qquad (3.23)$$

The *halomethanes* may also add to alkenes via radical intermediates; the reaction shown is of relevance to the combination of metabolites of carbon tetrachloride with unsaturated lipid *in vivo* as is discussed in Chapter 10 (Section 10.5).

$$CCl_4 + CH_2{=}CH{-}C_6H_{13} \longrightarrow \underset{\displaystyle CCl_3}{CH_2}{-}\underset{\displaystyle Cl}{CH}{-}C_6H_{13} \qquad (3.24)$$

3.2 Stable radicals

Despite the high general reactivity of free radicals described above, and the large number of reactions that free radicals may undergo, some free radicals are extraordinarily unreactive and stable even in solution at room temperatures. This greatly increased stability may result from both steric factors and resonance effects. For example in triphenylmethyl (Figure 3.2a) the bulky aromatic rings shield the central carbon atom and also provide a number of possible resonance structures that delocalise the unpaired free electron. A well-known stable free radical is **diphenylpicrylhydrazyl** (Figure 3.2b). This radical is stable in the solid state for long periods and in solution has a deep violet colour which fades on the reaction of diphenylpicrylhydrazyl with other radicals. Diphenylpicrylhydrazyl may thus be used in a spectrophotometric test for the detection of free radicals in solution. Figure 3.3 gives the absorption spectrum of the compound before and after reaction with benzosemiquinone.

A stable inorganic free radical that is water soluble is Fremy's salt (Figure 3.2c; peroxylamine disulphonate; potassium nitrosodisulphonate), which is used as a mild oxidising agent capable of oxidising phenols to quinones or aryl amines to quinone-imines. In the solid state Fremy's salt may exist as the dimer (Yamada and Tsuchida, 1959):

$$(SO_3K)_2{-}N{-}O{-}O{-}N{-}(SO_3K)_2$$

This is an interesting possibility in view of the speculative N—O—O—N peroxide discussed in Chapter 13 in relation to the carcinogenicity of polycyclic hydrocarbons. Like diphenylpicrylhydrazyl, Fremy's salt is

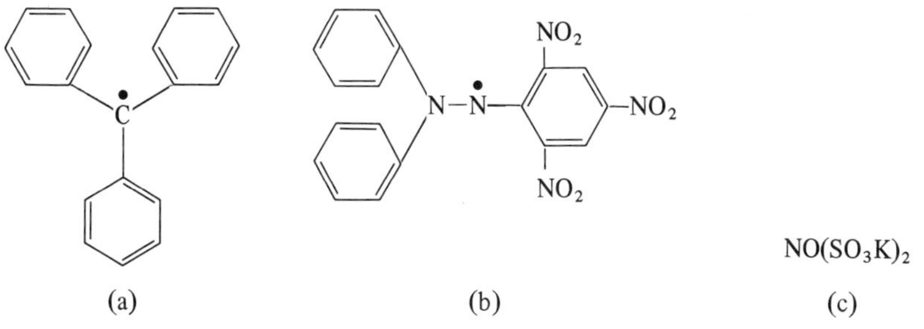

Figure 3.2. Structures of (a) triphenylmethyl, (b) diphenylpicrylhydrazyl and (c) potassium nitrosodisulphonate (Fremy's salt).

often used as a calibration standard for electron spin resonance studies for it gives three single lines separated by 13 gauss.

Semiquinone radicals may also be relatively stable particularly as the anionic form in alkaline solution. The participation of these radical species in flavin-linked enzyme reactions is discussed in other Sections of this book (2.5; 10.12.1).

The phenomenon of radicals trapped in an organic matrix is one that occurs frequently during polymerisation reactions or during pyrolysis (see Section 2.2), when the radicals formed during combustion may be trapped in soot particles. The biological activity of such trapped radicals is discussed in Chapter 13.

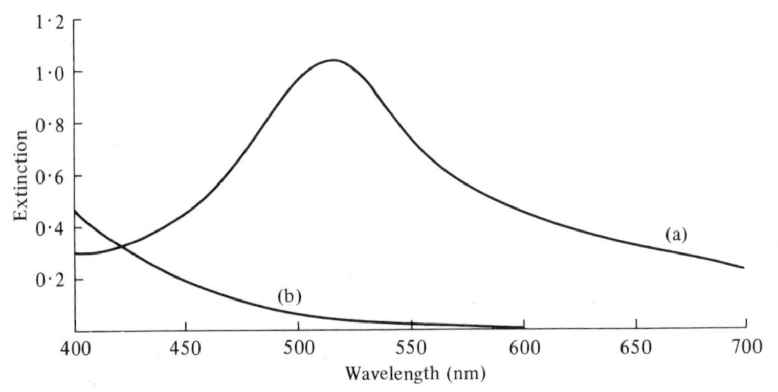

Figure 3.3. Absorption spectrum of an ethanolic solution of the stable free radical diphenylpicrylhydrazyl (DPPH, a). Spectrum (b) was obtained by reducing DPPH with an excess of hydroquinone to produce diphenylpicrylhydrazine. The concentration of DPPH in (a) was approximately 0·11 mM.

3.2.1 Spin labelling

Stable free radicals have recently been used in a most interesting manner, mainly by McConnell's group (Ohnishi and McConnell, 1965; Griffith and McConnell, 1966), to obtain information about the structure and behaviour of biological membranes. The technique is known as **spin labelling**.

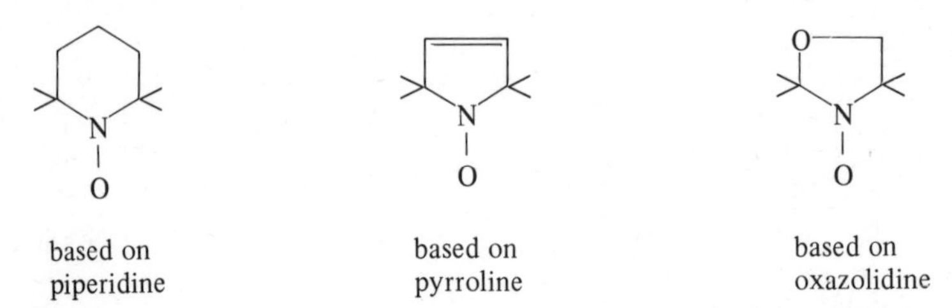

Figure 3.4. Examples of nitroxide ring structures that have been used in the production of spin labels. The carbon atoms adjacent to the ring nitrogen are generally substituted with methyl groups. Structure (a) based on piperidine, (b) based on pyrroline and (c) based on oxazolidine. For further details see Griffiths and Waggoner (1969).

A stable free radical is introduced into a membrane system and under certain conditions electron spin resonance spectroscopy of the modified system can give data on membrane structure in the neighbourhood of the free radical. For this purpose it is essential that the electron spin resonance spectrum obtained from the stable free radical alters with the *orientation* of the radical species. Studies with this technique have so far concentrated on the **nitroxyl**-radical (>N•—O), in which the nitrogen atom is joined to tertiary carbon atoms. In most cases the tertiary carbon atoms have been constituents of five- or six-membered rings such as pyrroline, piperidine or oxazolidine (see Figure 3.4). The 1-oxy substituent giving the nitroxyl radical has the unpaired electron mainly associated with a p-orbital of the ring nitrogen. The free electron distribution is thereby asymmetrically orientated with respect to the ring components; altering the orientation of the ring results in changes in the electron spin resonance spectrum obtained. By attaching the nitroxyl radical to a variety of membrane components it is possible to obtain information about the rotational mobility of the radical in its locality, the polarity of the neighbourhood in which the radical resides, as well as the relative orientation of the stable free radical. The structures of two nitroxyl-labelled compounds that have been used in such studies are shown in Figure 3.5. A short discussion of the technique is given by Ingram (1969); a more comprehensive account is by Griffith and Waggoner (1969).

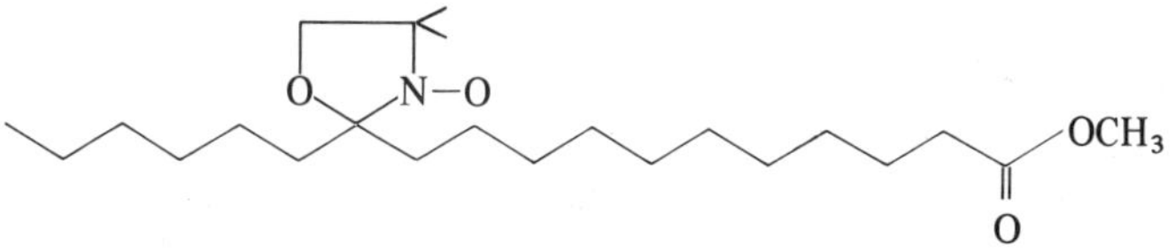

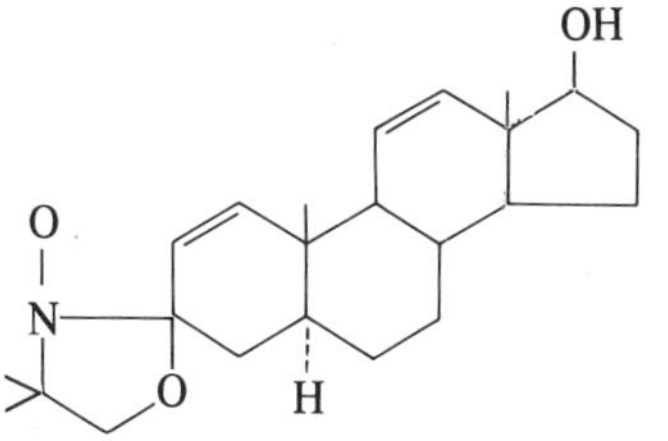

Figure 3.5. Examples of nitroxide derivatives that have been used in spin labelling experiments concerning membrane structure; (a) methyl stearate nitroxide and (b) a nitroxide derivative of a steroid. See Waggoner *et al.* (1969) and Hubbell and McConnell (1969).

Although the nitroxyl radical is basically very stable, it may act as a powerful lipid antioxidant under appropriate conditions. Weil *et al.* (1968) have shown that nitroxyl radicals based on piperidine will compete very favourably with oxygen for alkyl radicals

$$R_1^{\bullet} + R_2NO^{\bullet} \longrightarrow R_1{-}NO{-}R_2 \quad (3.25)$$

and are highly efficient in inhibiting squalene autoxidation.

3.3 Chain reactions

One very important feature of free radicals that has already been referred to briefly is their ability to enter into sequential reactions often of considerable length. Such chain reactions may result in the decomposition of a large number of molecules of the starting material from a single free-radical initiation step. The classical example of this is the photochemical combination of dry chlorine and hydrogen discussed in Section 2.3. If the chain process entails a net increase in the number of free radicals per cycle (a branched chain reaction) then, as in nuclear fission reactions above the critical mass limit, the overall situation becomes explosive. This sort of mechanism operates in the lighting of a safety match or in the detonation of an unstable peroxide.

3.3.1 Plastics

Economically important illustrations of chain reactions arise in several polymerisation reactions which, with free-radical intermediates, usually proceed through an addition polymerisation mechanism:

$$nX \longrightarrow X_n \quad (3.26)$$

Some examples of polymers that are formed by such a mechanism are given:

Polyethylene

Ethylene is heated under pressure in the presence of a free-radical initiator

$$CH_2{=}CH_2 \longrightarrow -(CH_2{-}CH_2)_n{-}CH_2{-}CH_2^{\bullet} \quad (3.27)$$

$$\downarrow$$

$$-(CH_2{-}CH_2)_n{-}CH_2{-}CH_2-$$

Polyethylene plastic is relatively flexible and inert and is used for containers and electric insulation.

Teflon

This is produced by the condensation of tetrafluoroethylene ($CF_2{=}CF_2$). It is an inert material with a high melting point and a low coefficient of friction and is used, for example, in the biochemical laboratory to make homogeniser pistons to prevent effects of frictional heat on labile enzymes during the preparation of the tissue suspension.

Vinyl plastics
The rigid polymers of vinyl chloride ($CH_2{=}CHCl$) and vinylidene chloride ($CH_2{=}CCl_2$) are of considerable importance in the construction of pipes and insulation material.

Perspex
This very clear plastic is formed by the polymerisation of methyl methacrylate:

$$-\left(\begin{array}{l} COOCH_3 \\ | \\ \dot{C}-CH_2 \\ | \\ CH_3 \end{array}\right)_n-$$

These are but a few examples of the ever-increasing number of plastics that are manufactured by free-radical chain mechanisms. For a comprehensive discussion of polymerisation reactions see Flory (1953) and Stille (1962).

3.3.2 Lipid peroxidation
An example of a free-radical chain process that is deleterious rather than economically valuable is that of **lipid peroxidation**. Biological material contains a wide variety of unsaturated lipid materials, particularly in the form of membrane lipid. Rat liver mitochondrial membranes, lysosomal membranes and endoplasmic reticulum contain unsaturated fatty acids in considerable amounts (see Section 12.2). Food products such as milk, cheese and meat fat also contain large amounts of unsaturated fatty acids. A characteristic property of such unsaturated fatty acids is that in the presence of a free-radical initiator and oxygen they undergo oxidative deterioration. This is usually accompanied by a decidedly rancid smell and the change makes the food unpalatable. Since the process is of such economic importance to the food (particularly the dairy) industry, a large number of investigations of the mechanisms underlying this oxidative deterioration has been carried out. The process is called by the general name of lipid peroxidation and, as we shall see in later Chapters, is of considerable importance in our attempts to understand the mechanisms underlying cellular injury.

A simplified scheme for the reactions involved in lipid peroxidation is shown in Figure 3.6 in which $R^{\bullet}$ is the radical initiator. In the first reaction, $R^{\bullet}$ abstracts hydrogen from the unsaturated fatty acid (I) to yield the free radical (II). This is followed by the uptake of oxygen by (II) to produce the fatty acid peroxy radical (III). This radical (III) reacts with another molecule of (I) to yield a further molecule of (II) and an unsaturated hydroperoxide (IV).

The process of lipid peroxidation in biological samples is associated with the loss of unsaturated bonds in lipid that is normally quite rich in materials such as arachidonic acid, linoleic acid and linolenic acid. There is also an uptake of oxygen and in general a production of a material that reacts with thiobarbituric acid similarly to malonaldehyde. The **thiobarbituric acid** reaction is discussed in detail in Section 4.2.3 and the discussion here will be restricted to a scheme (Figure 3.7) that is consistent with the major events stated above. The scheme is that outlined by Dahle *et al.* (1962) and shows how an unsaturated fatty acid hydroperoxide undergoes hydrogen abstraction through interaction with the radical R• (reaction I); this is followed by a rearrangement of the double bonds (reaction II) and uptake of oxygen (reaction III). The formation of a five-membered oxygen ring (reaction IV) leads to bond fission with formation of malonaldehyde (reaction VII). It can also be seen that the rearrangement of double bonds in reaction II leads to a conjugated diene arrangement which is responsible for the increased absorption observed at 233 nm in peroxidising lipids (Figure 3.7).

It can readily be imagined that the process of lipid peroxidation can cause extensive disturbance to the finely ordered structure of biological membranes which are rich in unsaturated fatty acids. The experimentally observed disturbances in intracellular membranes resulting from lipid peroxidation are discussed in Section 7.3; practical aspects of the estimation of lipid peroxides, of malonaldehyde and the conjugated diene shift are described in Section 4.2; the metabolism of lipid peroxides and malonaldehyde are outlined in Section 4.2.3 and in Section 10.8.1.

Finally in this Section on peroxidation it is important to consider the kinetics of chain propagation by an initiator. The general kinetic behaviour

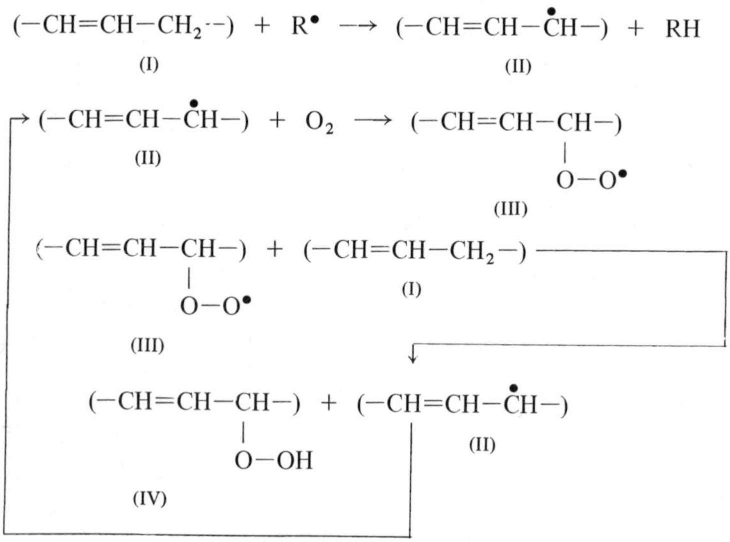

$$\underset{\text{(I)}}{(-CH{=}CH{-}CH_2{-}{-})} + R^\bullet \longrightarrow \underset{\text{(II)}}{(-CH{=}CH{-}\dot{C}H{-})} + RH$$

$$\underset{\text{(II)}}{(-CH{=}CH{-}\dot{C}H{-})} + O_2 \longrightarrow \underset{\text{(III)}}{(-CH{=}CH{-}CH(O{-}O^\bullet){-})}$$

$$\underset{\text{(III)}}{(-CH{=}CH{-}CH(O{-}O^\bullet){-})} + \underset{\text{(I)}}{(-CH{=}CH{-}CH_2{-})} \longrightarrow$$

$$\underset{\text{(IV)}}{(-CH{=}CH{-}CH(O{-}OH){-})} + \underset{\text{(II)}}{(-CH{=}CH{-}\dot{C}H{-})}$$

Figure 3.6. A simplified scheme for the chain reaction involved in the formation of unsaturated lipid hydroperoxide (IV).

of chain reactions is described in detail by Walling (1957) and by Waters (1948). The discussion here is restricted to a particular example that is of relevance to liver injury produced by carbon tetrachloride (sec Chapters 9 and 10). Evidence will be presented in Chapters 9 and 10 that carbon tetrachloride produces several of its damaging actions on liver cells by a homolytic cleavage in the endoplasmic reticulum to trichloromethyl and chlorine. This is accompanied by an increased lipid peroxidation and the

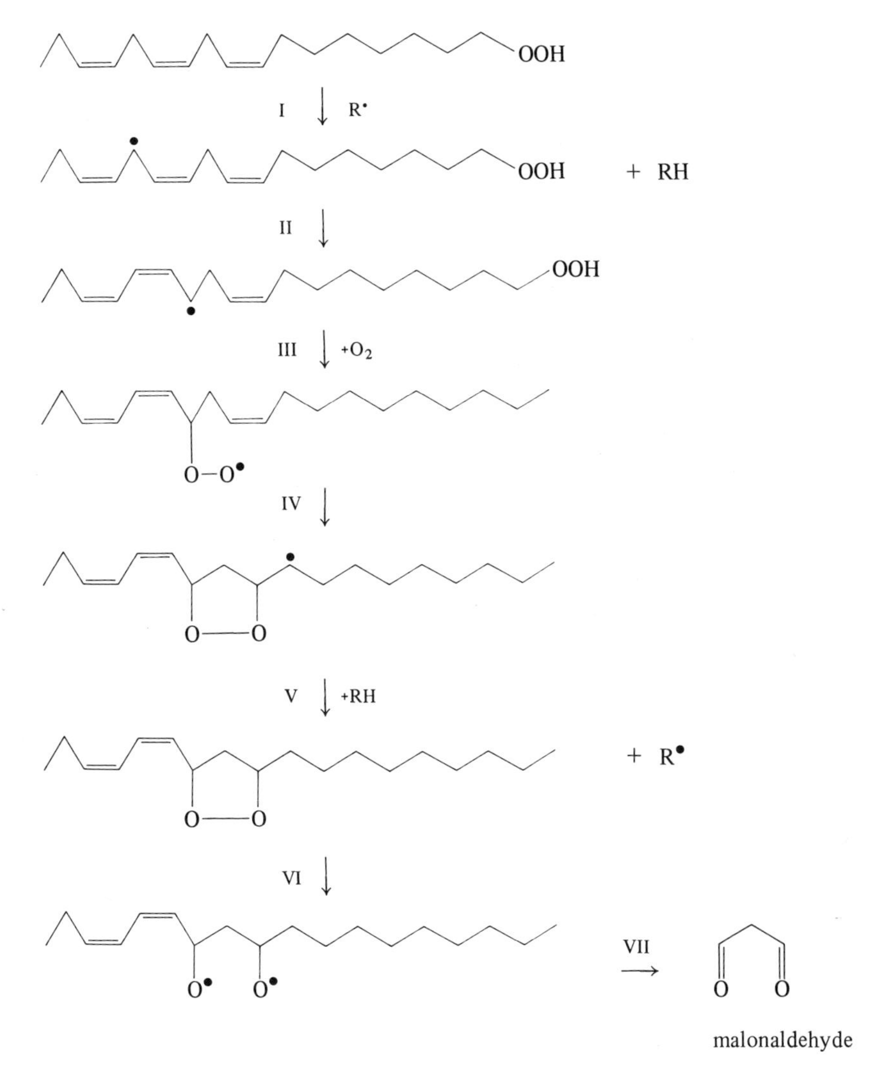

Figure 3.7. Scheme for the oxidative breakdown of an unsaturated fatty acid hydroperoxide to malonaldehyde (after Dahle *et al.*, 1962).

general behaviour of the system can be represented as shown in Figure 3.8. In that diagram the reactive species is shown as $CCl_3^\bullet$ but it could equally well be $Cl^\bullet$.

$$(1)\quad FPH_3 + CCl_4 \xrightarrow{K_F} CCl_3^\bullet + Cl^- + FPH_2^\bullet + H^+$$

$$(2)\quad CCl_3^\bullet + USL \xrightarrow{K_I} USL^\bullet + CHCl_3$$

$$(3)\quad USL^\bullet + O_2 \xrightarrow{K_o} USLO_2^\bullet$$

$$(4)\quad 2USL^\bullet \xrightarrow{K_y} Y$$

$$(5)\quad 2USLO_2^\bullet \xrightarrow{K_x} X$$

$$(6)\quad USL^\bullet + USLO_2^\bullet \xrightarrow{K_z} Z$$

$$(7)\quad USLO_2^\bullet + USL \xrightarrow{K_p} USLO_2H + USL^\bullet$$

In general $K_o \gg K_p$ and $K_x > K_z, K_y$

In the steady state there is a constant production of $USL^\bullet$; thus

$$K_I[CCl_3^\bullet] = K_x[USLO_2^\bullet]^2$$

$$-\frac{dO_2}{dt} = K_o[USL^\bullet][O_2] = K_p[USLO_2^\bullet][USL] = K_p[USL]\left(\frac{K_I}{K_x}[CCl_3^\bullet]\right)^{1/2}$$

since

$$[CCl_3^\bullet] = K_F[CCl_4][FP^\bullet]$$

then

$$-\frac{dO_2}{dt} = \text{constant}[CCl_4]^{1/2} \ .$$

Figure 3.8. Scheme for the relationship between oxygen uptake and the concentration of carbon tetrachloride in a peroxidising lipid environment. USL, unsaturated lipid; FP, flavoprotein; X, Y and Z, unspecified breakdown products of $USLO_2^\bullet$, $USL^\bullet$ and $USL^\bullet$ + $USLO_2^\bullet$ respectively. For comments on the scheme see the text.

Two basic assumptions are made in deriving the kinetic relationship. First, that the concentration of the endogenous reduced flavoprotein (see Section 10.5) with which carbon tetrachloride interacts is constant. Secondly that the rate of reaction of the unsaturated fatty acid free radical $USL^\bullet$ with oxygen is very much faster than reactions K_y or K_z; this is generally true in situations that have been examined (see Walling, 1957). With these assumptions it can be seen that the rate of uptake of oxygen, which measures the rate of lipid peroxidation, is proportional to the square root of the steady-state concentration of trichloromethyl. In experimental situations studied *in vitro* where low concentrations of carbon tetrachloride are diffused into contact with the NADPH–flavoprotein (see Chapter 10) then the concentration of trichloromethyl radicals is given by:

$$[CCl_3^\bullet] = K_F[CCl_4][FP^\bullet] \tag{3.28}$$

Since the assumption has been made that the concentration of FP$^{\bullet}$ is constant then it follows that the rate of lipid peroxidation is proportional to the square root of the concentration of carbon tetrachloride. Experimental data are in accord with this finding (Section 10.6).

References

Caldin, E. F., 1964, *Fast Reactions in Solution* (Blackwell, Oxford).
Dahle, L. K., Hill, E. G., Holman, R. T., 1962, *Archs. Biochem. Biophys.*, **98**, 253.
Florey, P. J., 1953, *Principles of Polymer Chemistry* (Cornell University Press, New York).
Griffith, O. H., McConnell, H. M., 1966, *Proc. natn. Acad. Sci. U.S.A.*, **55**, 8, 708.
Griffith, O. H., Waggoner, A. S., 1969, *Acct. Chem. Res.*, **2**, 17.
Hubbell, W. L., McConnell, H. M., 1969, *Proc. natn. Acad. Sci. U.S.A.*, **63**, 16.
Ingold, C. K., 1970, *Structure and Mechanism in Organic Chemistry*, second edition (Cornell University Press, New York).
Ingram, D. J. E., 1969, *Biological and Biochemical Applications of Electron Spin Resonance* (Adam Hilger, London).
Linnett, J. W., 1964, *The Electronic Structure of Molecules* (Methuen, London).
Ohnishi, S., McConnell, H. M., 1965, *J. Am. chem. Soc.*, **87**, 2293.
Pryor, W. A., 1966, *Free Radicals* (McGraw-Hill, New York).
Stille, J. K., 1962, *Introduction to Polymer Chemistry* (John Wiley, New York).
Waggoner, A. S., Kingzett, T. J., Rottschaefer, S., Griffith, O. H., Keith, A. D., 1969, *Chem. Phys. Lipids*, **3**, 245.
Walling, C., 1957, *Free Radicals in Solution* (John Wiley, New York).
Waters, W. A., 1948, *Chemistry of Free Radicals*, second edition (Oxford University Press, Oxford).
Weil, J. T., Van der Veen, J., Olcott, H. S., 1968, *Nature, Lond.*, **219**, 168.
Williams, G. H., 1960, *Homolytic Aromatic Substitution* (Pergamon Press, Oxford).
Yamada, S., Tsuchida, R., 1959, *Bull. chem. Soc. Japan*, **32**, 721.

4

Detection and measurement of free radicals

4.1 Introduction

The unpaired electron that characterises a substance as a free radical confers high chemical reactivity in general, as well as a resultant magnetic moment. Various methods of detection and measurement of free-radical concentrations based on these two features have been devised; these may be described as (a) chemical methods and (b) physical methods of detection.

4.2 Chemical methods of detection

The earliest demonstration of the highly reactive property of free radicals was by Paneth and Hofeditz (1929), who demonstrated that the thermal decomposition of metal-alkyls gave very unstable products that would react with and remove metallic mirrors. The disappearance of the metallic mirror could be followed visually, spectrophotometrically or by the analysis of the products formed, which are also metal-alkyls. The reaction is applicable to gaseous free radicals and is relatively adaptable by the use of different metals to form the mirror; further, the use of metal radioisotopes for the mirror greatly improves the sensitivity of product identification. Figure 4.1 illustrates the apparatus used by Paneth to demonstrate this effect.

The occurrence of free-radical intermediates in chemical reactions may be inferred from the formation of products that would not be expected if the reaction followed an ionic pathway. Examples of this behaviour occur in aromatic substitution reactions as already described in Section 3.1.3,

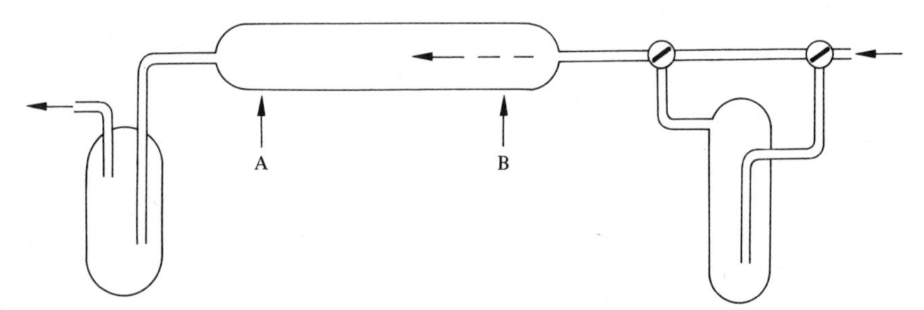

Figure 4.1. Simplified representation of the apparatus used by Paneth and Hofeditz (1929) to demonstrate formation of reactive products after heating metal alkyls. Tetramethyl lead was passed into the horizontal tube in the direction shown by the arrows. The horizontal tube contained a metallic deposit ('mirror') at A. On heating the horizontal tube with a moveable furnace at B the metal alkyl decomposed and the reactive products formed then caused a destruction of the metallic mirror at A and the formation of a new mirror at the site of heating B. The interpretation is that the metal alkyl is decomposed by heat, giving a deposition of a metallic mirror and reactive alkyl free radicals that interact with the metallic mirror further along the tube at A, thereby destroying it.

and in anti-Markovnikov addition to unsaturated alkenes (see Section 3.1.6). A further feature of many free-radical reactions is chain propagation in which a single free-radical initiation step may result in the sequential decomposition of thousands of molecules (see Section 3.3). The recognition of such a process provides a strong indication that radical intermediates are involved in the system under study.

The ability of a radical initiation step to propagate a chain reaction has been used as a sensitive method for free-radical detection by Fridovich and Handler (1961). Sulphite is oxidised by a free-radical chain mechanism that can be followed by the uptake of oxygen. Addition of free radicals to a stabilised sulphite solution results in a rapid increase in oxygen uptake that can be monitored in a Warburg apparatus or by the use of an oxygen electrode. Sato and Iwaizumi (1969) have used this technique to establish the formation of triphenyltetrazolium free radicals during microsomal electron transport. The triphenyltetrazolium free radical is then believed to react with oxygen as shown in Figure 4.2.

Work by McCord and Fridovich (1969) has provided an elegant procedure for the detection of reactions involving the superoxide anion radical ($O_2^{\bullet -}$). They have found that copper–proteins previously isolated by other workers and named haemocuprein and erythrocuprein have very high superoxide dismutase activity for Reaction (4.1).

$$O_2^{\bullet -} + O_2^{\bullet -} + 2H^+ \longrightarrow O_2 + H_2O_2 \tag{4.1}$$

Similar enzymic activity was found in many tissues examined. The interesting observation was made that the superoxide dismutase strongly inhibited sulphite oxidation in the presence but not in the absence of EDTA, indicating that EDTA altered the reaction mechanism. The oxidation of adrenalin to adrenochrome catalysed by xanthine oxidase was also studied by McCord and Fridovich (1969), who found that the

$$(NADPH + FP) + TTC \longrightarrow TTC^{\bullet}$$

$$TTC^{\bullet} + O_2 \longrightarrow O_2^{\bullet -} + TTC$$

$$O_2^{\bullet -} + H^+ \longrightarrow O_2H^{\bullet}$$

$$O_2H^{\bullet} + O_2H^{\bullet} \longrightarrow H_2O_2 + O_2$$

$$O_2H^{\bullet} + TTC^{\bullet} \longrightarrow O_2H^- + TTC$$

$$O_2H^- + H^+ \longrightarrow H_2O_2$$

$$TTC^{\bullet} + e^- \longrightarrow F$$

$$TTC^{\bullet} + TTC^{\bullet} \longrightarrow TTC + F$$

Figure 4.2. Participation of triphenyltetrazolium (TTC) free radicals in NADPH-oxidase of liver endoplasmic reticulum (see Sato and Iwaizumi, 1969). The triphenyltetrazolium free radicals are formed by an interaction at the NADPH–flavoprotein site and subsequently react with molecular oxygen as shown.

reaction involved the superoxide anion and was inhibited by 3×10^{-10} M superoxide dismutase.

Many reactions that involve free-radical intermediates are inhibited by the addition to the system of free-radical scavengers. Such behaviour can be used as a means of detecting the radical nature of the initial mechanism. Well-known scavengers in this respect are oxygen, iodine, quinones and antioxidants such as propyl gallate. Inhibition of a particular reaction by a number of such scavengers is suggestive evidence that a free-radical mechanism is involved. However, the situation is much less clear in biological systems *in vivo*, for here the individual scavengers may possess a variety of other properties that confuse the picture. For example, many antioxidants act *in vivo* as inducers of the microsomal cytochrome P_{450} system (Gilbert *et al.*, 1969); some, for example, propyl gallate (Torrielli and Slater, 1970), greatly decrease the activity of enzymes such as NADPH-cytochrome *c* reductase. Oxygen, which may inhibit free-radical reactions *in vitro* by quenching, has numerous effects on biological material; when present in high concentration it causes increased metabolic activity and membrane changes in the liver and is, of course, necessary as a substrate in peroxidative changes. As a consequence of such side reactions it is often difficult to separate effects *in vivo* that result from a free-radical scavenging action from other diverse properties of the scavenger.

It is possible to detect and measure some free radicals by ordinary light **absorption spectrophotometry**. For example, triphenylmethyl in equilibrium with hexamethylethane is yellow whereas the hexamethylethane alone is colourless. The stable free radical diphenylpicrylhydrazyl is intensely violet and has a well-defined absorption spectrum, as shown in Figure 3.3. This free radical can be used as a detector of other free-radical reactions as the formation of diphenylpicrylhydrazine results in a loss of violet colour. The reaction is complicated, however, and difficult to quantitate.

The use of absorption spectrophotometry for free-radical detection in solution at room temperature is limited to stable free radicals unless very fast techniques for obtaining absorption spectra can be used. However, relatively unstable radicals can be stabilised by trapping in a relatively rigid glass by freezing so that their lifetime is prolonged. Striking demonstrations of the increases in lifetime of excited molecules trapped in such frozen lattices are given by Szent-Györgyi (1957). The use of very fast techniques[(1)] for studying the absorption spectra of unstable reaction intermediates has rapidly developed after early studies by Norrish and Porter (1949) with **flash photolysis** in the gas phase. The flash of incident light (of suitable wavelength) is first directed on the reaction mixture, leading to homolytic dissociation. A second flash of light some 10^{-3}–10^{-6} s later enables absorption spectra to be obtained

[(1)] For a short review see Hammes (1966).

of the free-radical intermediates that were formed by the initial photochemical reaction. The development of lasers has enabled an extension of this technique by providing a concentrated coherent transient source of monochromatic radiation. Modifications of the flash-photolysis technique to allow the use of high-energy radiation **(pulse radiolysis)** have also been developed. For example, Land and Swallow (1969) have studied the reduction of riboflavin to the semiquinone by high-energy pulsed *electrons* (0·5 or 5 μs pulses of 8–14 MeV).

The importance of **lipid peroxidation** in tissue disturbances has already been outlined in Section 3.3.2, where the mechanism underlying peroxidation and the subsequent breakdown to malonaldehyde was described. It is evident that the process of lipid peroxidation in tissue samples containing unsaturated lipid bathed in an oxygen-rich medium is potentially of common occurrence. Moreover, lipid peroxidation has been suggested to be of major importance in the development of several types of tissue injury so that methods for the estimation and detection of lipid peroxidation have been the subject of considerable research. The occurrence and estimation of lipid peroxidation has been tackled in various ways: (a) by the measurement of the steady-state concentration of the *lipid peroxide* itself; (b) by the measurements of the *uptake of oxygen* associated with peroxidation and by the *decrease* in the content of *unsaturated fatty acid*; (c) by the measurement of a decomposition product of lipid peroxide that reacts with thiobarbituric acid to yield a red colour, and which is generally considered to be *malonaldehyde*; (d) by the measurement of *diene conjugation*. These procedures will be examined in turn.

4.2.1 Determination of lipid peroxides

The methods used are generally based on the reaction of the lipid peroxide with iodide in acid solution, which results in the liberation of iodine; the iodine is titrated with thiosulphate. There are many difficulties in such an estimation, particularly with biological material, since artefactual peroxidation can occur during the isolation of the lipid peroxide unless strict anaerobic precautions are observed. With such strict precautions the steady-state concentration of lipid peroxides in normal tissue is generally very low[(2)]. Problems associated with the measurement of lipid peroxides in biological material are discussed by Bunyan *et al.* (1967). Advantage can be taken of the ability of EDTA to prevent a metal-catalysed non-enzymic lipid peroxidation going on in tissue extracts during the extraction procedures (see Ghoshal and Recknagel, 1965). Methods used to determine the iodine liberated have included micro-titration, reaction with cadmium acetate and spectrophotometry of the

[(2)] This does not mean that lipid peroxides are not produced in significant amounts *in vivo* but that the steady-state concentration is low. Lipid peroxides readily undergo metabolism in tissue suspensions, as will be discussed in Section 4.2.3.

product (Svoboda and Lea, 1958) and an amperometric method (Oette *et al.*, 1963). For a comprehensive review of methods of determination of lipid peroxides see Johnson and Siddiqi (1970).

4.2.2 Measurement of oxygen uptake and loss of unsaturated fatty acid

The oxygen uptake coupled to peroxidation can be readily followed by using an oxygen electrode (Hochstein and Ernster, 1963). This is a relatively sensitive procedure but, of course, in biological systems oxygen uptake is associated with a variety of enzymic steps and it is important to establish the contribution due to lipid peroxidation itself. An example of this technique is shown in Figure 4.3, where the stimulation in oxygen uptake by liver microsomes on addition of ferrous ions and ADP is dramatic. This stimulation in oxygen uptake is accompanied by an increased production of malonaldehyde and by a decreased content of polyunsaturated fatty acids (see Hochstein and Ernster, 1963; May and McCay, 1968). The loss of unsaturated polyenoic fatty acids can be followed by gas chromatography (May and McCay, 1968).

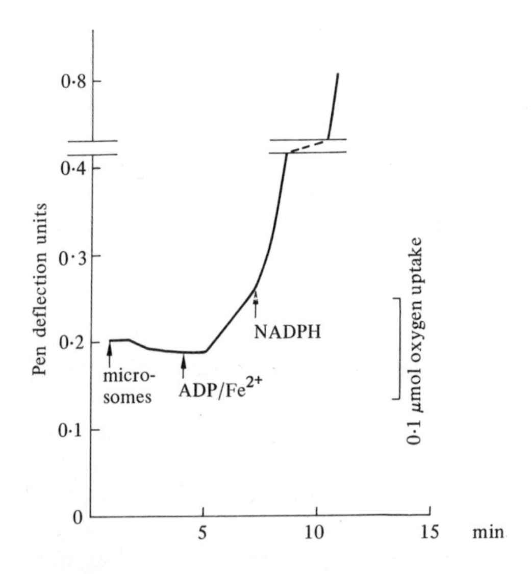

Figure 4.3. Oxygen uptake by a liver microsome suspension. The increased rate in the presence of NADPH (0·36 μmol), ADP (2·3 μmol) and $FeSO_4$ (4 μg), final volume 2·77 ml, can be clearly seen. For other details see Slater (1968).

4.2.3 Thiobarbituric acid reaction

This reaction is concerned with the formation of a coloured product when malonaldehyde is heated in the presence of thiobarbituric acid. The structures involved are shown in Figure 4.4. Since this reaction is used extensively in studies of lipid peroxidation in biological material, and results obtained by its use are involved in much of the discussion in Chapter 10, it will be described in some detail.

Dox and Plaisance (1916) showed that aromatic aldehydes gave a red colour on heating with thiobarbituric acid. In 1944 Kuhn and Liversedge found that the aerobic incubation of brain suspensions (and some other tissues) produced a material that on heating with thiobarbituric acid gave a red colour similar to that reported by Dox and Plaisance; the reactive material was derived from the lipid fraction of the tissue, as shown by Bernheim *et al.* (1948). Since the reaction of the material with thiobarbituric acid was decreased on the addition to the mixture of phenylhydrazine or semicarbazide it was considered that the reactive material contained a carbonyl group.

Wilbur *et al.* (1949) tested a number of aliphatic aldehydes and found that glyoxylic acid gave a red colour on heating with thiobarbituric acid but the maximum absorption of the colour was at 525–550 nm rather than at 535 nm as was found with tissue suspensions by Kuhn and Liversedge (1944). Patton and Kurtz (1951) tested a number of other materials and found that malonaldehyde produced a colour on heating with thiobarbituric acid (Figure 4.5) identical with that obtained from oxidised milk fat or from oxidised methyl linoleate. Sidwell *et al.* (1954) investigated the use of the thiobarbituric acid reaction for following the oxidation of fats in foodstuffs and found that in general the thiobarbituric acid index (i.e. a measure of the amount of material present that reacts with thiobarbituric acid to give a red colour) was proportional to the concentration of lipid peroxide and to the carbonyl

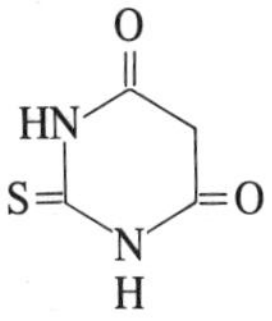

(a) Thiobarbituric acid

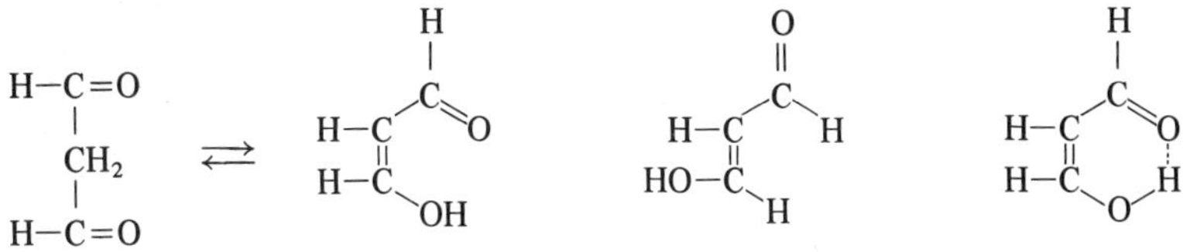

I II III IV

(b) Malonaldehyde

Figure 4.4. Structures of (a) thiobarbituric acid and (b) malonaldehyde. The keto form of malonaldehyde (I) is in equilibrium with the enol forms (S-*cis*, II; S-*trans*, III) of β-hydroxyacrolein. Structure (IV) is a chelated ring form (see Kwon *et al.*, 1964).

content of the tissue. However, when the storage temperature was increased the thiobarbituric acid index increased faster than peroxide concentrations. Tarladgis and Watts (1960) correlated the oxygen consumption of autoxidising fatty acid suspensions with the thiobarbituric acid reaction; linolenic acid and arachidonic acid produced more thiobarbituric acid colour in a given period of incubation than did linolenic acid. The correlation between the thiobarbituric acid index and peroxide production was confirmed by Kenaston *et al.* (1955) with purified samples of methyl linolenate and methyl linoleate, ultraviolet irradiation being used as the stimulus for autoxidation; the thiobarbituric acid index was found to be proportional to the amount of peroxide and to the conjugated diene content as measured by absorption at 233 nm.

Sinnhuber *et al.* (1958) isolated the substance responsible for the red colour formed from the reaction between thiobarbituric acid and a rancid oil suspension and found that it was similar to that produced by heating thiobarbituric acid with pure malonaldehyde. They suggested that the reaction involved the condensation of two molecules of thiobarbituric acid with one molecule of malonaldehyde. Bieri and Anderson (1960) found that homogenates of various tissues when incubated *in vitro* gave a thiobarbituric acid colour that was decreased in extent after predosing the rats with DL-α-tocopheryl acetate; this effect was found in liver, heart and muscle suspensions, but not in brain or kidney.

Tarladgis *et al.* (1962) studied the reaction variables of the thiobarbituric acid–malonaldehyde reaction and found that the reaction product gave

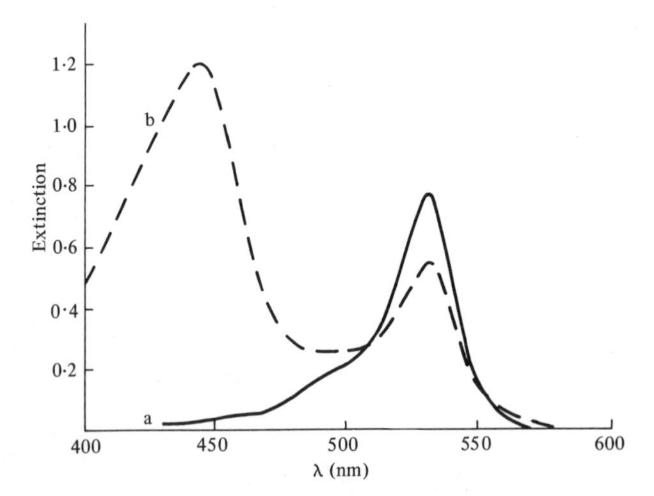

Figure 4.5. Absorption spectra of (a) the coloured product formed by heating malonaldehyde with a thiobarbituric acid–trichloroacetic acid mixture, and (b) the product formed by heating sucrose with a thiobarbituric acid–trichloroacetic acid mixture. The absorption maximum for (a) is 531 nm. Final concentrations of malonaldehyde and sucrose were approximately 1 mM and 62·5 mM respectively.

several spots on paper chromatography. This was confirmed by work of Saslow *et al.* (1963; 1966), who performed ultraviolet photolysis of linolenic acid suspensions and then carried out countercurrent extraction of the reaction product after heating with thiobarbituric acid. They obtained several fractions, indicating a complex reaction. Franz and Cole (1962) irradiated methyl linolenate suspensions and found that the product on heating with thiobarbituric acid gave several spots on thin-layer chromatography.

The effects of inorganic metal ions on the reaction between thiobarbituric acid and tissue aldehydes has been studied by Wills (1964; 1965), McKnight and Hunter (1965) and by Schoenmaker and Tarladgis (1966). *High* concentrations of inorganic iron were found to stimulate the reaction considerably. This work is relevant to the observation of Claisen (1903) that ferric chloride produces a coloured product on mixing with malonaldehyde *in vitro*.

It can be seen from the above discussion that the material produced by aerobic incubation of tissue suspensions is related to, if not identical with, malonaldehyde. The reaction with thiobarbituric acid is, however, complex and a number of coloured products are formed. Nevertheless, in the situations that have been studied at constant temperature the material giving a positive reaction with thiobarbituric acid[(3)] showed a constant relationship to lipid peroxide, to oxygen uptake and to diene conjugation. Under other conditions where the incubation temperature is varied such a relationship no longer holds[(4)]. It may be concluded that the thiobarbituric acid reaction is directly related to lipid peroxidation in certain well-defined circumstances but that its application to other situations requires careful preliminary investigation. Malonaldehyde is a reactive molecule that can cause cross-linking of protein molecules and disturbances of enzyme properties *in vivo* and *in vitro* (Kwon *et al.*, 1965; Crawford *et al.*, 1967). The chemistry of malonaldehyde is discussed by Kwon and Watts (1964). Malonaldehyde is rapidly metabolised *in vivo* (Placer *et al.*, 1965) and by tissue suspensions *in vitro*. Recknagel and Ghoshal (1966) have shown that rat liver mitochondria will metabolise malonaldehyde *in vitro* in the presence of ATP and magnesium ions. This reaction has been re-investigated by Horton and Packer (1970), who have found that liver tissue contains a mitochondrial aldehyde oxidase of low specificity that metabolises malonaldehyde, methylglyoxal, butyraldehyde and several other low molecular weight aldehydes. The rapid metabolism of these aldehydes is of relevance to the views of Szent-Györgyi (1965) that keto-aldehydes of low molecular weight play an important role in the control of cell division. Further, the ready

[(3)] *Malonaldehyde* will be used synonymously with *'material(s) giving a positive reaction with thiobarbituric acid'* throughout the rest of this book.

[(4)] See Table 4.1 and discussion at the end of this Section.

metabolism of malonaldehyde by liver mitochondria probably explains the difficulty that many investigators have had in showing that an increased concentration of malonaldehyde is present in various types of liver injury supposedly associated with an increased lipid peroxidation.

In connection with the metabolism of malonaldehyde mentioned above it is important to note that lipid peroxides themselves can be readily metabolised by tissue suspensions at 37°C. If a freshly prepared liver homogenate in ice-cold medium is immediately analysed for lipid peroxides by an iodometric procedure (Section 4.2.1) then a relatively high content is found (Table 4.1); the malonaldehyde concentration, however, is quite low in such suspensions. If the homogenate is warmed to 37°C then the content of lipid peroxide decreases and the concentration of malonaldehyde rises (Table 4.1). These results suggest that the lipid peroxide is metabolised by the liver homogenate and this is accompanied by a production of malonaldehyde. Further, if a lipid peroxide (for example lauryl peroxide) is added to a liver suspension at 37°C then the peroxide grouping can be shown to be destroyed very quickly. The mechanism of lipid peroxide metabolism has been studied by Christopherson (1968, 1969) and O'Brien and Little (1969).

Christopherson (1968) showed that linoleic hydroperoxide is metabolised by a rat liver enzyme system, located in the soluble compartment, and involving the oxidation of glutathione. The products were oxidised glutathione and a hydroxy-dienoic acid. The enzyme involved is glutathione peroxidase; the reaction with the hydroperoxide is analogous with that utilising hydrogen peroxide [Reactions (4.2) and (4.3)].

$$\mathrm{ROOH + 2GSH \longrightarrow R{-}OH + GSSG + H_2O} \tag{4.2}$$

$$\mathrm{H{-}OOH + 2GSH \longrightarrow 2H_2O + GSSG} \tag{4.3}$$

Table 4.1. Relationship between malonaldehyde production and lipid peroxide concentrations in suspensions of liver microsomes plus cell sap. A rat liver homogenate in 0·25 M-sucrose (1:5 w/v) was centrifuged at 11 700 *g* for two separate periods of 10 min at 0°C. The final supernatant was used for analysis. Samples were stored at 0°C or incubated at 37°C with and without the addition of ADP (0·73 mM) and $FeSO_4$ (0·65 mM). Malonaldehyde was determined by the thiobarbituric acid reaction and lipid peroxide by an iodometric procedure modified by Swoboda and Lea (1958). Data were obtained by T. F. Slater and B. C. Sawyer.

Incubation	Malonaldehyde ($\mathrm{nmol\ ml^{-1}}$)	Lipid peroxide ($\mathrm{nmol\ ml^{-1}}$)
60 min, 0°C	2·2	870
60 min, 37°C	21·7	250
60 min, 37°C plus ADP/Fe^{2+}	81·7	720

In 1969 Christopherson extended these observations to linolenic acid hydroperoxide and identified five isomeric hydroxy-trienoic acids. He suggested that the reaction was a general one for fatty acid hydroperoxides, which converted them into easily metabolisable products. In this manner a destructive sequence of lipid peroxidation could be stopped. O'Brien (1969) found that linoleic acid hydroperoxide was decomposed by a radical mechanism in the presence of metal ions, but through a non-radical mechanism to the unsaturated hydroxy acid in the presence of nucleophiles such as ascorbic acid. With tissue suspensions O'Brien and Little (1969) found that reduced glutathione decomposed the linoleic acid hydroperoxide essentially by the nucleophilic pathway, giving products similar to those described by Christopherson. Green and O'Brien (1970) have studied the subcellular localisation of the glutathione peroxidase that is involved in the tissue-catalysed metabolism of fatty acid hydroperoxides. The enzyme occurred mainly in the soluble compartment but some also was localised in the mitochondrial matrix. In the absence of added hydroperoxide, Jocelyn (1970) has found that reduced glutathione is oxidised by a rather complex route involving the soluble enzyme xanthine oxidase [Reaction (4.4)], the soluble enzyme glutathione peroxidase [Reaction (4.6)] and the peroxisomal enzyme urate oxidase [Reaction (4.5)]. The process is dependent on an endogenous source of xanthine or hypoxanthine.

$$\text{Xanthine} + H_2O + O_2 \longrightarrow \text{urate} + H_2O_2 \qquad (4.4)$$

$$\text{Urate} + O_2 \longrightarrow \text{products} \qquad (4.5)$$

$$H_2O_2 + 2\text{GSH} \longrightarrow 2H_2O + \text{GSSG} \qquad (4.6)$$

4.2.4 Diene conjugation

The process of lipid peroxidation is accompanied by a rearrangement of the double bonds in natural unsaturated fatty acids leading to diene conjugation (see Figure 3.7 and Section 3.3.2). This diene conjugation results in an increase in absorption at 233 nm. Spectrophotometric examination at this wavelength of lipid extracts will therefore indicate whether diene conjugation (and, by inference, lipid peroxidation) has occurred.

4.3 Physical methods of detection

Physical methods for the detection of free radicals rely almost exclusively on the presence of a magnetic moment resulting from the unpaired electron. A relatively insensitive method for detection has been devised based on the measurement of **magnetic susceptibility**, since the unpaired electron gives rise to a paramagnetic contribution. This method, however, is liable to give serious errors resulting from contamination of the cavity by ferromagnetic impurities. The use of susceptibility measurements for the detection of free radicals is discussed in detail by Ingram (1958).

Another technique for free-radical detection is based on **mass spectrometry**. In this procedure free radicals entering the mass spectrometer chamber are ionised by electron impact and the sizes of the ions so produced are determined by spectrometer measurements. By far the most investigations on free-radical reactions, however, have been based on **electron spin resonance**. This technique detects the presence of the unpaired electron in the free radical by observing the resonance absorption of the microwave radiation required to reverse the direction of the magnetic dipole.

Each electron in a chemical structure can be imagined to be spinning about its axis and this imparts to it a small intrinsic magnetic moment (a dipole) that is subject to quantum conditions. In the presence of an applied d.c. magnetic field, H, the electron dipoles will line up in the direction of the magnetic field in either a parallel or antiparallel position (Figure 4.6). The antiparallel position involves a higher energy level than the alternative position. The difference in energy between the two possible positions is $g\beta H$, where β is the Bohr magneton, a constant involved in converting angular momentum to magnetic moment; g is a spectroscopic splitting factor that has a value of 2·0023 for a free electron possessing only spin angular momentum. When the unpaired electron interacts appreciably with its surroundings the value of g may depart significantly from 2·0023.

If an electromagnetic field is applied to such an arrangement of orientated electrons, then at a certain frequency, ν, of the incident radiation the energy supplied will be sufficient to cause a transition from the lower to the higher energy states. At this frequency Equation (4.7) holds, where h is Planck's constant.

$$h\nu = g\beta H \qquad (4.7)$$

At these particular values of ν and H there will be absorption of the incident electromagnetic radiation that can be detected electronically. Laboratory facilities and practical advantages normally restrict the value of H to a maximum of approximately 10000 G and this places the required frequency for energy transition into the microwave region of the electromagnetic spectrum. The X-band of radar detection is commonly

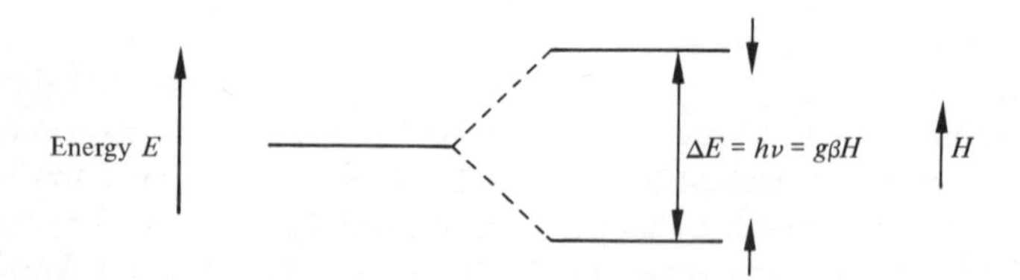

Figure 4.6. Splitting of electronic energy levels by a magnetic field H. The difference in energy between the levels with opposed electron spin is $\Delta E = h\nu = g\beta H$, where ν is the frequency of adsorbed radiation, g is the spectroscopic splitting factor and β is the Bohr magneton. For further details see Section 4.2.4.

used for electron spin resonance measurements on biological samples. X-band detection requires a magnetic field strength of approximately 3·3 kG for *g* 2·0. In general the microwave frequency is kept constant while *H* is varied and the resultant resonance absorption of microwave energy is recorded as a first derivative signal to indicate the presence of unpaired electrons in the material under study (Figure 4.7). The position (e.g. *g* value) and size of the microwave absorption peaks can give information about the type and quantity of free radicals present in the sample.

Because of the reactivity of the free-radical structures in general, and in an effort to lengthen their transitory existence, measurements on biological samples are often done at very low temperatures. This increases the ratio of signal height to noise[(5)] and increases the lifetime of free

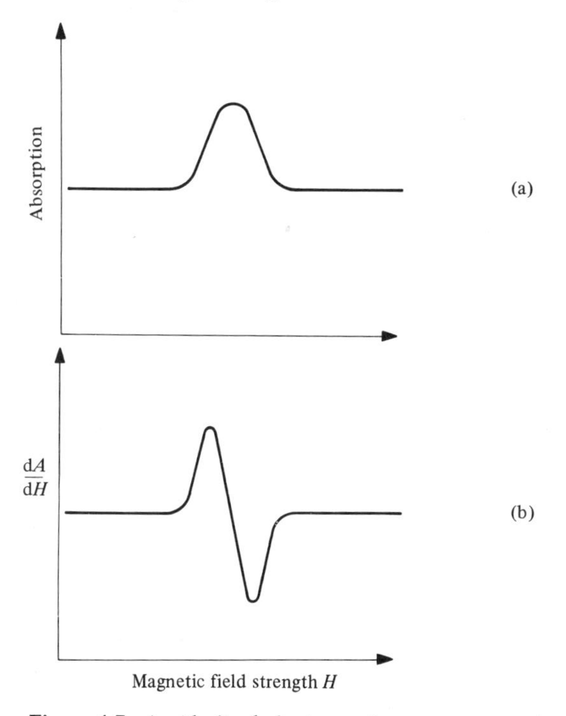

Figure 4.7. An idealised electron spin resonance spectrum showing (a) the variation in the absorption of microwave power by the sample as the magnetic field is varied. In (b) the first derivative of the microwave power absorption is plotted against *H* to give the representation usually depicted for such spectra.

[(5)] The sensitivity of the technique is dependent on the difference in the numbers of molecules (n_1 and n_2) occupying the two energy levels. In most cases the ratio n_1/n_2 is given by $n_1/n_2 = e^{-h\nu/kT}$. It can be seen that the ratio is increased as the temperature (T) is decreased and as the frequency (ν) is increased.

radicals trapped in the frozen lattice. A further increase in sensitivity may be obtained by coupling the spectrometer to a computer that will average the data from repeated scans of a sample; after many successive scans random noise tends to cancel out. The ratio of signal/noise is increased by such a procedure in proportion to $\sqrt{n}$ where n is the number of repetitive scans through the sample. Detailed explanations of the theory and practice of electron spin resonance, particularly in relation to biological studies, can be found in Ingram (1958, 1969) and Wyard (1969).

References

Bernheim, F., Bernheim, M. L. C., Wilbur, K. M., 1948, *J. biol. Chem.,* **174**, 257.
Bieri, J. G., Anderson, A. A., 1960, *Archs. Biochem. Biophys.,* **90**, 105.
Bunyan, J., Murrell, E. A., Green, J., Diplock, A. T., 1967, *Br. J. Nutr.,* **21**, 475.
Christopherson, B. O., 1968, *Biochim. biophys. Acta,* **164**, 35.
Christopherson, B. O., 1969, *Biochim. biophys. Acta,* **176**, 463.
Claisen, L., 1903, *Ber.,* **36**, 3664.
Crawford, D. L., Yu, T. C., Sinnhuber, R. O., 1967, *J. Fd. Sci.,* **32**, 332.
Dox, A. W., Plaisance, G. P., 1916, *J. Am. chem. Soc.,* **38**, 2156.
Franz, J., Cole, B. T., 1962, *Archs. Biochem. Biophys.,* **96**, 382.
Fridovich, I., Handler, P., 1961, *J. biol. Chem.,* **236**, 1836.
Ghoshal, A. K., Recknagel, R. O., 1965, *Life Sc.,* **4**, 2195.
Gilbert, D., Martin, A. D., Gangolli, S. D., Abraham, R., Golberg, L., 1969, *Fd. cosmetic Toxic.,* **7**, 603.
Green, R. C., O'Brien, P. J., 1970, *Biochim. biophys. Acta,* **197**, 31.
Hammes, G. G., 1966, *Science,* **151**, 1507.
Hochstein, P., Ernster, L., 1963, *Biochem. biophys. Res. Commun.,* **12**, 388.
Horton, A. A., Packer, L., 1970, *Biochem. J.,* **116**, 19P.
Ingram, D. J. E., 1958, *Free Radicals as Studied by Electron Spin Resonance* (Butterworths, London).
Ingram, D. J. E., 1969, *Biological and Biochemical Applications of Electron Spin Resonance* (Adam Hilger, London).
Jocelyn, P. C., 1970, *Biochem. J.,* **117**, 951.
Johnson, R. M., Siddiqi, I. W., 1970, *The Determination of Organic Peroxides* (Pergamon Press, Oxford).
Kenaston, C. B., Wilbur, K. M., Ottolenghi, A., Bernheim, F., 1955, *J. Am. Oil Chem. Soc.,* **32**, 33.
Kuhn, H. I., Liversedge, M., 1944, *J. Pharmac. exp. Ther.,* **82**, 292.
Kwon, T.-W., Menzel, D. B., Olcott, H. S., 1965, *J. Fd. Sci.,* **30**, 808.
Kwon, T.-W., Watts, B. M., 1964, *J. Fd. Sci.,* **29**, 294.
Land, E. J., Swallow, A. J., 1969, *Biochemistry, Easton,* **8**, 2117.
May, H. E., McCay, P. B., 1968, *J. biol. Chem.,* **243**, 2296.
McCord, J. M., Fridovich, I., 1969, *J. biol. Chem.,* **244**, 6049, 6056.
McKnight, R. C., Hunter, F. E., 1965, *Biochim. biophys. Acta,* **98**, 640.
Norrish, R. G. W., Porter, G., 1949, *Nature, Lond.,* **164**, 658.
O'Brien, P. J., 1969, *Can. J. Biochem. Physiol.,* **47**, 485.
O'Brien, P. J., Little, C., 1969, *Can. J. Biochem. Physiol.,* **47**, 493.
Oette, K., Peterson, M. L., McAuley, R. L., 1963, *J. Lipid Res.,* **4**, 212.
Paneth, F., Hofeditz, W., 1929, *Ber.,* **62**, 1335.
Patton, S., Kurtz, G. W., 1951, *J. Dairy Sci.,* **34**, 669.
Placer, Z., Veselkova, A., Rath, R., 1965, *Experientia,* **21**, 19.
Recknagel, R. O., Ghoshal, A. K., 1966, *Expl. molec. Path.,* **5**, 108.

Saslow, L. D., Anderson, H. J., Waravdekar, V. S., 1963, *Nature, Lond.,* **200**, 1098.
Saslow, L. D., Corwin, L. M., Waravdekar, V. S., 1966, *Archs. Biochem. Biophys.,* **114**, 61.
Sato, S., Iwaizumi, M., 1969, *Biochim. biophys. Acta,* **172**, 30.
Schoenmaker, A. W., Tarladgis, B. G., 1966, *Nature, Lond.,* **210**, 1153.
Sidwell, C. G., Salwin, H., Benca, M., Mitchell, J. H., 1954, *J. Am. Oil Chem. Soc.,* **31**, 603.
Sinnhuber, R. O., Yu, T. C., 1958, *Fd. Res.,* **23**, 626.
Slater, T. F., 1968, *Biochem. J.,* **106**, 155.
Swoboda, P. A. T., Lea, C. H., 1958, *Chem. Ind.,* 1090.
Szent-Györgyi, A., 1957, *Bioenergetics* (Academic Press, New York).
Szent-Györgyi, A., 1965, *Science,* **149**, 34.
Tarladgis, B. G., Pearson, A. M., Dugan, L. R., 1962, *J. Am. Oil Chem. Soc.,* **39**, 34.
Tarladgis, B. G., Watts, B. M., 1960, *J. Am. Oil Chem. Soc.,* **37**, 403.
Torrielli, M. V., Slater, T. F., 1971, *Biochem. Pharmac.,* **20**, 2027.
Wilbur, K. M., Bernheim, F., Shapiro, O. W., 1949, *Archs. Biochem. Biophys.,* **24**, 305.
Wills, E. D., 1964, *Biochim. biophys. Acta,* **84**, 475.
Wills, E. D., 1965, *Biochim. biophys. Acta,* **98**, 238.
Wyard, S. J., 1969, *Solid State Biophysics* (McGraw-Hill, London), p.3.

5

Free-radical scavengers

5.1 Introduction

The damaging effects that free radicals may exert on biological membranes have been mentioned already; other deleterious actions of free radicals on tissue components are dealt with in Chapter 7. In view of such potent biological activity, it is not surprising to find that cells contain a number of substances that are capable of removing (i.e. scavenging) free radicals normally produced in the cells' environment. It is only when these natural defence mechanisms are overwhelmed that significant cellular disturbance results from free-radical processes.

In this Chapter we shall examine very briefly the main classes of free radical scavenger that occur naturally, or that are ingested frequently in food, or occasionally as drugs. Free-radical scavengers are closely related to the antioxidants[(1)], and in biological systems involving unsaturated lipids and oxygen the two terms are, to some extent, synonymous. There is a voluminous literature on the chemistry and industrial applications of antioxidants and obviously this Chapter can serve only as a very general background to a complex area. For general reviews of antioxidants and autoxidation the reader is referred to Lundberg (1961).

5.2 Naturally occurring free-radical scavengers

There is a large variety of substances that occur naturally in animal tissues and which may function as free-radical scavengers or antioxidants under certain experimental conditions, particularly when used in high concentration. The number of endogenous materials that can function as antioxidants and free-radical scavengers at *physiological* concentrations, however, is certainly much more restricted. Such materials may be subdivided into (a) water-soluble materials, (b) fat-soluble materials and (c) substances bound to materials of high molecular weight.

5.2.1 Water-soluble scavengers

The main substances that may be included under this heading in animal tissues are ascorbate (vitamin C), glutathione, ergothioneine and purine

[(1)] In the broadest sense, antioxidants are substances that inhibit the oxidative deterioration of a wide variety of materials (for example, of unsaturated fats in food). Such reactions involve free-radical stages in which a radical initiation step is followed by a chain mechanism. Antioxidants may inhibit the production of the free-radical initiator (for example, by chelation of an essential trace metal) or they may decrease the efficiency of chain propagation. In the latter event they act as radical scavengers by lowering the concentration of the reactive free-radical intermediates. Antioxidants may thus include substances that are not free-radical scavengers (for example, inhibitors of free-radical initiation stages); conversely, free-radical scavengers may not act in a particular system as an antioxidant (for example, the system may be a non-oxidative one).

bases or related materials. The relative efficiencies of these different scavengers in mammalian blood and in liver have been assessed by Glavind and Faber (1966); their data on tissue concentrations are summarised in Table 5.1.

Table 5.1. Concentrations of various antioxidants in rat liver and spleen. Results for ergothioneine are from Melville *et al.* (1954); other data are from Glavind and Faber (1966). The values in parentheses were obtained after exposing rats to 3×10^6 rd.

	Concentration (μequiv./g of tissue)	
	Liver	Spleen
Uric acid		0·4 (0·1)
Glutathione	8·1 (4·2)	3·9 (2·1)
Ascorbic acid	2·6 (0·7)	3·5 (1·5)
Ergothioneine	0·6	0·05

Glavind and Faber's results were obtained with acid-soluble extracts of rat liver and spleen; it can be seen that the major acid-soluble antioxidants in liver are glutathione and ascorbic acid. Whole animal irradiation produced a marked decrease in the concentrations of all of the antioxidants studied.

Vitamin C (**ascorbic acid**, A) was first isolated by Szent-Györgyi in 1927; it is a mild reducing agent and readily undergoes oxidation to dehydroascorbic acid (DHA) through the intermediate formation of the *monodehydroascorbic acid* radical (MDHA):

$$\mathrm{A} \longrightarrow \mathrm{MDHA}^{\bullet} + \mathrm{H}^{\bullet} \longrightarrow \mathrm{DHA} + 2\mathrm{H}^{\bullet} \qquad (5.1)$$

The properties of monodehydroascorbic acid have been described by Levandoski *et al.* (1964) and by Kluge *et al.* (1967); its structure is shown in Figure 5.1.

The two-step oxidation can be easily demonstrated by titration with the stable free radical diphenylpicrylhydrazyl as reported by Blois (1958), who found that two molecules of diphenylpicrylhydrazyl (DPPH) were utilised per molecule of ascorbic acid [Reactions (5.2) and (5.3)].

$$\begin{cases} \mathrm{DPPH}^{\bullet} + \mathrm{A} \longrightarrow \mathrm{DPPH}_2 + \mathrm{MDHA}^{\bullet} & (5.2) \\ \mathrm{DPPH}^{\bullet} + \mathrm{MDHA}^{\bullet} \longrightarrow \mathrm{DPPH}_2 + \mathrm{DHA} & (5.3) \end{cases}$$

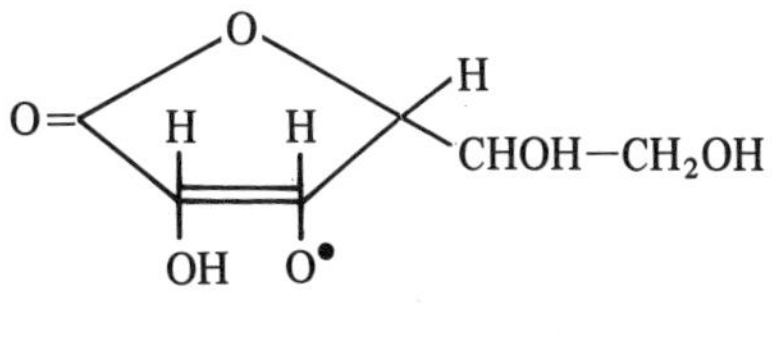

Figure 5.1. Structure of the monodehydroascorbic acid radical.

In the reaction ascorbic acid is acting as a scavenger of the (stable) diphenylpicrylhydrazyl radical. A biological example of the scavenging action of vitamin C was found by Slater and Riley (1966) in connection with photosensitisation: vitamin C inhibits the porphyrin-catalysed oxidative degradation of lysosomal membranes in rat tail epidermis (see Section 14.3.1). Because of its usefulness as a free-radical scavenger ascorbic acid is often added to food preparations to prevent oxidative deterioration.

Ascorbic acid readily oxidises in air and this reaction is increased by the presence of metal ions (e.g. Cu^{2+}, Fe^{2+}). In Section 2.4 it was mentioned that a mixture of ascorbic acid and a metal complex functions as a source of $O_2H^{\bullet}$ radicals. Ascorbic acid also *stimulates* a non-enzymic lipid peroxidation process in mitochondria and in microsomal fractions *in vitro*. This process is accompanied by substantial decreases in the enzyme activities of these fractions (see Nordenbrand *et al.*, 1964) and is strongly inhibited by metal chelators. Under conditions *in vivo* the metal ions that catalyse the lipid peroxidation reaction are presumably in an inaccessible or bound form; preparation of relatively dilute tissue homogenates, however, permits peroxidation to proceed. In this context it is of interest that liver homogenates prepared from scorbutic guinea pigs undergo lipid peroxidation at a slower rate than homogenates from control guinea pigs (Abramson, 1949).

Thiol compounds (including glutathione, ergothioneine and cysteine) are well-known scavengers of free radicals and several are reasonably effective as radiation protectors. **Cysteamine** (Figure 5.2) has been extensively studied in this respect (see Bacq, 1965); although it is a powerful scavenger its action *in vivo* is limited by its very short half-life owing to rapid excretion.

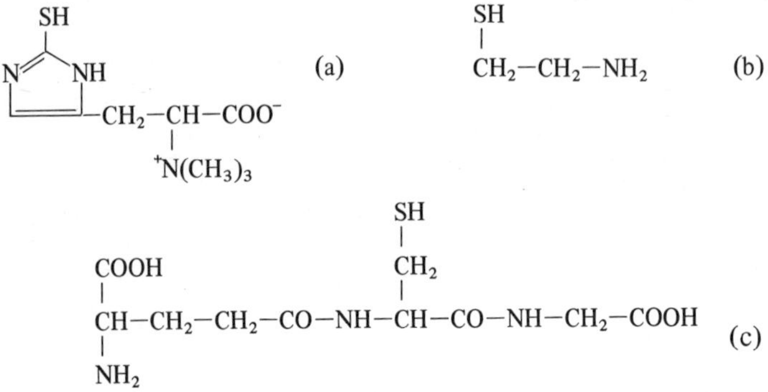

(a) Ergothioneine: imidazole ring (N, NH) bearing SH, with side chain $-CH_2-CH(^{+}N(CH_3)_3)-COO^-$

(b) $HS-CH_2-CH_2-NH_2$

(c) $HOOC-CH(NH_2)-CH_2-CH_2-CO-NH-CH(CH_2SH)-CO-NH-CH_2-COOH$

Figure 5.2. Structures of (a) ergothioneine, (b) cysteamine and (c) glutathione.

Thiol compounds probably inhibit free-radical reactions by a process of hydrogen donation [Reaction (5.4)]:

$$R_3-SH + R_2^{\bullet} \longrightarrow R_3-S^{\bullet} + R_2-H \tag{5.4}$$

and it is important for a scavenging action that the newly formed species $R_3{-}S^\bullet$ does not catalyse the initial step [Reaction (5.5)].

$$R_2{-}X \longrightarrow R_2^\bullet + X^\bullet \tag{5.5}$$

The radical species $R_3{-}S^\bullet$ may often undergo dimerisation to a stable form [Reaction (5.6)]

$$2R_3{-}S^\bullet \longrightarrow R_3{-}S{-}S{-}R_3 \tag{5.6}$$

thus effectively terminating the chain.

A mechanism for radical scavenging of hydroxyl radicals by a disulphide has been discussed by Foye (1969) [Reactions (5.7)-(5.9)].

$$R{-}S{-}S{-}R + OH^\bullet \longrightarrow R{-}\underset{\displaystyle OH}{\overset{\bullet}{\underset{|}{S}}}{-}S{-}R \tag{5.7}$$

$$R{-}\underset{\displaystyle OH}{\overset{\bullet}{\underset{|}{S}}}{-}S{-}R + O_2 \longrightarrow R{-}\underset{\displaystyle OH}{\underset{|}{\overset{\displaystyle O_2^\bullet}{\overset{|}{S}}}}{-}S{-}R \tag{5.8}$$

$$R{-}\underset{\displaystyle OH}{\underset{|}{\overset{\displaystyle O_2^\bullet}{\overset{|}{S}}}}{-}S{-}R \longrightarrow R{-}SO_3H + RS^\bullet \tag{5.9}$$

Glutathione (Figure 5.2) occurs widely in animal tissues but clear evidence for its intracellular role has proved difficult to obtain. However, Kosower and Kosower (1969) have devised a method that gives results indicating that "glutathione is an essential cellular constituent, absolutely required for maintenance and survival". Their conclusion is based on the effects of a group of thiol-oxidising agents which can convert glutathione (GSH) into oxidised glutathione (GSSG) and at the same time produce harmful free radicals. The thiol-oxidisers are derivatives of **diazine carboxylic acids** (Figure 5.3) and react with GSH as shown in Reactions (5.10)-(5.14).

$$2GSH + C_6H_5N{=}NCOOCH_3 \longrightarrow GSSG + C_6H_5NHNHCOOCH_3 \tag{5.10}$$

$$C_6H_5N{=}N{-}COOCH_3 + H_2O \longrightarrow CH_3OH + C_6H_5{-}N{=}NCOOH \tag{5.11}$$

$$C_6H_5N{=}NCOOH \longrightarrow C_6H_5N{=}NH + CO_2 \tag{5.12}$$

$$C_6H_5N{=}NH + OH^- \longrightarrow C_6H_5^\bullet + H_2O + N_2 \tag{5.13}$$

$$C_6H_5N{=}NH + O_2 \longrightarrow C_6H_5^\bullet + HO_2^\bullet + N_2 \tag{5.14}$$

Kosower and Kosower found that erythrocytes suspended in oxygen-rich buffer and exposed to azo-ester suffer *intracellular* damage with Heinz-body formation. When the cells are treated with azo-ester in the absence of intracellular oxygen (i.e. pretreated with carbon monoxide) then *cell membrane* damage is produced with subsequent lysis. GSH appears to be concerned with the stabilisation of cell membranes against free-radical attack. In this respect it is interesting that O'Brien and Little (1969) have found that fatty acid hydroperoxides are decomposed by GSH in the presence of glutathione peroxidase, probably via a non-radical mechanism that requires NADPH for regeneration of the GSH from GSSG. Glutathione is not, however, very effective in protecting animals against high doses of ionising radiation (see Hope, 1959) in which the production of very reactive radicals of the hydroxyl type overcomes the normal body defence mechanisms.

		GSH reaction	Hydrolytic stability	Free-radical formation
(a)	$C_6H_5-N=N-C(=O)-O-CH_3$	10 s	1200 s	++
(b)	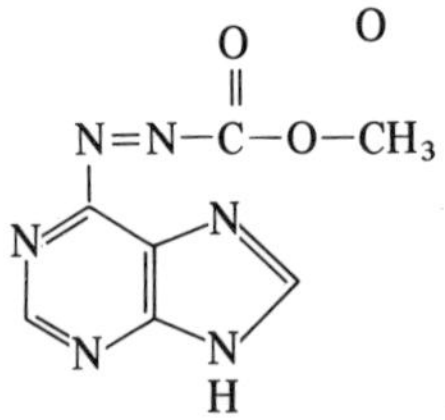 (purinyl)$-N=N-C(=O)-O-CH_3$	2 s	70 s	—

Figure 5.3. Properties of the thiol-oxidising agents described by Kosower and Kosower (1969). The trivial names for substances (a) and (b) are azo-ester and adenine ester respectively. The glutathione reaction time is a measure of the half-life of the reagent in the presence of glutathione at pH 7·4, both reactants 1 mM. The hydrolytic stability is the half-life of the hydrolysis in aqueous buffer at pH 7·4. For other details see Kosower and Kosower (1969).

The role of **purine bases**, nucleosides and nucleotides as antioxidants has been investigated by several groups of workers: for example, Matsushita *et al.* (1963) and Melzer and Tomlinson (1966). The effects of a number of bases, nucleosides and nucleotides on the autoxidation of linoleic acid at pH 7 and pH 9 were studied by Matsushita and colleagues; an abbreviated summary of their data is given in Table 5.2. They found that variation in the concentration of some materials in the incubation medium altered the observed effect from a *pro*-oxidant to an *anti*-oxidant effect. Figure 5.4 and Table 5.2 illustrate this statement for α-tocopherol, uric acid, adenine, hypoxanthine and ribonucleic acid. Even in the presence of added metal ions there was a pronounced antioxidant effect by purine bases, of which uric acid was particularly effective. This behaviour contrasted with that shown by α-tocopherol, which was ineffective as an antioxidant in the presence of added copper ions.

It is known that the pyrimidine deoxynucleotides are more susceptible to peroxidative attack by hydrogen peroxide than are the corresponding purine bases. However, in DNA the difference in susceptibility found *in vitro* is not so apparent. Melzer and Tomlinson (1966) found that mixtures of bases are oxidised to a lesser extent than are the individual components: the purine bases in the DNA inhibit the peroxidative destruction of pyrimidine bases by hydrogen peroxide. In fact, in this context, Machover and Duchesne (1965) have suggested that one

Table 5.2. Effects of various antioxidants *in vitro* on the autoxidation of linoleic acid at pH 9·0 for 48 h at 37°C (data of Matsushita *et al.*, 1963). At the end of the incubations, the malonaldehyde content was determined by the thiobarbituric acid reaction. Values are given as percentages of that for the control incubation in the absence of added antioxidant.

Antioxidant	Concentration (mM)			
	5	1	0·25	0·1
Control (100%)				
α-Tocopherol	59	53	146	135
Uric acid	40	20	16	16
Adenine	36	36	63	72
Hypoxanthine	30	25	60	72
RNA[a]	34	28	43	55

[a] RNA concentration of 2 mg ml^{-1} taken as 5 mM.

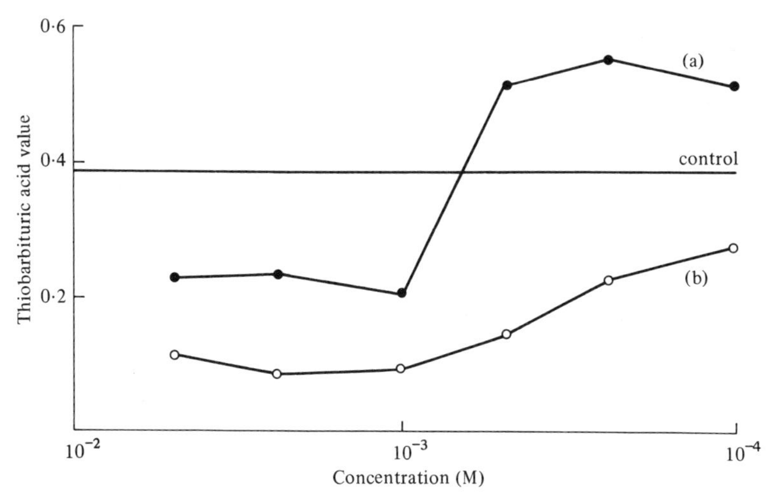

Figure 5.4. Effect of different concentrations of (a) α-tocopherol or (b) hypoxanthine on the malonaldehyde production from autoxidising linoleic acid (48 h at 37°C). Data are from Matsushita *et al.* (1963).

advantage of the Watson–Crick pairing in the DNA helix is that a purine base is always in a suitable position to inhibit oxidative damage to the neighbouring pyrimidine.

In view of the marked antioxidant activity of purine bases it is noteworthy that tissues contain a wide variety of such compounds in the acid-soluble fraction. Table 5.3 gives some data for rat liver acid-soluble nucleotides, which, as well as carrying out specialised coenzyme activities, may function also as internal free-radical scavengers. The nucleotides occur to some extent in association with intracellular structures, such as the mitochondria, and some of this nucleotide-pool is tightly bound and removed only with difficulty. An interesting finding in this context is that of Siekevitz (1955), who reported that rat liver *endoplasmic reticulum* contained a high concentration of **inosine** (0·5–1·0 μmol/10 g wet wt. of liver). This relatively high concentration suggests that inosine has a role in protecting the endoplasmic reticulum (or a particular part of the endoplasmic reticulum) against peroxidative attack. In fact, as will be discussed in Section 10.8.2, inosine is decreased in concentration during the metabolism *in vivo* of carbon tetrachloride to free-radical products by the endoplasmic reticulum. Figure 5.5 shows the metabolic relationships between inosine and adenine nucleotides.

A number of other naturally occurring materials have been shown to have antioxidant activity *in vitro* but, in general, it seems likely that the concentrations of these compounds *in vivo* are too low for significant activity of this nature. For example, 10^{-5} M-thyroxine has been reported to protect vitamin E-deficient rats against dialuric acid-induced haemolysis (Bunyan *et al.*, 1961). Thyroxine (10^{-5} M) was also found to be effective *in vitro* against erythrocyte haemolysis and against lipid peroxidation in liver homogenates prepared from vitamin E-deficient rats. The concentration of thyroxine in rat body water (approximately 5×10^{-8} M) is, however, considerably less than that used in the above-mentioned investigations. Cash *et al.* (1966) have shown that the antioxidant action

Table 5.3. Acid-soluble nucleotides of rat liver. The nucleotides were extracted into cold perchloric acid and then separated by Dowex-1 chromatography. Values are from Wang *et al.* (1963) except for NAD^+ and $NADP^+$, which are from Slater *et al.* (1964) who used direct enzymic assays on liver extracts.

Nucleotide	Normal liver (μmol/g of wet liver)	Nucleotide	Normal liver (μmol/g of wet liver)
AMP	0·77	GDP	0·03
ADP	0·79	UMP	0·04
ATP	0·93	NAD^+	0·50
GMP	0·01	$NADP^+$	0·04

of thyroxine is complicated by the presence of metal ions. The hormone protects *in vitro* against lipid peroxidation and mitochondrial swelling produced by ferrous ions, but *potentiates* swelling and lipid peroxidation in the presence of very low concentrations of calcium ions. The concentration of calcium ions required for this effect is so low that contamination of many commonly used reagents by calcium would satisfy the requirement.

In addition to the water-soluble antioxidants mentioned above there are also a number of similarly effective substances found in plants that may be ingested in food; these include gallic acid, eugenol and **nordihydroguarietic acid** (NDHG). The last-named material has found wide use in the food industry as an antioxidant in lards, oils, baking mixes, etc. commonly at 0·01–0·02% concentration. Its formula is shown in Figure 5.6. Nordihydroguarietic acid has a low toxicity *in vivo* (LD_{50} 800 mg kg^{-1} in mice by parenteral administration) and has been reported by Burk and Woods (1963) to inhibit anaerobic glycolysis in tumour cells *in vitro* at concentrations of approximately 100 μM.

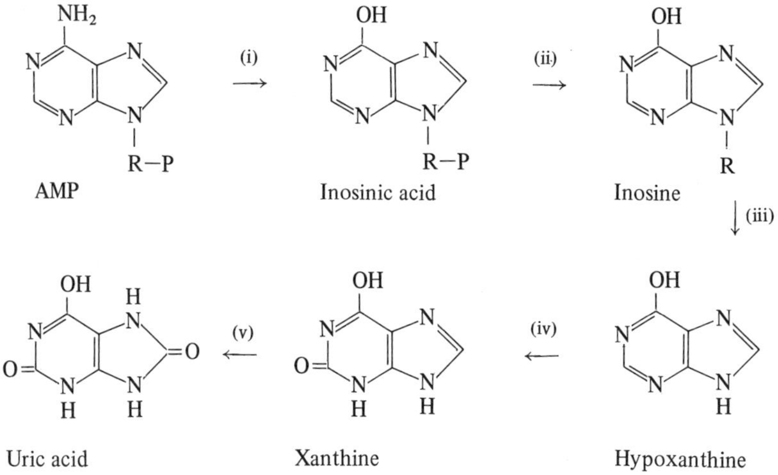

Figure 5.5. Metabolic transformations of purine nucleotides, nucleosides and free bases. Abbreviations: R, β-D-ribose; P, phosphate.

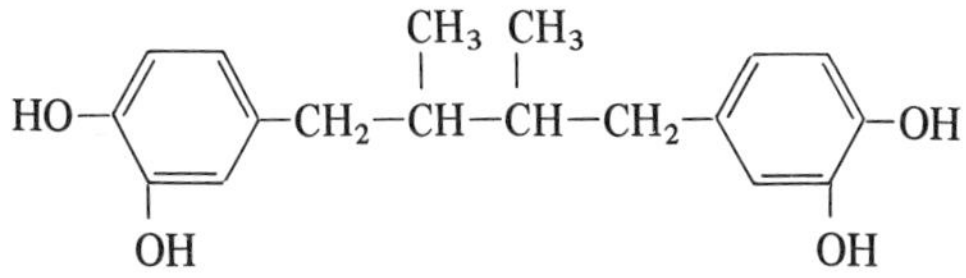

Figure 5.6. Structure of nordihydroguaiaretic acid (NDGA).

5.2.2 Fat-soluble scavengers

The main animal fat-soluble scavengers are vitamin E, vitamin A and carotenes, and ubiquinones (see Figure 5.7). The literature on the biological activity of vitamin E is voluminous (see: *Annotated Bibliography of Vitamin E*, National Vitamin Foundation Inc., New York). Reviews of interest are by Tappel (1962), Roels (1967) and Draper and Csallany (1969).

There has been no difficulty in demonstrating an antioxidant action for **vitamin E** in many diverse conditions *in vitro*. For example, it readily inhibits lipid peroxidation in microsomal suspensions (Hochstein and Ernster, 1963). During such an interaction the vitamin is oxidised to the tocopherolquinone (Figure 5.8). It has been very difficult, however, to

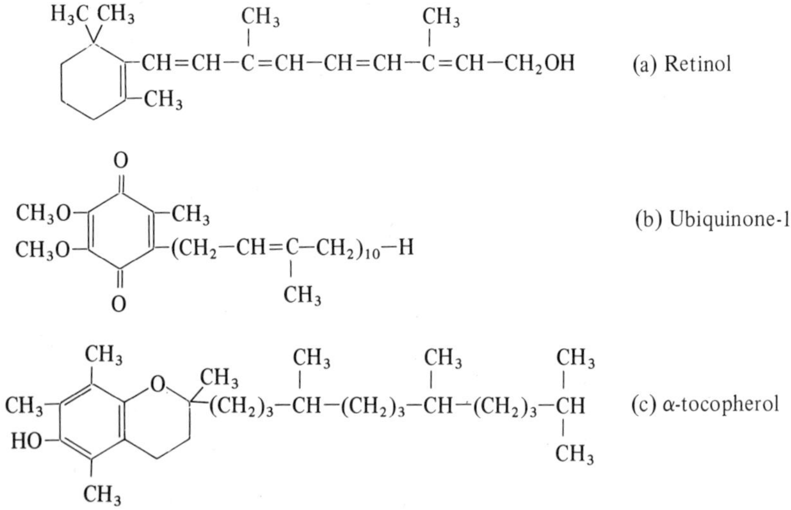

H_3C CH_3 CH_3 CH_3
$-CH{=}CH{-}C{=}CH{-}CH{=}CH{-}C{=}CH{-}CH_2OH$
$-CH_3$
(a) Retinol

O
CH_3O- $-CH_3$
CH_3O- $-(CH_2{-}CH{=}C{-}CH_2)_{10}{-}H$
O CH_3
(b) Ubiquinone-10

CH_3 CH_3 CH_3 CH_3
O CH_3
CH_3- $-(CH_2)_3{-}CH{-}(CH_2)_3{-}CH{-}(CH_2)_3{-}CH$
HO- CH_3
CH_3
(c) α-tocopherol

Figure 5.7. Structures of (a) vitamin A, retinol, (b) ubiquinone-10, (c) vitamin E, α-tocopherol.

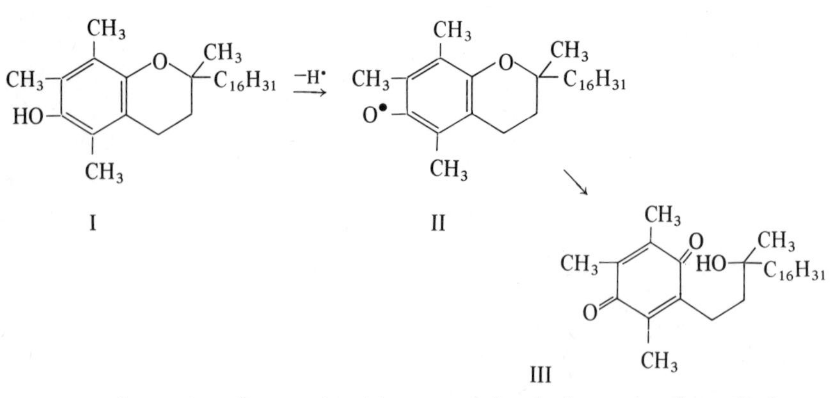

CH_3 O CH_3 $-C_{16}H_{31}$ $\xrightarrow{-H^\bullet}$ CH_3 O CH_3 $-C_{16}H_{31}$
CH_3- HO- CH_3 CH_3- $O^\bullet$ CH_3
I II

CH_3 O CH_3
CH_3- HO- $C_{16}H_{31}$
O CH_3
III

Figure 5.8. Conversion of α-tocopherol into α-tocopherolquinone via a free-radical intermediate (see Tappel, 1962).

demonstrate unequivocally a similar role for vitamin E under conditions *in vivo* where lipid peroxidation is suspected to occur (e.g. in certain types of acute liver injury). This problem is discussed in recent contributions from Green's laboratory (Green *et al.*, 1969; Bunyan *et al.*, 1969). In a few examples *in vivo*, a deficiency of vitamin E has been shown to be associated with increased peroxidation: in erythrocytes subjected to hyperoxia (Mengel and Kann, 1966) and in adipose tissue and adrenals. Such examples are discussed in Sections 12.3 and 12.7.2.

Ubiquinone is a component of the mitochondrial respiratory chain in rat liver although it is difficult to decide whether it functions on the main electron-transport chain or on a side route, perhaps involving oxidative phosphorylation, for which several mechanisms involving quinone phosphate intermediates have been suggested. Like many quinones (see Section 5.2.1) it can function as an antioxidant and Di Luzio (1967) has used ubiquinol-4 as a protective agent against ethanol-induced fatty liver (Chapter 12). Under conditions of low oxygen partial pressure equivalent to those existing in tissues, Mellors and Tappel (1966) found that ubiquinol-6 was as effective as α-tocopherol in inhibiting several free-radical motivated reactions.

Carotenes have been shown to participate in free-radical reactions; Hove (1953) showed that β-carotene was destroyed on incubation with linoleate hydroperoxide. The breakdown was accelerated by low concentrations of carbon tetrachloride and rather less by chloroform. Forbes and Taliaferro (1945) found that a diet rich in carrots protected rats against liver injury due to carbon tetrachloride which, as discussed in Chapter 10, is most probably mediated through a free-radical mechanism. Mathews (1964) reported that β-carotene was effective in protecting mice against an otherwise lethal photosensitisation due to haematoporphyrin; porphyrin photosensitisation is believed to involve a free-radical attack on lysosomal membranes in the epidermis (see Section 14.3.1). β-Carotene is known to form free radicals on irradiation with visible light (Smaller, 1960), and the light-induced changes in the biologically active retinal in the retina are also believed to be free radical in nature (Grady and Borg, 1968).

5.2.3 Antioxidants bound to material of high molecular weight

Here it is only necessary to draw attention to the widespread occurrence of free thiol groups in proteins that may function in a scavenging role; the antioxidative effect of high molecular weight RNA has already been mentioned in Section 5.1.

5.3 Exogenous antioxidants

The main sources of exogenous antioxidants are foodstuffs and therapeutic drugs.

5.3.1 Food antioxidants

It is common to find that tinned foods have low concentrations of antioxidants such as butylated hydroxytoluene (BHT), propyl gallate or ethoxyquin etc. (see Figure 5.9). As already mentioned, nordihydroguarietic acid is added widely to lards and similar products. Detailed regulations have been devised to restrict the administration of harmful quantities of antioxidants to foodstuff; in England, for example, see the *Statutory Instruments, Antioxidants in Food Regulation,* 1966, number 1500, *Food and Drugs Regulations* (HMSO). The concentrations permitted are low; for example, with ethoxyquin in apples and pears, the maximum permitted concentration is 3 p.p.m. The toxic manifestations arising from the ingestion of very large amounts of antioxidants will be discussed briefly in Section 12.7.1; the speculations regarding the beneficial effect of lower doses of antioxidants on generalised ageing phenomena are reviewed in Section 12.6. The mechanism of antioxidant action of the phenolic antioxidants is discussed by Stockey (1962), who also reviews the synergistic action of certain plant acids such as citric acid on antioxidant effectiveness. Phenols (and amines) act as antioxidants by hydrogen donation [Reaction (5.15)].

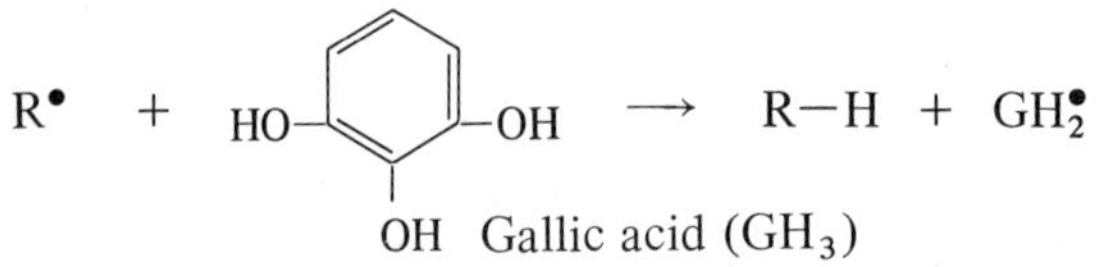

$$R^{\bullet} + \text{HO–C}_6\text{H}_2\text{(OH)–OH} \longrightarrow \text{R–H} + \text{GH}_2^{\bullet} \qquad (5.15)$$

Gallic acid (GH_3)

In Reaction (5.15) the effectiveness of the antioxidant is dependent on the newly formed radical ($GH_2^{\bullet}$) being inactive in chain propagation or being destroyed by side reactions. An example of the inhibition of a radical reaction through hydrogen donation producing a very stable free radical is that shown in Reaction (5.16), involving diphenylpicrylhydrazine ($DPPH_2$):

$$R^{\bullet} + DPPH_2 \longrightarrow R\text{–}H + DPPH^{\bullet} \qquad (5.16)$$

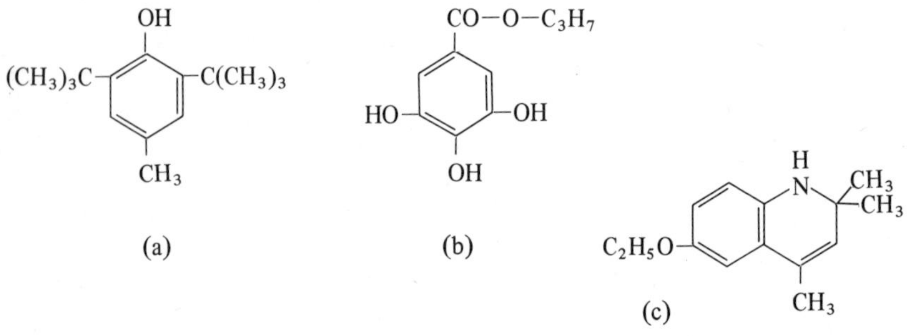

OH, $(CH_3)_3C$–, –$C(CH_3)_3$, CH_3 (a)

$CO–O–C_3H_7$, HO–, –OH, OH (b)

H, N, CH_3, CH_3, C_2H_5O–, CH_3 (c)

Figure 5.9. Structures of (a) butylated hydroxytoluene (BHT), (b) propyl gallate and (c) ethoxyquin.

In the inhibition of lipid peroxidation by an antioxidant (AH_2) it is generally suggested that the radical destroyed by reactions indicated above is the peroxy-free radical [Reactions (5.17)–(5.19)].

$$R{-}O{-}O^{\bullet} + AH_2 \longrightarrow R{-}OOH + AH^{\bullet} \quad (5.17)$$

$$2AH^{\bullet} \longrightarrow AH_2 + A \quad (5.18)$$

$$ROO^{\bullet} + AH^{\bullet} \longrightarrow ROOH + A \quad (5.19)$$

The inhibition of such chain reactions by metal salts utilises a different mechanism known as ligand transfer [Reaction (5.20)].

$$R^{\bullet} + \text{Metal}^{n+}\text{Anion}^{n-} \longrightarrow R{-}\text{Anion} + \text{Metal}^{(n-1)+}\text{Anion}^{(n-1)-} \quad (5.20)$$

5.3.2 Drug antioxidants

Many heterocyclic compounds with a ring nitrogen atom or sulphur atom function as free-radical scavengers. A particularly well-known example is the **phenothiazine** group (Figure 5.10), used commercially in the oil industry as antioxidants (Murphy *et al.*, 1950) and in the drug industry as antihistamines and tranquillisers. Piette *et al.* (1964) have demonstrated the existence of the chlorpromazine free radical by using the electron-spin-resonance technique. The chlorpromazine free radical was found to be exceptionally stable; this, together with the solubility properties of phenothiazines that allows them to penetrate into lipoprotein membranes, probably accounts for the effectiveness of some phenothiazines *in vivo* as free-radical scavengers, e.g. as protectors against liver injury produced by carbon tetrachloride (see Section 10.9).

Promethazine also illustrates a point of considerable biological importance that can be called the **cross-over effect**. Briefly, this phenomenon concerns the action of a particular drug that may be protective at one concentration

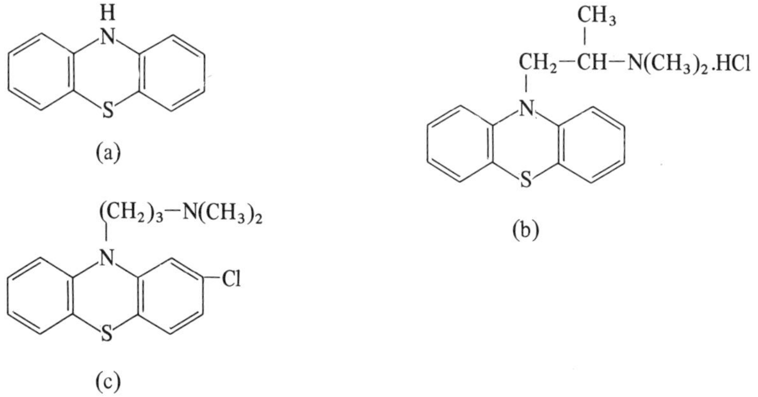

(a)

(b)

(c)

Figure 5.10. Structures of (a) phenothiazine, (b) promethazine hydrochloride and (c) chlorpromazine.

and deleterious at another. For example, promethazine stabilises the lysosomal membrane at low concentrations (about μM) but at mM concentrations causes lysosomal lysis (Slater and Riley, 1966). Similar observations have been made by Seeman and Weinstein (1966) for red blood cells. In some instances, such behaviour may represent a concentration-dependent change over from the drug acting mainly as a radical scavenger to a major action as a radical initiator [Reactions (5.21) and (5.22)].

$$R^{\bullet} + Ph \longrightarrow R\!:^{-} + Ph^{\bullet +} \quad (5.21)$$

$$Ph \longrightarrow e^{-} + Ph^{\bullet +}(+\ R{-}R) \longrightarrow R^{\bullet} + R^{+} + Ph \quad (5.22)$$

However, it is probable that in most cases, this effect is due to the operation of two different properties of the molecule that require different concentration ranges for their disclosure. With promethazine these properties are (i) antioxidant action at low concentrations and (ii) a *surface-active* detergent action at high concentrations. Whatever type of mechanism is operative in a given compound, however, the phenomenon is of considerable pharmacological importance since in such examples the dose–response characteristics are decidedly non-linear.

An interesting reaction *between* two agents that have been dealt with here as antioxidants has been described by Hoffman and Discher (1968). It is known that patients receiving prolonged therapy with high doses of chlorpromazine have a tendency to develop ocular disturbances such as corneal and lens opacities. Hoffman and Discher have shown that in the presence of light and oxygen chlorpromazine produces a decrease in glutathione thiol groups *in vitro* [(2)]; a similar decrease in thiol groups was found to occur in serum albumin exposed to chlorpromazine free radicals.

Other commonly used materials that have been reported to yield free-radical intermediates that may be of relevance to their biological action are salicylates and imipramine (Borg, 1965a), adrenalin, insulin and oestrogens (Borg, 1965b).

References

Abramson, H., 1949, *J. biol. Chem.*, **178**, 179.

Bacq, Z. M., 1965, *Chemical Protection Against Ionizing Radiation* (Charles C. Thomas, Springfield, Illinois).

Blois, M. S., 1958, *Nature, Lond.*, **181**, 1199.

Borg, D. C., 1965a, *Biochem. Pharmac.*, **14**, 115, 627.

Borg, D. C., 1965b, *Proc. natn. Acad. Sci., U.S.A.*, **53**, 633, 829.

Bunyan, J., Cawthorne, M. A., Diplock, A. T., Green, J., 1969, *Br. J. Nutr.*, **23**, 309.

Bunyan, J., Green, J., Edwin, E. E., Diplock, A. T., 1961, *Biochim. biophys. Acta*, **47**, 401.

(2) Ohnishi *et al.* (1969) have also studied this reaction. They find that the chlorpromazine radical acts as a one-electron oxidant converting glutathione into the monodehydro form; this rapidly dimerises to oxidised glutathione. The glutathione content of the lens has been found to be lower than normal in all forms of cataract. For a discussion of glutathione metabolism in the lens see van Heyningen (1962).

Burk, D., Woods, M., 1963, *Radiat. Res. Supplements,* **3**, 212.
Cash, W. D., Gardy, M., Carlson, H. E., Ekong, E. A., 1966, *J. biol. Chem.,* **241**, 1745.
Di Luzio, N. R., 1967, *Progr. Biochem. Pharmac.,* **3**, 325.
Draper, H. H., Csallany, A. S., 1969, *Fedn Proc. Fedn Am. Socs exp. Biol.,* **28**, 1690.
Forbes, J. C., Taliaferro, I., 1945, *Proc. Soc. exp. Biol. Med.,* **59**, 27.
Foye, W. O., 1969, *J. Pharmac. Sci.,* **58**, 283.
Glavind, J., Faber, M., 1966, *Int. J. Radiat. Biol.,* **11**, 445.
Grady, F. J., Borg, D. C., 1968, *Biochemistry, Easton,* **7**, 675.
Green, J., Bunyan, J., Cawthorne, M. A., Diplock, A. T., 1969, *Br. J. Butr.,* **23**, 297.
Heyningen, R. van, 1962, in *The Eye,* volume 2, Ed. H. Davson (Academic Press, London), p.213.
Hochstein, P., Ernster, L., 1963, *Biochem. biophys. Res. Commun.,* **12**, 388.
Hoffman, A. J., Discher, C. A., 1968, *Archs. Biochem. Biophys.,* **126**, 728.
Hope, D. B., 1959, "Glutathione", in *Biochemical Society Symposium number 17,* Ed. E. M. Crook (The Biochemical Society, London), p.93.
Hove, E. L., 1953, *J. Nutr.,* **51**, 609.
Kluge, H., Rasch, R., Brux, B., Frunder, H., 1967, *Biochim. biophys. Acta,* **141**, 260.
Kosower, E. M., Kosower, N. S., 1969, *Nature, Lond.,* **224**, 117.
Levandoski, N. G., Baker, E. M., Canham, J. E., 1964, *Biochemistry, Easton,* **3**, 1465.
Lundberg, W. O. (Ed.), 1961, *Autoxidation and Antioxidants,* volumes 1 and 2 (Interscience, New York).
Machover, P., Duchesne, J., 1965, *Nature, Lond.,* **206**, 618.
Mathews, M. M., 1964, *Nature, Lond.,* **203**, 1092.
Matsushita, S., Ibuki, F., Aoki, A., 1963, *Arch. Biochem. Biophys.,* **102**, 446.
Mellors, A., Tappel, A. L., 1966, *J. biol. Chem.,* **241**, 4353.
Melville, D. B., Horner, W. H., Lubschez, R., 1954, *J. biol. Chem.,* **206**, 221.
Melzer, M. S., Tomlinson, R. V., 1966, *Archs. Biochem. Biophys.,* **115**, 226.
Mengel, C. E., Kann, H. E., 1966, *J. clin. Invest.,* **45**, 1150.
Murphy, C. M., Ravner, H., Smith, N. L., 1950, *Chem. Ind.,* **42**, 2479.
Nordenbrand, K., Hochstein, P., Ernster, L., 1964, *Sixth International Congress of Biochemistry, New York,* volume 8 , p.76.
O'Brien, P. J., Little, C., 1969, *Can. J. Biochem. Physiol.,* **47**, 493.
Ohnishi, T,, Yamazaki, H., Iyanagi, T., Nakamura, T., Yamazaki, I., 1969, *Biochim. biophys. Acta,* **172**, 357.
Piette, L. H., Bulow, G., Yamazaki, I., 1964, *Biochim. biophys. Acta,* **88**, 120.
Roels, O. A., 1967, *Nutr. Rev.,* **25**, 33.
Seeman, P., Weinstein, J., 1966, *Biochem. Pharmac.,* **15**, 1737.
Siekevitz, P., 1955, *J. biophys. biochem. Cytol.,* **1**, 477.
Slater, T. F., Riley, P. A., 1966, *Nature, Lond.,* **209**, 151.
Slater, T. F., Sträuli, U. D., Sawyer, B. C., 1964, *Biochem. J.,* **93**, 260.
Smaller, B., 1960, *Free Radicals in Biological Systems,* Eds. M. Blois, H. W. Brown, R. M. Lemmon, R. O. Lindblom, M. Weissbluth (Academic Press, New York).
Stockey, B. N., 1962, *Lipids and Their Oxidation*, Ed. H. W. Schultz (Avi, Westport, Conn.), p.139.
Tappel, A. L., 1962, *Vitamins and Hormones: Advances in Research and Applications,* Volume 20, Ed. R. S. Harris (Academic Press, New York), p.493.
Wang, D. Y., Greenbaum, A. L., Harkness, R. D., 1963, *Biochem. J.,* **86**, 62.

Part 2

Free radicals in biological reactions

Occurrence of free radicals in biological material

6.1 Introduction

A variety of methods has been used successfully to demonstrate the occurrence of free-radical intermediates in diverse metabolic reactions in tissue and tissue extracts, thereby confirming the early proposals of Michaelis on the oxidation and reduction of organic materials by univalent electron transfer. Only a few examples of the occurrence of free-radical intermediates in biological systems will be given here[(1)]; these examples will be of particular relevance to later Sections of this book.

6.2 Observations on whole tissue samples

The majority of investigations that have been concerned with demonstrating the occurrence of free radicals in biological material have used electron spin resonance techniques. The study of tissue samples under physiological conditions with electron spin resonance is complicated, however, by the high absorbance of the microwave radiation by liquid water. As a consequence, and also to increase the ratio of signal to noise, most investigations have used frozen or freeze-dried material (for the underlying reasons, see Chapter 4). The use of very thin sample chambers, however, has enabled studies with tissue samples in *aqueous* solution at temperatures close to 37°C. The short dimension of the sample cell is parallel to the direction of propagation of the microwave radiation and the sample position is adjusted to occupy a position of minimum electric flux. This minimises the absorption of the electric field component by water (see Figure 6.1). Vithayathil *et al.* (1965) have studied the electron spin

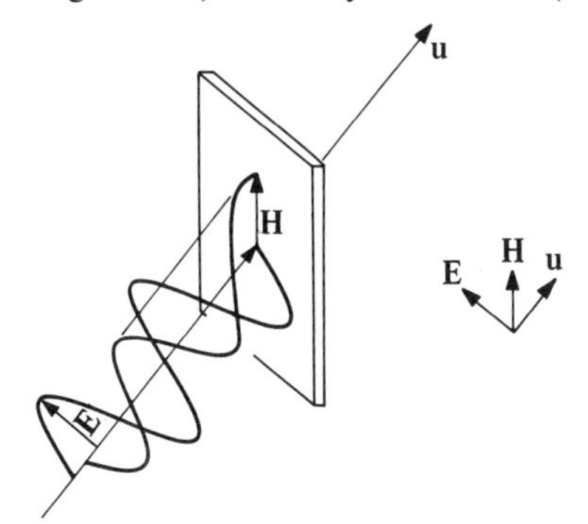

Figure 6.1. The microwave radiation is shown travelling in direction **u**. The electric (**E**) and magnetic (**H**) contributions to the electromagnetic radiation are shown as mutually perpendicular waveforms that are out of phase by 0·25λ. A thin sample holder is shown with the short dimension in the direction of **u**.

(1) The subject is well covered by the following reviews: Michaelis (1946); Commoner *et al.* (1954, 1957); Blois *et al.* (1961); Isenberg (1964); Ingram (1969); Mallard and Kent (1969); Wyard (1969).

resonance spectra of rat liver *slices* suspended in 5% glucose solution at 15°C, by using a computer of average transients to increase the ratio of signal to noise. They obtained good signals with 50–100 mg wet wt. of liver slices in the sample chamber. Although such conditions are closer to physiological conditions than those used in studies with frozen samples at −196°C, the difficulties of maintaining aerobic conditions in the thin sample chambers have been stressed (Isenberg, 1964).

Numerous studies have been made on *frozen powders* of a wide variety of tissues with or without freeze-drying. Wyard (1969) gives references to studies on adrenal, blood, brain, heart, intestine, kidney, lens, liver, lung, lymph node, muscle, skin, spleen and thymus. To this list we may add cervix, ovary and uterus (Slater and Cook, 1969). In general, complex spectra have been obtained with such samples. The precise identifications of the molecular species responsible for the signal absorptions have not been achieved in other than a very few instances. In several tissues (for example liver), however, the major part of the signal absorption in the $g = 2$ region may be correlated with mitochondrial components (Commoner and Ternberg, 1961).

Figure 6.2 illustrates, for example, the type of electron spin resonance spectrum obtained with frozen tissue powders; the spectrum illustrated was obtained by the author (in collaboration with J. W. R. Cook of Decca Radar Ltd.) with a sample of normal rat liver. In comparison with the signals found with frozen powders of liver, frozen *homogenates* of liver in 0·15 M potassium chloride give a much smaller signal response than would be expected simply from the dilution factor involved in the preparation of the homogenate. The mechanism of this quenching phenomenon is unknown but has been previously observed: Heckly and Dimmick (1967), for example, reported that the addition of EDTA, sucrose, glucose or sodium chloride to tissue before freezing decreased the electron spin resonance signals subsequently observed.

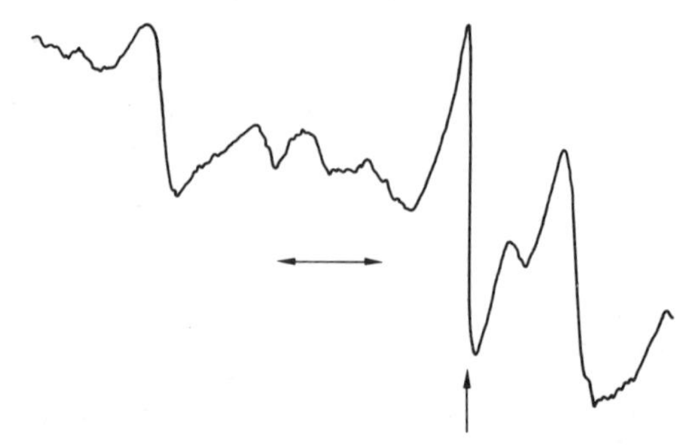

Figure 6.2. Electron spin resonance spectrum of a powdered sample of rat liver at −180°C obtained with a Decca X-1 spectrometer (data of Slater and Cook, 1969). The vertical line marks the position of the free electron spin at $g = 2{\cdot}0023$. The horizontal scale represents 100 gauss.

6.3 Observations on intracellular fractions

Despite the quenching effect observed with homogenates, strong signals have been obtained with suspensions of *intracellular* particles. Commoner and Hollocher (1960) observed an electron spin resonance signal in heart muscle **mitochondria** that was related to the activity of succinate dehydrogenase. Since then many studies (see Rieske *et al.*, 1964; van Voorst *et al.*, 1967; Hemmerich *et al.*, 1969) have been carried out on the electron spin resonance behaviour of submitochondrial particles and purified mitochondrial enzymes, mostly in relation to the enzymic activity of succinate dehydrogenase (a flavin-linked enzyme) and cytochrome oxidase (a copper- and iron-containing enzyme). The signal at $g = 2$, which was generally believed to originate largely from the flavin components of the mitochondrial respiratory chain, has now been shown to contain a substantial contribution (at least 50%) from the semiquinone form of ubiquinone (Bäckström *et al.*, 1970). In view of the demonstrated existence of free-radical species at various sites along the respiratory chain it is worth noting that Wang (1970) has outlined a scheme for coupled oxidative phosphorylation based on free-radical imidazolyl intermediates.

Hashimoto *et al.* (1962) reported the existence of a strong signal in rabbit liver **microsomes** at $g = 2{\cdot}24$ and smaller signals at $g = 2{\cdot}41$ and $g = 1{\cdot}91$. These signals all changed in unison when reaction conditions were varied and were ascribed to a single paramagnetic species designated microsomal Fe_x. The intensity of the signal was decreased by the addition of NADPH or NADH under anaerobic conditions and the introduction of oxygen resulted in the reappearance of the signals; NADPH was more effective than NADH in decreasing signal intensity. Several studies have since been made on this interesting material and microsomal Fe_x is now believed to be closely associated with the cytochrome P_{450} region of the microsomal drug-detoxifying sequence. Ichikawa and Yamano (1967) showed that a correlation existed between the cytochrome P_{450} content and microsomal Fe_x in a variety of tissues: pretreatment of rabbits with phenobarbitone or Sudan III increased both cytochrome P_{450} and microsomal Fe_x in the liver (Table 6.1). In the first few weeks after birth there are similar changes in cytochrome P_{450} and microsomal Fe_x

Table 6.1. Correlation between the cytochrome P_{450} content and the electron spin resonance signal at $g = 2{\cdot}25$ (Fe_x) in liver microsomes prepared from control and treated rabbits. Phenobarbital was given by intraperitoneal injection of 85 mg kg^{-1} body wt. daily for 4 days; with Sudan III the corresponding dose was 4 mg kg^{-1} body wt. Data of Ichikawa and Yamano (1967).

Microsome sample	Cytochrome P_{450}	Fe_x
Control	1·72	1·70
Phenobarbital	5·40	5·10
Sudan III	5·10	4·80

whereas cytochrome-b_5 showed a different pattern of change (Ichikawa and Yamano, 1967). A representative electron spin resonance spectrum of a rat liver microsomal suspension in tris buffer and examined at −196°C is illustrated in Figure 6.3.

Several other types of subcellular particle have been studied with the electron spin resonance technique. For example, Commoner *et al.* (1957) studied the reaction of **chloroplasts** to light. The signal obtained from chloroplasts suspensions was increased by illumination; the primary steps involved in photosynthesis are discussed in terms of the electron spin resonance evidence by Calvin and Androes (1962), Isenberg (1964) and by Weaver (1968) [see also Mallard and Kent (1969) for other references]. Although doubts have been raised about whether the electron spin resonance signal in photosynthetic systems has any direct relationship to photosynthesis, McElroy *et al.* (1969) give evidence that the signal in a bacterial photosynthetic system originates from oxidised bacteriochlorophyll. From an analysis of the kinetics they conclude that the free-radical species is associated with the primary step of bacterial photosynthesis. Mechanisms involved in photosynthesis are discussed in a review by Gibbs (1967). Other subcellular particles that have been studied in connection with free-radical processes occurring within them include the **melanosome** (see Section 14.3.1) and **lysosomes** in relation to lipofuschin formation (see Section 7.2).

The above discussion has indicated very briefly the wide variety of investigations carried out on the existence of free radicals in whole tissue samples and in subcellular fractions. The changes that have been observed in the electron spin resonance signal with change of metabolic state, or in different physiological states of the whole organism, strongly indicate that the signals observed have biological significance. The changes found in the

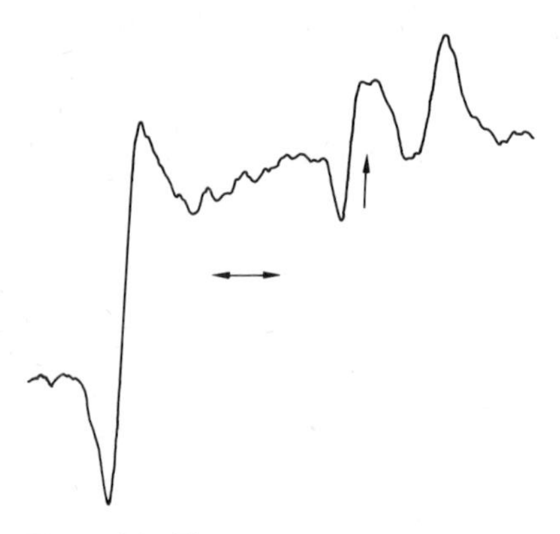

Figure 6.3. Electron spin resonance spectrum of a frozen suspension of rat liver microsomes in tris buffer, pH 7·4, at −180°C; protein concentration 68 mg ml^{-1} suspension (data of Slater and Cook, 1969). The horizontal line represents 100 gauss; the vertical line marks the position of the free electron spin at $g = 2{\cdot}0023$.

electron spin resonance spectra of tissues with disease[(2)] are further indications that the spectra are of relevance to biochemical mechanisms *in vivo* and are not artefacts of experimental technique. Nevertheless, it is important to note that several groups of investigators have stressed the pitfalls in applying electron spin resonance techniques to biological material, particularly freeze-dried material (for references see Dettmar *et al.*, 1967). The process of freeze-drying has been suggested to create new radical species[(3)] (see Kent and Mallard, 1967, and discussion in Wyard, 1969); this has been denied by Heckly and Dimmick (1967), who believe that the effects reported by Kent and Mallard and by others result from the presence of oxygen in contact with the freeze-dried material. The role of oxygen in altering the electron spin resonance spectrum of dried biological material has also been discussed by Isenberg (1964). Evidently, freeze-drying is a procedure that may well lead to complications in the interpretation of the data obtained by electron spin resonance examination of biological material. This problem has been well reviewed by Mallard and Kent (1969).

6.4 Observations on specific metabolic reactions

Turning now to the occurrence in biological material of *specific* free-radical intermediates it is possible to give unequivocal evidence for the existence of these in a variety of important metabolic reactions. The examples given are illustrative of a considerable number of related investigations.

It is known that irradiation of β-carotene with visible light produces free-radical structures (Smaller, 1960). Visible light also isomerises retinal (vitamin A aldehyde) to an equilibrium mixture of *cis*- and *trans*-isomers. In the **visual process** rhodopsin (a complex of retinal and a protein opsin) is converted by light into prelumirhodopsin; this is believed to involve the isomerisation of the retinal originally present as the 11-*cis* isomer into the all-*trans* form. The structures involved are shown in Figure 6.4. Grady

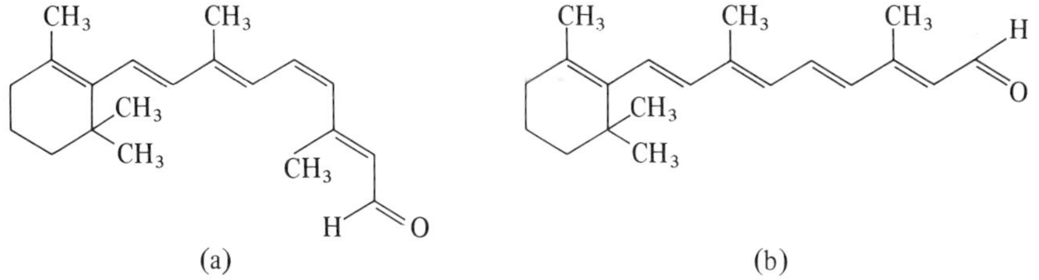

(a) (b)

Figure 6.4. Structures of the isomeric forms of retinal: (a) 11-*cis*-retinal; (b) all-*trans*-retinal.

(2) For example, in tumours (Section 13.2), in jaundice (Ternberg and Commoner, 1963) and in irradiation damage (Kenny and Commoner, 1969).

(3) Mechanical grinding of material has also been reported to increase the free-radical content of a sample (Urbanski, 1967).

and Borg (1968) have examined this reaction by using electron spin resonance techniques and have suggested that the reaction involves a triplet state of the retinal which possibly then transforms into a free-radical anion. In this example a free-radical intermediate participates at the primary stage of the conversion of *light-energy* into (eventually) an *electric impulse* in the neuron.

The role of free radicals in the initial reactions of **photosynthesis** has already been briefly mentioned above; here the conversion is of *light-energy* into *chemical free energy*. A further and interesting example of this is given by Eisenstein and Wang (1969), who have studied the porphyrin-sensitised photoreduction of ferredoxin by glutathione. Their results indicate that light raises the porphyrin to an excited state and it then abstracts an electron from reduced glutathione. The porphyrin radical anion then transfers its electron to ferredoxin, a redox component of photosynthetic systems containing non-haem iron. A similar example is the chlorophyllin-sensitized photoreduction of $NADP^+$ by cytochrome *c* (Tu and Wang, 1969).

The conversion of *chemical free energy* into *light-energy* under physiological conditions is attracting much current interest in terms of evaluating the mechanisms of **bioluminescence**. Reviews on this subject are by Hastings (1968) and McCapra (1966). Many examples of chemiluminescence under physiological conditions are known which involve very large changes in free energy; it is probable that the excited states are free radical in nature. For example, Cormier and Prichard have studied the peroxidation of luminol (5-amino-2,3-dihydro-1,4-phthalazinedione, Figure 6.5) by peroxidase in the presence of hydrogen peroxide. Electron spin resonance studies indicated the formation of luminol free radicals as intermediates. A possible mechanism is shown in Figure 6.6. Steele and

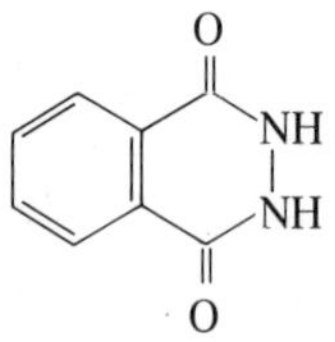

O
NH
NH
O

Figure 6.5. Structure of luminol.

$$E + H_2O_2 \longrightarrow C_1$$

$$C_1 + LH_2 \longrightarrow C_2 + LH^\bullet$$

$$C_2 + LH_2 \longrightarrow E + LH^\bullet$$

$$2LH^\bullet + H_2O_2 \longrightarrow \text{aminophthalate}^* + \text{products}$$

$$\text{Aminophthalate}^* \longrightarrow \text{aminophthalate} + \text{light}$$

Figure 6.6. Suggested mechanism for the chemiluminescent reaction involving luminol (LH_2) and peroxidase (E). C_1 and C_2 are complexes I and II involving peroxidase and H_2O_2 (see Cormier and Prichard, 1968).

colleagues (Williams and Steele, 1965; Vorhaben and Steele, 1967) have investigated the chemiluminescence of mixtures of riboflavin, hydrogen peroxide, Cu^{2+} and vitamin C and conclude that free-radical intermediates are involved. The free-energy changes involved in such reactions are very large (60–80 kcal mol^{-1}) and may be trapped to perform aromatic hydroxylation reactions. Williams and Steele (1965) have drawn a very interesting analogy between the photosynthetic reaction, in which light-energy is *utilised* to reduce NADPH, and the NADPH–oxygen–metal ion-hydroxylating system of endoplasmic reticulum, which, under appropriate conditions, *releases* light-energy with consumption of NADPH.

An interesting example of the formation of free radicals by light-irradiation in biological material concerns **melanin**[(4)]. This compound is a polymer of DOPA-quinone, 5,6-dihydroxyindole and 5,6-dihydroxyindole-2-carboxylic acid and exhibits a stable free-radical signal probably due to the free radicals being trapped in the polymer cage (Commoner *et al.*, 1954; Mason *et al.*, 1960; see also Pathak and Stratton, 1968). The intensity of the electron-spin-resonance signal was increased by irradiating the melanin with light of wavelength greater than 300 nm; light with a wavelength less than 320 nm not only increased the melanin signal but also produced additional signals with new *g* values, as might be expected from the high energy of the short wavelength ultraviolet radiation. The photoenhanced melanin radical was unstable, in contrast to the endogenous melanin signal (Pathak and Stratton, 1968). It is probable that the melanin-rich zone in the epidermis protects the underlying dermis from damage that could result from exposure to short wavelength sunlight. During irradiation the melanin molecules would be raised to an excited state and the excess of energy is presumably dissipated in a controlled manner to prevent damage to surrounding structures. Slater and Riley (1966) have suggested that the localisation of melanin in a membrane-linked particle (the melanosome) prevents the random distribution of melanin free radicals throughout the cytoplasm, which would be expected to lead to deleterious results.

Several studies have demonstrated free-radical intermediates in reactions involving commonly used reagents. Yuferov *et al.* (1970) have investigated the reaction between amino acids and ninhydrin in alcoholic solution at room temperature. Electron spin resonance signals were obtained that increased in intensity when the reaction was carried out at pH values greater than 7·0 under anaerobic conditions and in the presence of vitamin C.

Yamazaki and Tolbert (1970) have studied the photoreduction of free or bound flavin mononucleotide by zwitterionic buffers. It is known that the flavin group can be photoreduced under anaerobic conditions by nitrogen-containing compounds such as amino acids or EDTA, and also

[(4)] Further discussion of melanin and of the production of free radicals in skin during irradiation is given in Section 14.3.1.

by an intramolecular reaction with the ribityl side chain. Yamazaki and co-workers have now extended the list of reactive compounds to commonly used buffers such as Tricine[5] and Bicine. Under anaerobic conditions the reaction can be followed spectrophotometrically by measuring the production of reduced flavin; under aerobic conditions the reduced flavin autoxidises with production of hydrogen peroxide and destruction of the buffer. Under anaerobic conditions the reduced flavin can be coupled directly to a dye such as dichlorophenolindophenol. The authors comment that such photoreductions occur with light-intensities commonly encountered under laboratory conditions and may complicate kinetic analyses. Tris, phosphate and HEPES buffers were much less reactive than Tricine and Bicine.

Kimura and Szent-Györgyi (1969) have studied the reaction between *N*-methylphenazonium sulphate (PMS) and electron donors including imidazole, indole, naphthylamine, cyclohexylamine, glycine, glutamine, lipoic acid and acetylacetone. Electrons were found to be transferred most readily from compounds containing a nitrogen atom with a lone-pair of electrons; the transfer was less intense with sulphur and still less intense from oxygen. The product of the reaction was the monovalent PMS free-radical anion ($PMS^{\bullet-}$), which was detected by the electron spin resonance technique.

This Chapter has indicated very briefly the evidence obtained from electron spin resonance showing that free radicals normally occur as intermediates in diverse metabolic reactions. If free-radical intermediates were formed in a relatively diffusible form in the cell not only would their reaction with the subsequent component of the particular enzyme sequence be inefficient but the free radicals might well initiate deleterious reactions in a variety of cell components, particularly intracellular membranes containing unsaturated lipid. In general, free-radical intermediates involve components that are built into an organised 'insoluble' structural system (e.g. the respiratory chain), where they can react efficiently with a neighbouring acceptor but are constrained from deleterious reactions that would arise were they freely diffusible. In this sense the free-radical intermediates of *normal* metabolism are '*frozen radicals*'. In the following Chapter we will consider how the uncontrolled production of *diffusible* free radicals can damage cellular processes and initiate cell injury.

References

Bäckström, D., Norling, B., Ehrenberg, A., Ernster, L., 1970, *Biochim. biophys. Acta,* **197**, 108.

Blois, M. S., Brown, H. W., Lemmon, R. M., Lindblom, R. O., Weissbluth, M. (Eds.), 1961, *Free Radicals in Biological Systems* (Academic Press, New York).

[5] Tricine, Bicine and HEPES buffers are respectively: *N*-tris(hydroxymethyl)methyl-glycine; *NN*-bis-(2-hydroxyethyl)glycine; *N*-2-hydroxyethylpiperazine-*N*′-2-ethane-sulphonic acid.

Calvin, M., Androes, G. M., 1962, *Science,* **138**, 867.
Commoner, B., Heise, J. J., Lippincott, B. B., Norberg, R. E., Passonneau, J. V., Townsend, J., 1957, *Science,* **126**, 57.
Commoner, B., Hollocher, J. C., 1960, *Proc. natn. Acad. Sci. U.S.A.,* **46**, 405.
Commoner, B., Ternberg, J. L., 1961, *Proc. natn. Acad. Sci. U.S.A.,* **47**, 1374.
Commoner, B., Townsend, J., Pake, G. E., 1954, *Nature, Lond.,* **174**, 689.
Cormier, M. J., Prichard, P. M., 1968, *J. biol. Chem.,* **243**, 4706.
Dettmer, C. M., Discoll, D. H., Wallace, J. D., Neaves, A., 1967, *Nature, Lond.,* **214**, 492.
Eisenstein, K. K., Wang, J. H., 1969, *J. biol. Chem.,* **244**, 1720.
Gibbs, M., 1967, *A. Rev. Biochem.,* **36**, 757.
Grady, F. J., Borg, D. C., 1968, *Biochemistry, Easton,* **7**, 675.
Hashimoto, Y., Yamamo, T., Mason, H. S., 1962, *J. biol. Chem.,* **237**, PC3843.
Hastings, J. W., 1968, *A. Rev. Biochem.,* **37**, 597.
Heckly, R. J., Dimmick, R. L., 1967, *Nature, Lond.,* **216**, 1003.
Hemmerich, P., Ehrenberg, A., Walker, W. H., Eriksson, L. E. G., Salach, J., Bader, P., Singer, T. P., 1969, *FEBS Letters,* **3**, 37.
Ichikawa, Y., Yamano, T., 1967, *Archs. Biochem. Biophys.,* **121**, 742.
Ingram, D. J. E., 1969, *Biological and Biochemical Applications of Electron Spin Resonance* (Adam Hilger, London).
Isenberg, I., 1964, *Physiol. Rev.,* **44**, 487.
Kenny, P., Commoner, B., 1969, *Nature, Lond.,* **223**, 1229.
Kent, M., Mallard, J. R., 1967, *Nature, Lond.,* **215**, 736.
Kimura, J. E., Szent-Györgyi, A., 1969, *Proc. natn. Acad. Sci. U.S.A.,* **62**, 286.
Mallard, J. R., Kent, M., 1969, *Physics Med. Biol.,* **14**, 373.
Mason, H. S., Ingram, D. J. E., Allen, B. T., 1960, *Archs. Biochem. Biophys.,* **86**, 225.
McCapra, F., 1966, *Q. Rev. chem. Soc.,* **20**, 485.
McElroy, J. D., Feher, G., Mauzerall, D. C., 1969, *Biochim. biophys. Acta,* **172**, 180.
Michaelis, L., 1946, *Currents in Biochemistry,* Ed. D. E. Green (Interscience, New York).
Pathak, M. A., Stratton, K., 1968, *Archs. Biochem. Biophys.,* **123**, 468.
Rieske, J. S., Zaugg, W. S., Hansen, R. E., 1964, *J. biol. Chem.,* **239**, 3017, 3023.
Slater, T. F., Cook, J. W. R., 1969, *Cytology Automation,* Ed. D. M. D. Evans (Livingstone, Edinburgh), p.108.
Slater, T. F., Riley, P. A., 1966, *Nature, Lond.,* **209**, 151.
Smaller, B., 1960, *Free Radicals in Biological Systems,* Eds. M. Blois, H. W. Brown, R. M. Lemmon, R. O. Lindblom, M. Weissbluth (Academic Press, New York).
Ternberg, J. L., Commoner, B., 1963, *J. Am. med. Ass.,* **183**, 339.
Tu, S.-I., Wang, J. H., 1969, *Biochemistry, Easton,* **8**, 2970.
Urbanski, T., 1967, *Nature, Lond.,* **216**, 577.
Vithayathil, A. J., Ternberg, J. L., Commoner, B., 1965, *Nature, Lond.,* **207**, 1246.
Voorst, J. D. W. van, Veeger, C., Dervartanian, D. V., 1967, *Biochim. biophys. Acta,* **146**, 367.
Vorhaben, J. E., Steele, R. H., 1967, *Biochemistry, Easton,* **6**, 1404.
Wang, J. H., 1970, *Science,* **167**, 25.
Weaver, E. C., 1968, *A. Rev. Pl. Physiol.,* **19**, 283.
Williams, J. R., Steele, R. H., 1965, *Biochemistry, Easton,* **4**, 814.
Wyard, S. J., 1969, *Solid State Biophysics,* Ed. S. J. Wyard (McGraw-Hill, London), p.263.
Yamazaki, R. K., Tolbert, N. E., 1970, *Biochim. biophys. Acta,* **197**, 90.
Yuferov, V. P., Froncish, W., Kharitonenkov, I. G., Kalmanson, A. E., 1970, *Biochim. biophys. Acta,* **200**, 160.

7

Deleterious reactions of free radicals

7.1 Introduction

The generally high chemical reactivity of free radicals has been stressed in Chapter 3. It is evident from the discussion given there that the production of extraneous free radicals *in vivo* will have a multiplicity of effects on natural components; it may be confidently predicted that many of these effects will be deleterious to normal metabolism. For example, free radicals such as trichloromethyl produced in the membranes of the endoplasmic reticulum can be expected to react with neighbouring thiol groups, with the unsaturated bonds that occur in fatty acids associated with membranes, and with nucleic acid bases. We can consider such reactions as illustrative of the damage produced by free-radical attack on proteins that are dependent for activity on thiol groups, on membranes that involve a closely ordered arrangement of lipoprotein molecules with additional components, and on nucleotide coenzymes and polynucleotides that play an essential role in metabolism and in heredity.

7.2 Destruction of thiol groups

The reaction of thiol groups with free radicals underlies the biological action of radiation protectors such as cysteamine. Such compounds scavenge radicals very effectively, as discussed in Section 5.1.1.

Many enzymes depend on thiol groups for functional activity. With such enzymes it can be expected that significant intracellular production of reactive free radicals will lead to the inactivation of the enzymes provided that the essential thiol groups are close to the site of free-radical formation. In this context several studies have been made of the effects of lipid peroxidation on, or of the actual addition of organic peroxides to, isolated enzymes. Wills (1959) has shown that enzyme inactivation by **succinoyl peroxide** was approximately related to the thiol content of the enzyme. A similar relationship was found when free radicals were produced in the enzyme suspension by the addition of autoxidising unsaturated fatty acid (Wills, 1961; Lewis and Wills, 1962). More recent studies from Tappel's laboratory (Chio and Tappel, 1969) show that although lipid peroxidation decreases the activity of thiol-dependent enzymes, changes in the polypeptide structure itself were also involved. With ribonuclease A, for example, which is not a thiol-dependent enzyme, the loss of activity was correlated with intra- and inter-molecular cross-linking that was probably produced by reaction with malonaldehyde. High molecular weight complexes of protein and lipid resembling lipofuschin may also be formed *in vitro* during the peroxidation of lipid components of intracellular organelles (Chio *et al.*, 1969); this is of interest in view of many previous speculations that lipofuschin (or 'age-pigment') represents the chronic accumulation *in vivo* of autoxidising protein–lipid mixtures (Pearse, 1960).

7.3 Lipid peroxidation

The mechanisms underlying the peroxidative decomposition of unsaturated lipids have been described in Chapter 3. Since appreciable lipid peroxidation of lipoprotein membranes is accompanied by extensive disturbance of the organised structure of the membrane it is not surprising that the process results, in general, in the loss of specialised functions of the membrane, be it the mitochondrial, lysosomal or cell membrane itself. Three examples of such effects may be cited. (1) Hunter *et al.* (1963) showed that the initiation of lipid peroxidation in mitochondrial suspensions by a mixture of ferrous ions and vitamin C was accompanied by swelling, and by decreased P/O ratios. (2) Tappel and colleagues (Desai *et al.*, 1964) produced lipid peroxidation in purified suspensions of lysosomes and found that as the reaction proceeded there was a liberation of intralysosomal enzymes into the surrounding medium. A similar mechanism based on lipid peroxidative damage to the lysosomal membrane was proposed by Slater and Riley (1966) to explain the epidermal damage that occurs in photosensitisation. (3) Mengel and Kann (1966) have shown that lipid peroxidation may be initiated in suspensions of erythrocytes, particularly when the erythrocytes were taken from animals that were deficient in vitamin E. Peroxidation of the erythrocyte membrane was accompanied by lysis. All of these effects are described in more detail elsewhere (see Chapter 12, Section 12.3) in this Monograph and are cited here only to illustrate damage to intracellular membrane systems that results from lipid peroxidation.

Lipid peroxidation is not only associated with disorganisation of lipoprotein membranes; primary effects on other cellular components may also occur as already mentioned above for ribonuclease A. Desai and Tappel (1963) have studied the effects of lipid peroxidation on cytochrome *c* and have found that the peroxy-radicals produced during lipid peroxidation caused substantial alterations in the compound's physical properties. Roubal and Tappel (1967) have shown that lipid peroxidation can also affect nucleotides by studying the effect of an autoxidising suspension of ethyl arachidonate on ATP concentration. The concentration of ATP was decreased by lipid peroxidation and adenosine, AMP and ADP were identified among the reaction products. In contrast to the destructive action of ionising radiation on ATP, however, the efficiency of ATP breakdown by lipid peroxidation in autoxidising suspensions of ethyl arachidonate was low: 2×10^{-5} mol of ATP was destroyed/mol of peroxy-radical produced. The effects of lipid peroxidation on the destruction of α-tocopherol have been studied by O'Brien and Titmus (1967), who found that in the haematin-catalysed reaction between linoleate hydroperoxide and α-tocopherol, the main product was an α-tocopherone with some dimer and trimer. In the same system α-tocopheryl acetate was not oxidised.

A consequence of lipid peroxidation is the production of malonaldehyde (see Section 4.2.3), which may result in an extension of the damage already produced by the peroxidative reactions in the membrane. Malonaldehyde reacts with amino acids to yield **Schiff bases**. This reaction has been studied in detail by Chio and Tappel (1969), who found that two molecules of amino acid react with one molecule of malonaldehyde to produce *NN*′-disubstituted 1-amino-3-iminopropene derivatives (Figure 7.1). Malonaldehyde also reacts with proteins, to produce inter- and intra-molecular cross-linking, and with DNA (Brooks and Klamerth, 1968).

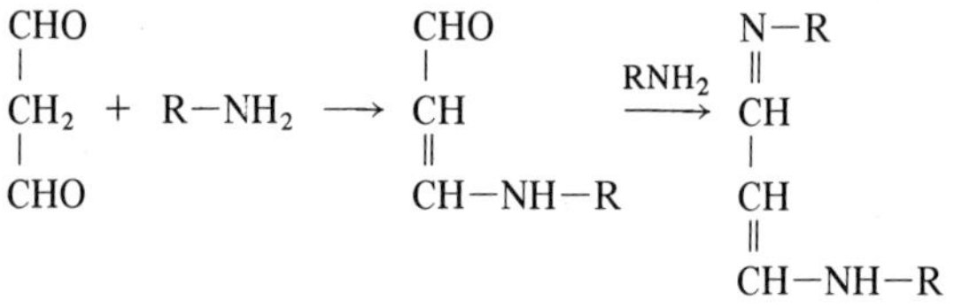

$$\begin{array}{ccccc} \mathrm{CHO} & & & \mathrm{CHO} & & \mathrm{N{-}R} \\ | & & & | & & \| \\ \mathrm{CH_2} & + & \mathrm{R{-}NH_2} & \longrightarrow & \mathrm{CH} & \xrightarrow{\mathrm{RNH_2}} & \mathrm{CH} \\ | & & & \| & & | \\ \mathrm{CHO} & & & \mathrm{CH{-}NH{-}R} & & \mathrm{CH} \\ & & & & & \| \\ & & & & & \mathrm{CH{-}NH{-}R} \\ \text{(i)} & & \text{(ii)} & \text{enamine (iii)} & & \text{(iv)} \end{array}$$

Figure 7.1. Reaction of malonaldehyde with an amino acid to produce an enamine. This then interacts with a further molecule of amino acid to yield the *NN*′-disubstituted 1-amino-3-iminopropene (see Chio and Tappel, 1969).

7.4 Nucleic acids and nucleotides

The effects of extraneous free radicals on nucleic acid structure and properties have been repeatedly studied in connection with the known effects of radiation on mutation rate and carcinogenesis. The radiation energy used in such experiments may be classified under three main headings: (1) high-energy ionising radiation, such as γ- or X-radiation; (2) ultraviolet irradiation; (3) radiation by long wavelength ultraviolet or visible radiation in the presence of a photosensitiser.

7.4.1 Ionising radiation

X-Radiation of nucleic acids in aqueous solution produces peroxidative changes mainly in the **pyrimidine** base components. The mechanisms are discussed by Latarjet *et al.* (1963). The extent of the reaction is decreased by lowering the oxygen partial pressure of the solution, and protection was also obtained by the addition of high concentrations of cysteamine *in vitro*. A mechanism suggested by Latarjet and colleagues is shown in Figure 7.2.

The protective action of cysteamine was taken to result from radical scavenging by the *cystamine* form of the protective agent, as well as by a reaction with the peroxide radical III in Figure 7.2 to give the glycol (Figure 7.3). Daniels and Schweibert (1967) showed that γ-radiation of cytosine could yield corresponding products to those shown above for thymine thereby indicating the formation of cytosine hydroperoxide.

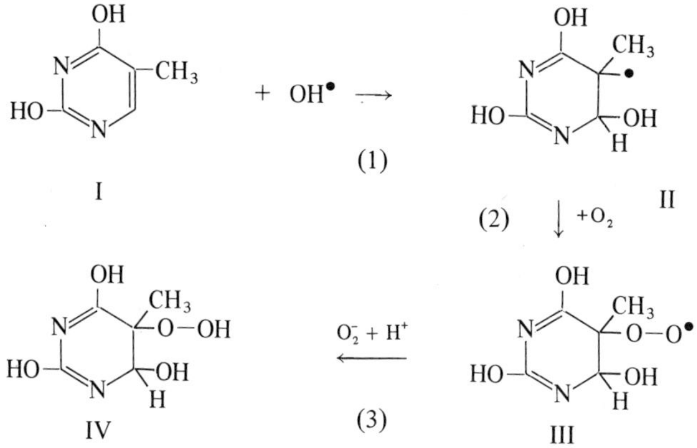

$$\text{I} + OH^{\bullet} \xrightarrow{(1)} \text{II} \xrightarrow[(2)]{+O_2} \text{III} \xrightarrow[(3)]{O_2^- + H^+} \text{IV}$$

Figure 7.2. Reaction of hydroxyl radical with a pyrimidine base (thymine) to produce the substituted hydroxyhydroperoxide (see Latarjet, 1963).

$$R{-}S{-}S{-}R + OH^{\bullet} \longrightarrow R{-}S^{\bullet} + R{-}SOH \quad \text{(a)}$$

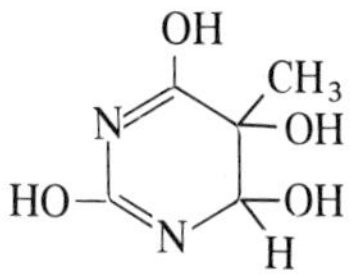

(b) V

Figure 7.3. Protective actions of a disulphide (R—S—S—R) on the hydroxyl attack on pyrimidine bases (e.g. thymine). In (a) the disulphide competes with the pyrimidine base (see Figure 7.2, reaction 1) for hydroxyl radicals. In (b) the thiyl radical formed from the disulphide reacts with the peroxide radical (3, Figure 7.2) and reduces it to the glycol, V. Mechanism (a) probably operates when the protective disulphide is present in high excess (see Latarjet, 1963).

7.4.2 Ultraviolet irradiation

Ultraviolet irradiation of frozen solutions of thymine and of DNA yields **thymine dimers** (see Ben-Hur *et al.*, 1967). The damage to DNA produced by direct ultraviolet irradiation seems to be mainly due to such changes in the thymine components. Thymine dimers may exist in several isomeric forms (Wulff and Fraenkel, 1961) and illustrations of two of these are shown in Figure 7.4. It is interesting that irradiation of the pyrimidines in combination with adenine gave photo-products not found by the irradiation of the pyrimidine alone (Smith, 1966). Rhaese and Freese (1968) have studied the action of hydroxyl radicals on DNA; the radicals were formed by ultraviolet irradiation of hydrogen peroxide or by the addition of transitional metals to hydrogen peroxide. The major destruction action was a decrease in unsaturated carbon double bonds, mainly in pyrimidine bases, with some breakage of the sugar-phosphate backbone and some liberation of free bases from the nucleotides.

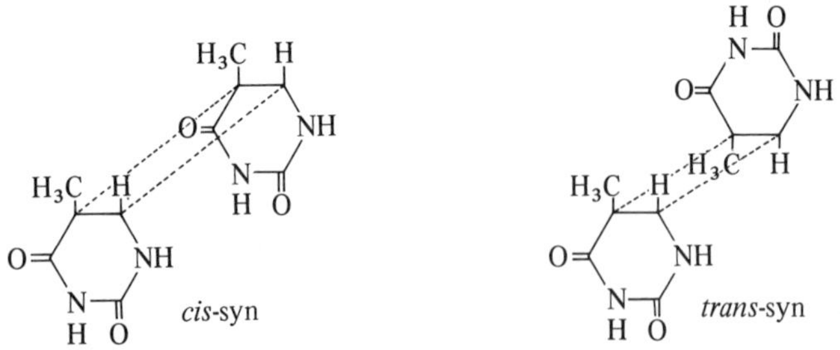

cis-syn *trans*-syn

Figure 7.4. Two of the possible thymine dimers are illustrated. For other isomeric structures see Ben-Hur *et al.* (1967). The bonds joining the thymine bases are shown dotted and have been lengthened to aid clarity.

7.4.3 Radiation in the presence of photosensitisers

In contrast to direct ultraviolet irradiation and ionising radiation damage that is largely confined to the pyrimidine bases, Simon and Vunakis (1962) found that visible light in the presence of methylene blue and oxygen primarily destroyed the **guanine** residues of DNA. In an analysis of the comparative lability of various purine and pyrimidine bases with visible radiation in the presence of methylene blue Simon and Vunakis (1964) found that uric acid was destroyed most rapidly, followed, in order of reactivity, by xanthine, guanosine and thymine.

This work was extended by Singer and Fraenkel-Conrat (1966), who showed that radiation of tobacco mosaic virus in the presence of several dyestuffs, particularly thiopyronine, caused loss of infectivity and a decrease in the guanine content of the irradiated virus.

The role of free-radical intermediates in dye-catalysed nucleotide reactions has been reviewed by Delmelle and Duchesne (1968), who have shown the appearance of electron spin resonance spectra identifiable with the base radical after visible light irradiation of the nucleotide in the presence of acridine dyes. The protective action of cysteamine added before irradiation was found to be associated with the appearance of the sulphur-containing free radical and with the non-appearance of the free radical form of the base. The results of Gräslund *et al.* (1969) suggest that intercalation of the acridine dye is necessary for the production of the base radicals which arise principally in regions rich in adenine–thymine pairs.

It is noteworthy that photosensitisation does not require conventional dyestuffs; for example, dimerization of thymine can be accomplished in aqueous solution with light of greater than 3000 nm provided that **acetone** is present as a photosensitiser.

Balazs *et al.* (1968) have reported the interesting finding that complexes of nucleic acids with dyestuffs (e.g. crystal violet) exhibited **paramagnetism** with a rapid increase in the electron spin resonance signal in the $g = 2$ region when illuminated with visible light. In this situation the photo-excited electron is associated with the dye rather than with the nucleic acid.

The effects of riboflavin on the photo-decomposition of nucleotides has been studied by Uehara *et al.* (1966). Riboflavin was found to catalyse the breakdown of NAD^+ and $NADP^+$ in solution when irradiated with visible light. Methylene blue was not effective in the system as a photosensitiser. The reaction involved the adenine moiety of NAD^+ and $NADP^+$, similar decreases in coenzyme A activity and in ATP concentration were also found under these conditions. The actual mechanism probably involves the deamination of the adenine to hypoxanthine. It was also noted that irradiation of mixtures of thymine and adenine, which, of course, hydrogen bond to each other in the double helical form of DNA, gave a greater destruction of adenine than the irradiation of adenine alone (these results with riboflavin should be compared with the action of hydrogen peroxide on nucleic acid bases; see Section 5.2.1).

In some cases the damage to nucleotides and nucleic acids that is produced by irradiation has been found to involve the formation of peroxides. Peroxides themselves have been studied for direct action on nucleic acids in the absence of irradiation. **Disuccinoyl peroxide** (10^{-8} M) was shown to decrease the transforming activity of DNA from *Pneumococcus* (Latarjet *et al.*, 1958). This effect was found to be dependent on very low concentrations of transitional metals (10^{-11} M-copper or -iron). Cumene hydroperoxide (10^{-3} M) was inactive, however, under similar conditions.

Finally in this section on irradiation-produced damage we may refer briefly again to some results of Slater and Jose (1969). Irradiation of bromotrichloromethane or of carbon tetrachloride in ethanol or acetone with ultraviolet light produced the homolytic cleavage of the halogenated methane; this was accompanied by a rapid decrease in NADH and NADPH (but not NAD^+ or $NADP^+$) present in solution. Here the mechanisms probably involve abstraction of hydrogen from the reduced nucleotide with subsequent dimerisation of the nucleotide free radical [Reactions (7.1)–(7.3)]. A similar reaction has been reported by Land and Swallow (1968), who subjected NAD^+ in solution to pulsed irradiation with high-energy electrons.

$$CCl_3Br \xrightarrow{h\nu} CCl_3^\bullet + Br^\bullet \tag{7.1}$$

$$CCl_3^\bullet + NADH \longrightarrow CHCl_3 + NAD^\bullet \tag{7.2}$$

$$2NAD^\bullet \longrightarrow (NAD)_2 \tag{7.3}$$

Another possible mechanism to account for the loss of coenzyme activity in such a system upon irradiation involves the formation of the $C_2H_5O^\bullet$ radical from the ethanol present in the reaction mixture. This radical is strongly reducing in character and may *further* reduce NADH and NADPH to inactive products [Reactions (7.4)–(7.5)].

$$CCl_3^\bullet + C_2H_5OH \longrightarrow CHCl_3 + C_2H_5O^\bullet \tag{7.4}$$

$$2C_2H_5O^\bullet + NADH \longrightarrow 2C_2H_4O + NADH_2 + H^+ \tag{7.5}$$

The first mechanism seems more likely since ethanol does not greatly influence the breakdown of NADPH in solution; acetone, however, does have some stimulatory action. The reaction may be followed by the decrease in fluorescence of the reduced nucleotide (Figure 7.5) or by determining the coenzyme activity at intervals during irradiation. The destruction of NADPH by $CCl_3^\bullet$ is discussed in Chapter 10 in connection with the hepatotoxicity of carbon tetrachloride.

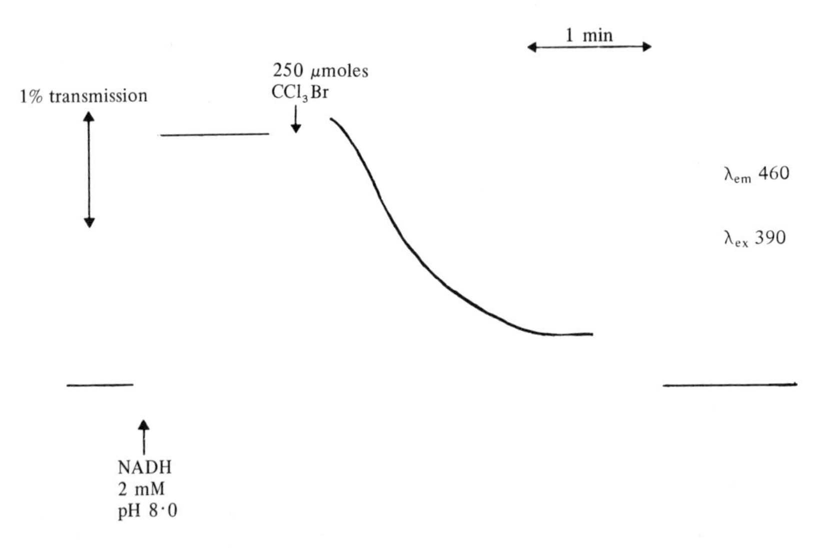

Figure 7.5. Decrease in fluorescence of NADP upon irradiation in presence of bromotrichloromethane. The nucleotide was dissolved in a mixture (2:1) of ethanol and tris buffer (pH 8·0, 0·1 M) and fluorescence at 460 nm measured in an Optica (Baird and Tatlock Ltd., London) spectrophotometer operating on single beam.

References

Balazs, E. A., Young, M. D., Phillips, G. O., 1968, *Nature, Lond.*, **219**, 154.
Ben-Hur, E., Elad, D., Ben-Ishai, R., 1967, *Biochim. biophys. Acta,* **149**, 355.
Brooks, B. R., Klamerth, O. L., 1968, *Eur. J. Biochem.*, **5**, 178.
Chio, K. S., Reiss, U., Fletcher, B., Tappel, A. L., 1969, *Science,* **166**, 1535.
Chio, K. S., Tappel, A. L., 1969, *Biochemistry, Easton,* **8**, 2821, 2827.
Daniels, M., Schweibert, M. C., 1967, *Biochem. biophys. Acta,* **134**, 481.
Delmelle, M., Duchesne, J., 1968, *Molecular Associations in Biology,* Ed. B. Pullman (Academic Press, New York), p.299.
Desai, I. D., Sawant, P. L., Tappel, A. L., 1964, *Biochim. biophys. Acta,* **86**, 277.
Desai, I. D., Tappel, A. L., 1963, *J. Lipid Res.*, **4**, 204.
Gräslund, A., Rigler, R., Ehrenberg, A., 1969, *FEBS Letters,* **4**, 227.
Hunter, F. E., Gebicki, J. M., Hoffsten, P. E., Weinstein, J., Scott, A., 1963, *J. biol. Chem.*, **238**, 828.
Land, E. J., Swallow, A. J., 1968, *Biochim. biophys. Acta,* **162**, 327.
Latarjet, R., Ekert, B., Demerseman, P., 1963, *Radiat. Res. Supplements,* **3**, 247.
Latarjet, R., Rebeyrotte, N., Demerseman, P., 1958, *Les peroxydes organiques en radiobiologie,* Ed. M. Haissinsky (Masson, Paris), p.61.

Lewis, S. E., Wills, E. D., 1962, *Biochem. Pharmac.*, **11**, 901.
Mengel, C. E., Kann, H. E., 1966, *J. clin. Invest.*, **45**, 1150.
O'Brien, P. J., Titmus, G., 1967, *Biochem. J.*, **103**, 33P.
Pearse, A. G. E., 1960, *Histochemistry: Theoretical and Applied* (Churchill, London).
Rhaese, H. J., Freese, E., 1968, *Biochim. biophys. Acta*, **155**, 491.
Roubal, W. T., Tappel, A. L., 1967, *Biochim. biophys. Acta*, **136**, 402.
Simon, M. I., Vunakis, H. V., 1962, *J. molec. Biol.*, **4**, 488.
Simon, M. I., Vunakis, H. V., 1964, *Archs. Biochem. Biophys.*, **105**, 197.
Singer, B., Fraenkel-Conrat, H., 1966, *Biochemistry, Easton*, **5**, 2446.
Slater, T. F.. Jose, P. J., 1969, *Biochem. J.*, **114**, 7P.
Slater, T. F., Riley, P. A., 1966, *Nature, Lond.*, **209**, 151.
Smith, K. L., 1966, *Biochem. biophys. Res. Commun.*, **23**, 426.
Uehara, K., Mizoguchi, T., Okada, Y., Kuwashima, J., 1966, *J. Biochem.*, **59**, 433.
Wills, E. D., 1959, *Biochem. Pharmac.*, **2**, 276.
Wills, E. D., 1961, *Biochem. Pharmac.*, **7**, 7.
Wulff, D. L., Fraenkel, G., 1961, *Biochim. biophys. Acta*, **51**, 332.

Part 3

Free radicals in tissue injury

8

Tissue injury: general comments

8.1 Introduction

In Part 3 of this Monograph the role of free-radical reactions in the initiation of cellular injuries is considered. The number of examples that could be cited in this respect is increasing rapidly, and Chapters 9–15 will describe a considerable variety of such processes in varying degrees of detail. It will be seen that many of the examples described concern the deleterious interactions between free-radical intermediates and *lipoprotein membranes*. We shall be concerned here only with examples of tissue injury that involve free-radical intermediates as important participants in the primary stages of the disturbances. More general aspects of cell injury are considered for examples in Cameron (1952, 1964), Dawkins and Rees (1959) and Cameron and Spector (1966).

A cellular or tissue *lesion* that is gross enough to be microscopically apparent is often the end-result of a most complex and intricate network of metabolic perturbations; for example, the developments of centrilobular necrosis of the liver, chemically produced malignant tumours or a severe inflammatory response. To study the primary mechanism(s) that is responsible for the development of the lesion it is usual to work backwards in time by analysing samples taken closer and closer to the time of administration of the toxic agent for significant alterations from the control situation. This approach is not without its hazards, however, for generally it would be wrong simply to equate an *earliness* of effect with its *importance* to the network of perturbations that culminate in the final lesion.

Two main examples will be dealt with in Part 3; these will illustrate the difficulties that underly the interpretation of pathological disturbances in chemical terms, and indicate the breadth and diversity of the experimental techniques required for such studies. In the first example (Chapters 9 and 10) the liver injury that results from the administration of carbon tetrachloride to rats will be discussed in considerable detail. The second example (Chapter 11) will deal with the effects of acute ethanol intoxication on the liver. First, however, it is necessary to define what is meant by 'cell injury'.

8.2 Cell injury

In this text **injury** will be taken to indicate all types of abnormal metabolic deviations whether apparent as a gross histopathological abnormality or detectable only by refined chemical analysis. The abnormal deviations included under this heading may affect, or be demonstrated at, a wide range of organisational levels, ranging from the whole body response, tissue abnormalities, cellular, subcellular or molecular disturbances. Rather than talk of 'injury' to a particular cell component or purified enzyme, however, the words '*damage*' or '*disturbance*' will be used synonymously with 'injury'. For example, an abnormal '*disturbance*' in the nucleotide concentration of a liver parenchymal cell may be considered as synonymous with its description as a cell *injury*.

The definition of injury given above is, of course, dependent on a knowledge of what the *normal* range is for a given property or component, so that an abnormal deviation may be recognised. In many instances, where the parameter under study can be quantitatively measured, statistical analysis can be used to estimate the probability that a particular value is abnormal. For example, it is common practice to consider as abnormal an experimental value that is more than two standard deviations away from the estimated mean value of the population. However, in other situations where subjective assessment is involved, or where the range of 'normal' values is large, the question of what is or what is not abnormal becomes more difficult to decide and the borderline between the two groups is indistinct. The examples discussed in this work will be restricted to disturbances where the assessment of abnormality by using standard statistical analysis is a relatively straightforward process.

The study of naturally occurring cell injuries is greatly facilitated by the use of model systems that enable particular reactions to be relatively well isolated from other interfering effects. For example, the mechanism underlying the action of carbon tetrachloride on rat liver may be studied at various levels of organisational complexity. The toxic agent can be given to a living rat and samples of whole liver analysed at sequential time-intervals. Alternatively, the whole liver may be removed and maintained in a *perfused* state free from hormonal or nervous influences whilst it is exposed to carbon tetrachloride. The liver may be cut into thin *slices* and its metabolism studied in this relatively intact state with or without exposure to carbon tetrachloride; very thin slices (*sections*) can be examined histochemically to see the changes in enzyme or coenzyme activities throughout a lobule. Microdensitometric techniques allow quantitative examination of *single cells* in such sections. The liver cell membranes may be disrupted by homogenisation and the behaviour of the *homogenate* or its constituent (intracellular) fractions examined. If the action of carbon tetrachloride can be located to involve a particular intracellular fraction (e.g. endoplasmic reticulum) then it may be convenient to extract and purify a single *enzyme* from the fraction for further study. Finally, having examined the interaction with a particular enzyme it may be possible to devise an even simpler model system involving only a part of the enzyme. Examples of the application of each of these levels of study to the investigation of the hepatotoxicity of carbon tetrachloride will be given in Chapters 9 and 10.

In such model experiments the use of inbred and genetically homogeneous animals is of considerable help as the standard deviation of each experimentally determined parameter is usually within a small percentage of the mean. It is important to note, however, that a description such as 'an inbred strain of white albino rat' hides a variety of differences in tissue composition dependent on the dietary background etc. of the rats concerned. For example, in the author's laboratory, female albino rats of the same body weight but from three different *sources* had liver ATP concentrations of $1{\cdot}85 \pm 0{\cdot}05$, $2{\cdot}44 \pm 0{\cdot}12$ and $2{\cdot}03 \pm 0{\cdot}13$ μmol/g wet wt. The same three groups of rats had microsomal glucose 6-phosphatase activities of 71 ± 5, 157 ± 7 and 127 ± 19 μg of phosphate/min/g of microsomal protein (Slater and Sawyer, 1969; Slater and Delaney, 1970). Similar variations in enzyme activities in rats of the same strain but from different sources have been reported by Eggleston and Krebs (1969).

For our purposes it can be stated that a cell or cell component has been 'injured' if its value or morphological character is significantly different from the range of values or characters existing in a normal population that is as homogeneous as possible. When the measurements are made on tissue or subcellular suspensions rather than, for example, with a purified enzyme preparation then certain difficulties arise in the sampling process. Most current biochemical techniques involve estimations of a minimum of 10 mg wet wt. of tissue per sample (or the equivalent of a subcellular fraction). Such a sample contains material from a large number of cells. Rat liver, for example, contains approximately 10^6 cells/10 mg wet wt. The average value obtained with such a sample therefore represents contributions from similar types of cell in various stages of growth or senescence, and also contributions from cells of different morphological and functional classes. The *proportions* of cells of different functional types may change quite substantially during the development of a particular lesion. This has been shown clearly by the studies of Daoust (1955) and Rubin *et al.* (1961) in which rats were fed with ethionine-rich diets; after 7 weeks of feeding the predominant cell type in the liver was the bile ductular cell. In normal liver, however, about 60% of the total cell population consists of hepatocytes; the remainder are mainly bile duct cells, Kupffer cells and blood vessel endothelial cells (Harkness, 1954). If values are required that are appropriate to a particular cell type in a determined metabolic status, then other and more complex techniques can be used; for instance, synchronous cell culture, quantitative cytochemistry or ultra-microdissection.

The deviation of a given property from a normal range, the deviation constituting the injury under investigation, may be reversed in time by **homeostatic** mechanisms that are capable of dealing with the perturbation. In contrast to such transient and reversible injuries, there can be deviations from the normal state so severe that they are essentially irreversible; in this case some permanent distortion of the cells' behaviour results. This may lead to a relatively stable though abnormal steady-state situation, or the disturbance may spread progressively through other interactions to produce a network of deleterious effects that culminate in premature cell death.

Living cells have finely balanced built-in regulatory mechanisms that favour in an appropriate time-scale the continuence of the *status quo*; this property leads to the establishment of steady-state situations (see Higgins, 1965). As a result of such homeostatic mechanisms it can be seen that a minor injury may be considered to be one that produces only a transient disturbance of the original metabolic pattern. Indeed, it is often profitable to study normal control mechanisms by just such a procedure: to apply a small disturbance and then to analyse the return of the system to its original state by procedures akin to those of chemical perturbation theory. By disturbing one point of a closely interdependent metabolic network it is possible to obtain information about dynamic interactions that are not normally easily observed by direct analysis of normal tissue. Although the last-named procedure is by far the most common practice in biochemical studies it suffers from a major disadvantage in being a time-frozen study on a time-dependent system. In this respect it can be seen that studies on cell injury, as well as providing information of value to the treatment of disease, may also uncover hitherto unrecognised features of normal metabolism.

8.3 Necrosis

Throughout Chapters 9–15 there will be frequent references to the development of necrosis. "Necrosis means the death of a group of cells as a result of injurious influences while still in contact with the living body" (Perry and Miller, 1961). The criteria used for assessing the presence of necrotic cells are histological and in general are reflections of changes *post-mortem*. It is important to realise that particular *functions* of the cell may be lost long *before* the histological appearance of necrosis is achieved. Conversely, the presence of necrosis does not mean that all metabolic processes in the affected tissue have ceased. Certain *reactions* in the cell are remarkably resistant to injury and may carry on more or less normally long after the cell has lost its last vestiges of *organised* behaviour. Although the term 'necrosis' is a convenient descriptive label for pathologists it has little to offer to a biochemical approach to cell injury. The recognition of necrosis indicates only that *fatal* disturbances to the cells have already occurred but gives no information about the dynamics of the processes involved. The end-result that is described broadly as necrosis may arise, of course, by any of a large number of very different metabolic disturbances.

8.4 Time-scale of changes

In any study of tissue injury it is as well to have an appreciation of the widely different time-scales of the events that occur at various organisational levels. Some changes may take hours, days or even weeks to become apparent by histological examination, thereby giving the impression that the basic disturbances were also slow in their development.

In general this is far from the truth, as a simple example concerning the effects of ischaemic anoxia of the liver will show.

A number of investigations have been made of the disturbances that result from tying off the vasculature supplying one lobe of the rat's liver (Gallagher *et al.*, 1956; De Duve and Beaufay, 1959; Bassi and Bernelli-Zazzera, 1964; Gaja *et al.*, 1965; Goldblatt *et al.*, 1965; Trump *et al.*, 1965). By histological criteria liver tissue appears to be capable of withstanding approximately 30 min of ischaemia without irreversible damage (i.e. the subsequent appearance of necrosis). This cannot be taken to imply that events potentially injurious to the liver cells have not occurred early on (whereas in fact they have) or that all tissues are equally resistant to ischaemic anoxia (whereas in fact they are not). Considering the latter point first, the range of susceptibilities to anoxia is very wide. For example the marked cerebral dysfunction produced by a few *minutes* of hypoxia is well known; by contrast, the newborn rabbit continues breathing for 30 min in an atmosphere of nitrogen (see Cameron, p.381, 1952).

A few illustrations of early and important metabolic perturbations in the anoxic liver can be given. First, if analyses are made of the total amounts of NAD^+ + NADH, $NADP^+$ + NADPH or ATP in the affected lobe it will be found that these decrease appreciably over the first few *minutes* of anoxia. Slater *et al.* (1964) reported that after 6 min of ischaemic anoxia the amounts of NAD^+ + NADH and $NADP^+$ + NADPH in liver samples had decreased by 12% and 21% respectively. ATP had decreased by approximately 20% by 2 min *post-mortem* (Delaney, 1969).

With the isolated perfused liver preparation Brauer (1965) has shown that total ischaemia rapidly affects bile flow. The average time of ischaemia required to reduce bile flow to one-half of its initial value was 46 s at 37·5°C. In this instance an effect was easily observable within a *minute* of inducing ischaemia.

With rapid freezing techniques for the sampling of rat liver tissues, and micro-methods for the estimations of substrate and coenzymes in the frozen liver powder, it has been found (Chance, 1965; Scholz and Bücher, 1965) that appreciable changes occur within *a few seconds* of imposing ischaemia. For example, after 20 s of ischaemia the *ratios* of ATP/ADP and of NAD^+/NADH changed from 5·7 and 7·0 to 0·55 and 2·6 respectively.

The discussion above illustrates the importance of rapidity in the sampling and fixation procedures used for tissue analysis if mechanistic details underlying any particular injury process are to be unravelled. Obviously, for a study of the kinetics of disturbances in biological systems to understand molecular perturbations involved in disease, it is essential to use specialised rapid sampling techniques of tissue *in situ*. A Symposium volume on this subject has been published (Chance *et al.*, 1964).

Consideration in Chapters 9–11 of disturbances to liver due to carbon tetrachloride and ethanol intoxications will enable a fairly comprehensive treatment to be given of injuries resulting in necrosis and fatty degeneration. Later Chapters will deal with injuries primarily of intracellular membrane systems (Chapter 12), of chemically induced cancer (Chapter 13) and of tissue damage resulting from radiation or excess metal accumulation (Chapters 14 and 15).

References

Bassi, M., Bernelli-Zazzera, A., 1964, *Expl. molec. Path.*, **3**, 332.
Brauer, R. W., 1965, *The Biliary System*, Ed. W. Taylor (NATO Advanced Study Group Symposium, Newcastle; Blackwell Scientific Publications, Oxford), p.58.
Cameron, G. R., 1952, *Pathology of the Cell* (Oliver and Boyd, Edinburgh).
Cameron, G. R., 1964, *Harben Lectures* (reprinted in *Jl. R. Inst. publ. Hlth. Hyg.*), **27**, 7, 33, and 65.
Cameron, G. R., Spector, W. G., 1966, *The Chemistry of the Injured Cell* (Charles C. Thomas, Springfield, Illinois).
Chance, B., 1965, *Control of Energy Metabolism*, Eds. B. Chance, R. W. Estabrook, J. R. Williamson (Academic Press, New York), p.415.
Chance, B., Eisenhardt, R. H., Gibson, Q. H., Lonberg-Holm, K. K. (Eds.), 1964, *Rapid Mixing and Sampling Techniques in Biochemistry* (Academic Press, New York).
Daoust, R., 1955, *J. Natn. Cancer Inst.*, **15**, 1447.
Dawkins, M. J. R., Rees, K. R., 1959, *A Biochemical Approach to Pathology* (Edward Arnold, London).
De Duve, C., Beaufay, H., 1959, *Biochem. J.*, **73**, 610.
Delaney, V. B., 1969, Ph. D. Thesis, University of London.
Eggleston, L. V., Krebs, H. A., 1969, *Biochem. J.*, **114**, 877.
Gaja, G., Bernelli-Zazzera, A., Sorgato, G., 1965, *Expl. molec. Pathol.*, **4**, 275.
Gallagher, C. H., Judah, J. D., Rees, K. R., 1956, *J. Path. Bact.*, **72**, 247.
Goldblatt, P. J., Trump, B. F., Stowell, R. E., 1965, *Am. J. Path.*, **47**, 183.
Harkness, R. D., 1954, *Liver Function: A Symposium on Quantitative Assessment*, Ed. R. W. Brauer (American Institute of Biological Sciences, Washington, DC), p.59.
Higgins, J. J., 1965, *Control of Energy Metabolism*, Eds. B. Chance, R. W. Estabrook, J. R. Williamson (Academic Press, New York), p.13.
Peery, T. M., Miller, F. N., 1961, *Pathology* (Churchill, London), p.10.
Rubin, E., Hutterer, F., Gall, E. C., Popper, H., 1961, *Nature, Lond.*, **192**, 886.
Scholz, R., Bücher, T., 1965, *Control of Energy Metabolism*, Eds. B. Chance, R. W. Estabrook, J. R. Williamson (Academic Press, New York), p.393.
Slater, T. F., Delaney, V. B., 1970, *Biochem. J.*, **116**, 303.
Slater, T. F., Sawyer, B. C., 1969, *Biochem. J.*, **111**, 317.
Slater, T. F., Sawyer, B. C., Sträuli, U. D., 1964, *Archs. int. Physiol. Biochim.*, **72**, 427.
Trump, B. F., Goldblatt, P. J., Stowell, R. E., 1965, *Lab. Invest.*, **14**, 343.

Hepatotoxicity of carbon tetrachloride: fatty degeneration

9.1 Introduction

In this and the following Chapter the hepatotoxicity of several halogen derivatives of methane: **chloroform** ($CHCl_3$), **fluorotrichloromethane** ($CFCl_3$), **bromotrichloromethane** ($CBrCl_3$) and **carbon tetrachloride** (CCl_4), will be discussed in general terms. By far the most attention will be paid to carbon tetrachloride, which damages the liver probably through mechanisms involving the formation of free-radical intermediates.

The halogeno-alkanes and -alkenes are industrially very important and many of them cause liver injury when ingested or inhaled in other than trace amounts. Unlike the halogeno-methanes, the halogen derivatives of the higher alkanes and alkenes have not been extensively studied biochemically, although considerable data are available for histopathological changes and dose–response characteristics. The reader is referred to comprehensive texts by van Oettingen (1955) and Browning (1965) for details of the toxicology of these particular halogeno-derivatives.

It is of value to list (Table 9.1) the main chemical and physical features of the halogenomethanes before discussing the general properties of these compounds.

Table 9.1. Properties of halogenomethanes.

Compound	*M*	*D* (g/cm³)	S_W (g/100 ml)	S_M (μg/ml)	ED_{50} mmol/kg body wt.	LD_{50} mmol/kg body wt.	H	*B*	b.p. (°C)
CCl_4	154	1·59	0·08	34	0·45	200	++++	68 (Cl)	77
$CHCl_3$	119	1·48	0·85	266	1·40	6	++++	90 (H)	61
$CFCl_3$	137	1·46	0·11	9			+	74 (Cl) 102 (F)	24
$CBrCl_3$	198	1·96	v.s.s.	2				49 (Br)	104

Abbreviations:

M Molecular wt.
D Density.
S_W Solubility in water.
S_M Solubility in a stock medium containing microsomes plus supernatant (see Slater and Sawyer, 1971).
H Hepatotoxicity assessed by study of conventionally stained sections.
B Bond dissociation energy (C—halogen or C—H with the 'leaving atom' given in parenthesis).
ED_{50} Minimum dose required to produce significant prolongation of sleeping time (see Plaa *et al.*, 1958).
LD_{50} Data for the mouse (Plaa *et al.*, 1958).
v.s.s. Very slightly soluble.

Chloroform, first synthesised by Liebig in 1831, is used industrially as a solvent and as a chemical intermediate; it has been used for over a hundred years as a general anaesthetic, as will be discussed later. Chloroform is readily oxidised (e.g. photochemically) to the highly poisonous phosgene ($COCl_2$); to inhibit this decomposition chloroform is usually mixed with a small quantity of ethanol and stored in the dark. It will be noticed from Table 9.1 that chloroform is considerably (approximately tenfold) more soluble in water or in a tissue suspension than is carbon tetrachloride. After the oral administration of equimolar amounts (260 *or* 2600 μmol/100 g body wt.) of chloroform and carbon tetrachloride to rats *in vivo*, however, the concentration of chloroform in the liver 1–3 h later was only approximately twice that found for carbon tetrachloride (Reynolds and Lee, 1967; Table 10.12).

The bond-dissociation energies shown in Table 9.1 are taken from Cottrell (1958) and Walling (1957). The homolytic fission of chloroform probably involves the C—H bond, for although the C—H bond-dissociation energy is greater than that for C—Cl the gain in resonance energy through the formation of trichloromethyl with trigonal symmetry is decisive (Walling, 1957). The situation is, however, fairly evenly poised for in $CDCl_3$ the homolytic fission yields $CDCl_2^\bullet$ [Reactions (9.1) and (9.2)].

$$CHCl_3 \longrightarrow CCl_3^\bullet + H^\bullet \qquad (9.1)$$

$$CDCl_3 \longrightarrow CDCl_2^\bullet + Cl^\bullet \qquad (9.2)$$

The bond-dissociation energy for D(C—H) in chloroform is some 20 kcal mol^{-1} greater than for carbon tetrachloride and by using Hirschfelder's (1941) rule it can be calculated that Reaction (9.3),

$$R^\bullet + CHCl_3 \longrightarrow RH + CCl_3^\bullet \qquad (9.3)$$

where $R^\bullet$ is an unspecified radical, proceeds about five times more slowly than the corresponding reaction involving carbon tetrachloride. This point will be of considerable importance in the argument outlined in Chapter 10.

Fluorotrichloromethane has a low freezing point (−168°C) and is used commercially as a refrigerant. It has a very low solubility in water in comparison with chloroform and is less soluble than carbon tetrachloride in tissue suspensions that contain lipid (Table 9.1). The bond-dissociation energy of the C—F bond in fluorotrichloromethane is 102 kcal $mole^{-1}$ (Cottrell, 1958), which compares with a bond-dissociation energy of 121 kcal mol^{-1} in carbon tetrafluoride. Although it might be expected that homolysis of fluorotrichloromethane favours the formation of $CFCl_2^\bullet$, the gain in energy through the resonance terms arising in $CCl_3^\bullet$ makes the prediction of cleavage of the compound to $CFCl_2^\bullet$ or to $CCl_3^\bullet$ rather difficult. It is likely that the bond-dissociation energy of $CFCl_2$—Cl is in the region of 75 kcal mol^{-1}, a value somewhat greater than the bond-dissociation energy of the C—Cl bond in chloroform: $D(CCl_2H—Cl) \leqslant 72$. A similar estimate of the bond-dissociation energy, D(C—Cl), of fluorotrichloromethane has been made by Gregory (1966).

Bromotrichloromethane is very slightly soluble in water and not much more so in tissue suspensions. The bond-dissociation energy of C—Br in bromotrichloromethane is low (49 kcal mol^{-1}), showing that homolysis of bromotrichloromethane will proceed more readily than is found with carbon tetrachloride. In fact bromotrichloromethane readily undergoes photochemical degradation in the presence of a reducing agent such as ethanol. The difference in reactivities between bromotrichloromethane and carbon tetrachloride is sufficiently great to allow homolytic reactions of the former to be studied in carbon tetrachloride as solvent (Sosnowsky, 1964). An indication of this difference in reactivities is the difference in absorption continua; for carbon tetrachloride the absorption continuum covers 2280–2800 Å whereas with bromotrichloromethane the range is around 3500 Å. The bond-dissociation energies discussed above show that the probable order of free-radical reactivity with a fixed radical species $R^\bullet$ is $CBrCl_3 > CCl_4 > CFCl_3$, $CHCl_3$.

9.2 Free-radical reactions involving carbon tetrachloride

The *homolysis* of the chlorinated methanes to yield **trichloromethyl**, particularly in the gas phase, has been the subject of many studies. Bromotrichloromethane and carbon tetrachloride, in particular, readily enter into homolytic reactions (see Walling, 1957). Four examples concerning carbon tetrachloride are described briefly.

1. Olefins containing a terminal double bond react readily with carbon tetrachloride upon irradiation [Reaction (9.4)].

$$R{-}CH{=}CH_2 + CCl_4 \xrightarrow{\text{radiation}} R{-}\underset{\displaystyle Cl}{\underset{|}{CH}}{-}CH_2CCl_3 \qquad (9.4)$$

2. The reaction of *N*-vinyl carbazole (NVC) with carbon tetrachloride in sunlight at 34°C yields high molecular weight polymers (Biswas and Ghoshal, 1966). This reaction is inhibited by oxygen and does not proceed in the dark [Reactions (9.5)–(9.7)].

$$\text{NVC} + CCl_4 \xrightarrow{\text{radiation}} [\text{NVC}, CCl_4]\text{ complex} \qquad (9.5)$$

$$[\text{NVC}, CCl_4]\text{ complex} \longrightarrow \text{NVC}^{\bullet+}Cl^- + CCl_3^\bullet \qquad (9.6)$$

$$\text{NVC} + CCl_3^\bullet \longrightarrow \text{NVC}{-}CCl_3^\bullet \longrightarrow \text{polymer} \qquad (9.7)$$

3. In the interesting reaction described by Urey and Eiszner (1952) in which diazomethane combines photochemically with carbon tetrachloride to yield pentaerythrityl tetrachloride [Reaction (9.8)]

$$4(\text{diazomethane}) + CCl_4 \xrightarrow{h\nu} C(CH_2Cl)_4 \qquad (9.8)$$

the initiation stages are those in Reactions (9.9) and (9.10).

$$CH_2N_2 \xrightarrow{\text{radiation}} CH_2\colon + N_2 \qquad (9.9)$$

$$CH_2\colon + CCl_4 \longrightarrow CH_2Cl^{\bullet} + CCl_3^{\bullet} \qquad (9.10)$$

4. The reaction between aliphatic amines and carbon tetrachloride gives a charge-transfer complex unstable to light. Decomposition to the aminium ion radical and trichloromethyl occurs (Biswas and Ghoshal, 1966).

$$\underset{\text{(DMA)}}{\textit{NN}\text{-Dimethylaniline}} + CCl_4 \xrightarrow{\text{radiation}} \underset{\text{radical}}{\text{DMA ion}} + CCl_3^{\bullet} \qquad (9.11)$$

9.3 Pharmacological effects of the halogenomethanes

Chloroform was introduced as a general anaesthetic by Simpson in 1847. An interesting account of the early uses of chloroform as an anaesthetic may be found in the book by Duncum (1947). Chloroform enjoyed considerable favour as a general anaesthetic for surgical purposes: it produces a more pleasant induction period than ether, and gives less irritation of the bronchi. This latter property is aided by the fact that chloroform is a more potent anaesthetic than ether and, consequently, a lower concentration is required to produce a given plane of anaesthesia. However, chloroform has several severe disadvantages compared with ether: it progressively lowers blood pressure and cardiac output, and may sensitise the heart to adrenalin and related agents. Sudden cardiac arrest with fatal results may occur during chloroform induction and this is associated with ventricular fibrillation. In fact, several such deaths were reported within a few years of the introduction of chloroform as an anaesthetic in 1847, but the mechanisms involved were not fully recognised for many years (see Goodman and Gilman, 1955). A further hazard of chloroform (and of carbon tetrachloride) anaesthesia is that prolonged inhalation can cause severe liver damage; this is more severe with carbon tetrachloride than with chloroform and the use of the former as a general anaesthetic was soon discontinued. Other general inhalation anaesthetics that are far superior to chloroform have been developed of which the most important is **halothane** (Fluothane; trifluorobromochloroethane, $CF_3-CHClBr$). The average concentrations of chloroform, ether and halothane in blood and liver during general anaesthesia are given in Table 9.2.

Halothane is very widely used in human surgery and has been extensively screened for toxicity, particularly for any deleterious effects on liver function. A recent survey involving almost a million patients showed that halothane produced about the same number of post-operative cases of massive hepatocellular necrosis as did ether[(1)]. The numbers of

[(1)] The incidence of acute liver necrosis after halothane was 1·02 in 10000 administrations; for ether and cyclopropane the corresponding figures were 0·49 and 1·70 respectively.

patients reacting in this way is so small that it is difficult to decide whether such patients were in the process of developing a viral hepatitis, for example, before halothane administration, or whether a genuine, abnormally high, sensitivity to halothane exists in an exceedingly small proportion of the population. A case has been reported of an anaesthetist who consistently became jaundiced on re-exposing himself to halothane (Klatskin and Kimberg, 1969). Since halothane is metabolised by enzyme systems in the liver endoplasmic reticulum it is possible that a bizarre metabolic route is available in a few individuals, resulting in the formation of toxic product(s) and subsequent liver damage. The literature concerning the possible existence of a hypersensitive state to halothane is voluminous and the interested reader is referred to the article summarising the National Halothane Study (1966)[2].

Table 9.2. Concentrations of ether, chloroform or halothane in blood, liver and adipose tissue during anaesthesia. The results for ether and chloroform refer to man, and are from Goodman and Gilman (1955). The results for halothane are for the rat (Duncan and Raventós, 1959); the rats were exposed to halothane + air (1·5:98·5, *v*/*v*).

Anaesthetic	Concentration		
	blood (mg/100 ml)	liver (mg/100 g)	adipose tissue (mg/100 g)
Ether (light anaesthesia)	50–130		
Ether (heavy anaesthesia)	130–150		
Chloroform (light anaesthesia)	2–10		
(heavy anaesthesia)	10–20		
Halothane 1 h exposure	16·9	22	100
2 h exposure	16·8	23	250
3 h exposure	22·0	35	450
6 h exposure	21·5	48	950

9.4 Liver damage due to carbon tetrachloride and chloroform

As already mentioned, high concentrations of chloroform sensitise the myocardium and fatal ventricular fibrillation can result. In fact, within 16 years of the introduction of chloroform as an anaesthetic the number of deaths assigned to its use had reached 123 (Chloroform Committee, 1864), a disturbing state of affairs that led to the setting up of a Committee of Investigation. The realisation of the *hepatotoxic* effects of chloroform came more slowly, partly since the consequences were less obviously dramatic than sudden cardiac arrest. The hepatotoxic consequences of inhalation of carbon tetrachloride were more readily

[2] Possible mechanisms to account for the very rare instances of acute liver damage that may result from halothane are discussed by Slater (1971).

apparent and its use as an anaesthetic, after its introduction by Nunnelly in 1849, was short-lived. The recognition of the hepatotoxicity of chloroform and of carbon tetrachloride led to direct experimentation on animals in attempts to understand the mechanisms involved in the tissue damage.

It was soon recognised that the administration of these compounds to rats or rabbits resulted in necrosis of the **centrilobular** parenchymal cells and that many cells throughout the lobules were overloaded with **fat**. The early literature is surveyed by Wells (1908) and by Graham (1915). A lucid account of the histopathological changes in experimental poisoning with carbon tetrachloride in the rat is given by Cameron and Karunaratne (1936), and their data with some later results form the basis of the following discussion on the time-sequence of events.

When carbon tetrachloride is given *orally* to rats at a dose of approximately 1 ml/kg body wt. fat begins to accumulate noticeably about 3 h after dosing. Chemical analysis of *liver triglyceride* shows that this increases linearly from the earliest stages of intoxication (Schotz and Recknagel, 1960): 1 h after dosing the increase was 34%, and 3 h after dosing the liver triglyceride had increased 195%.

In contrast, the histological signs indicative of necrosis are slower in development. A few isolated severely damaged cells are visible by 6 h but, in general, the main alteration in the centrilobular cells of the liver in the dosed rats during the early hours of intoxication is a loss of basophilia (Rosin and Doljanski, 1946; Leduc and Wilson, 1958). Later, cells near the central vein swell up and their nuclei become pycnotic; these balloon cells are very noticeable 18 h after dosing (Figure 9.1). By 24 h after dosing the full extent of centrilobular necrosis is evident (Figure 9.1); 2–3 days after dosing considerable regeneration and replacement of the dead cells has commenced. A detailed re-investigation of this time-sequence has recently been made by Wigglesworth (1964), with thin-sectioning techniques and osmic acid staining; his results are in agreement with those of Cameron and Karunaratne.

The time-scale of acute liver cell injury can be followed with some degree of precision by measuring the activity of a number of enzymes released by the damaged liver cells into the plasma. Enzymes normally located in the cell sap fraction of the liver cell (e.g. lactate and isocitrate dehydrogenases) increase in activity in the plasma within a few hours of administration of carbon tetrachloride. Enzymes normally located in the mitochondrial fraction (e.g. glutamate and malate dehydrogenases) leak out more slowly (see Rees and Sinha, 1960; Fox *et al.*, 1962). In a few cases the rise in plasma enzyme activity has been correlated with a decrease in the liver's content of the relevant enzyme. The release of liver protein into plasma has also been demonstrated by immunological procedures (Espinosa and Insunza, 1962). Some enzymes that increase in activity (e.g. acid ribonuclease) in the plasma during carbon tetrachloride

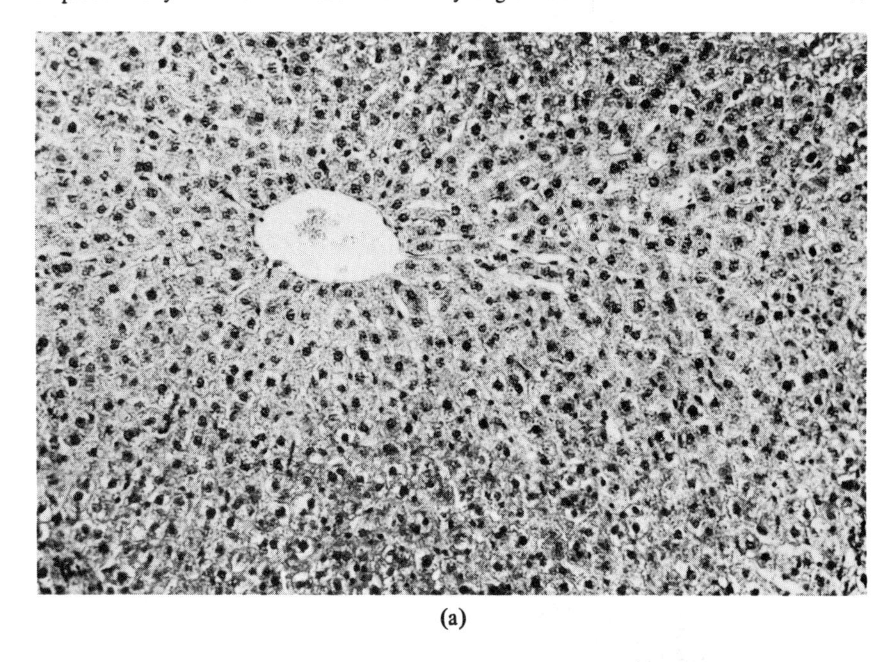

(a)

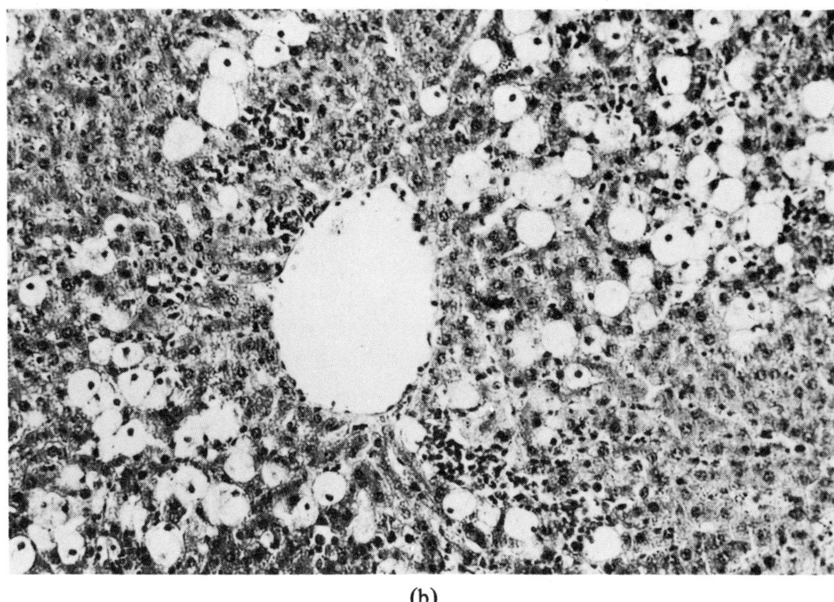

(b)

Figure 9.1. Photomicrographs of sections of rat liver.
(a) A liver from a control rat was perfused through the portal vein with cold 0·9% (*w/v*) NaCl solution and then with 4% glutaraldehyde in cacodylate buffer (0·1 M, pH 7·2). Sections were stained with haematoxylin and eosin. The picture shows an area around a central vein with uniformly sized nuclei and uniform cytoplasmic staining; (b) a liver was obtained from a rat poisoned orally 24 h previously with a mixture of carbon tetrachloride - liquid paraffin (1 : 3, *v/v*). The liver was processed as in (a). The picture shows an area around a central vein with loss of cytoplasmic staining, irregularly sized nuclei, and balloon cells.

intoxication do not appear, however, to originate from damaged liver cells; the tissue of origin in such cases is unknown (Slater and Greenbaum, 1965).

The *dose* of carbon tetrachloride may be quite critical for the demonstration and interpretation of certain metabolic disturbances. In the rat, increasing the dose much above $1 \cdot 5$ ml kg^{-1} does not cause an appreciable increase in the *extent* of centrilobular necrosis (similar results have been obtained by Gerhard *et al.*, 1970, in the mouse), yet many studies use doses as high as 5 ml kg^{-1}. At such high concentrations it is possible that secondary effects may cloud the picture, making it difficult to elucidate which changes are closely involved in the onset of centrilobular necrosis and which are secondary to the compound's action as a lipid solvent. Such high doses of carbon tetrachloride may, for example, increase the *rate* at which certain events progress; or, by virtue of the lipophilic nature of carbon tetrachloride the high doses may produce membrane changes (e.g. lysosomal) that are not essential for the production of necrosis. As the oral dose is increased there is the added danger of causing acute gastro-intestinal effects such as gastric haemorrhage. In such cases death may result not from any manifestation of liver injury but from intestinal shock and cardiovascular collapse. This suggests that caution is necessary when comparing widely differing mortality rates for carbon tetrachloride in rats previously subjected to varying experimental procedures: the alterations in the mortality rates may relate to different mechanisms responsible for death. Increasing the dose also increases the narcotic action of carbon tetrachloride and this is associated with a significant fall in *body temperature*; this latter response further complicates the interpretation of the experimental data (see Section 10.9). Some effects of carbon tetrachloride on the liver may be demonstrated with very small doses. Reynolds and Lee (1967), for example, have found that 9 μl of the compound/kg body wt. produces a detectable decrease in the enzyme glucose 6-phosphatase; liver necrosis was evident with a dose of 90 μl/kg body wt.

The *route* of administration of carbon tetrachloride is also of importance to the extent and nature of the liver injury that subsequently develops. Dodson *et al.* (1965) have reported that carbon tetrachloride given *subcutaneously* was less effective in producing liver injury than the same dose given *orally*; Cameron and Karunaratne (1936) reported that a small *intraportal* injection of carbon tetrachloride in the rat caused massive necrosis of the liver in contrast to the centrilobular type of necrosis normally observed after oral dosage.

Although chloroform gives a similar type of liver injury to that described for carbon tetrachloride it is less toxic in molar terms: Plaa *et al.* (1958) reported that the effective dose required to give a significant prolongation of sleeping time in 50% of the mice dosed (the ED_{50} dose) was $0 \cdot 45$ mmol kg^{-1} for carbon tetrachloride and $1 \cdot 40$ mmol kg^{-1} for chloroform. Similar results have been found by Reynolds and Lee (1967).

Chloroform is also less effective than carbon tetrachloride in decreasing total liver $NADP^+$ + NADPH (Slater and Sawyer, 1971; Table 9.3), and is virtually inactive in decreasing glucose 6-phosphatase (Reynolds, 1963). The comparison between chloroform and carbon tetrachloride is complicated, however, by the more profound anaesthetic properties of chloroform and by its higher solubility in aqueous media (Table 9.1).

A series of important discoveries in the period 1961–1966 suggested that the metabolic disturbances that result in the development of necrosis are different from those responsible for the increase in liver fat. The major studies leading to this conclusion were as follows: Leduc and Wilson (1958) showed that sulphaguanidine decreases the extent of liver necrosis but has little effect on the accumulation of fat resulting from the administration of carbon tetrachloride; a similar finding was reported by Rees *et al.* (1961) with the antihistamine drug promethazine. Other procedures that resulted in similar dissociations of centrilobular necrosis and fatty degeneration were adrenalectomy and the administration of adrenergic blocking drugs (Brody, 1963). Further, as will be discussed more fully in Section 10.4, McLean and McLean (1966) have found that rats placed on a protein-free diet develop markedly less liver necrosis than do rats on a normal diet, although triglyceride accumulation is unaffected. Finally, Alexander *et al.* (1967) have reported that dosing with asparagine decreases the accumulation of fat but has little effect on the development of necrosis. The findings described in the above-mentioned investigations will be discussed in detail in later Sections of this and the following Chapters. The hypothesis that fatty degeneration and necrosis develop along essentially different pathways will be accepted here, thereby enabling the two major events in carbon tetrachloride intoxication to be treated separately.

Table 9.3. Effects of chloroform *in vivo* on rat liver NADPH and glucose 6-phosphatase activity (data of T. F. Slater and B. C. Sawyer). Chloroform was administered by stomach tube at a dose of 1·25 ml/kg body wt. The numbers of estimations are in parentheses; values are given as percentages of the control group mean values.

Time (h)	Liver			
	$NADP^+$ + NADPH		Glucose 6-phosphatase	
	control	$CHCl_3$	control	$CHCl_3$
1[a]	100 (2)	114 (2)		
18[a]	100 (2)	52 (2)		
24			100 (3)	97 (3)

[a] Rats from different sources were used in these two experiments.

9.5 Mechanisms underlying fatty degeneration

The lipid that accumulates in the liver after rats are dosed with carbon tetrachloride is almost wholly triglyceride (Table 9.4). There are various ways in which an increased content of triglyceride can arise (see Figure 9.2): (1) by a *decreased oxidation* of fatty acids; (2) by an *increased synthesis* of fatty acids and of triglyceride in the liver; (3) by an *increased transport* of non-esterified fatty acids from the fat depots to the liver; (4) by a *decreased synthesis* of phospholipid, important for the excretion of triglyceride; (5) by a *decreased transport* of triglyceride from the liver into the plasma.

Table 9.4. Effect of carbon tetrachloride on the liver lipids in the rat (data from Rees and Shotlander, 1963). Values are given as percentages of the wet weight of the liver. Promethazine was given at time 0 and +6 h after carbon tetrachloride.

Time after CCl_4 (h)	Total lipid	Triglyceride	Phospholipid	Cholesterol
0	4·2	1·3	2·6	0·2
+5	6·2	3·2	2·7	0·3
+22	12·7	9·7	2·9	0·2
+22 + Promethazine	9·8	6·5	2·8	0·2
+ Promethazine [a]	4·8	1·6	3·0	0·2

[a] Promethazine given at time 0 and +6 h; rats were killed at +22 h; no carbon tetrachloride was administered.

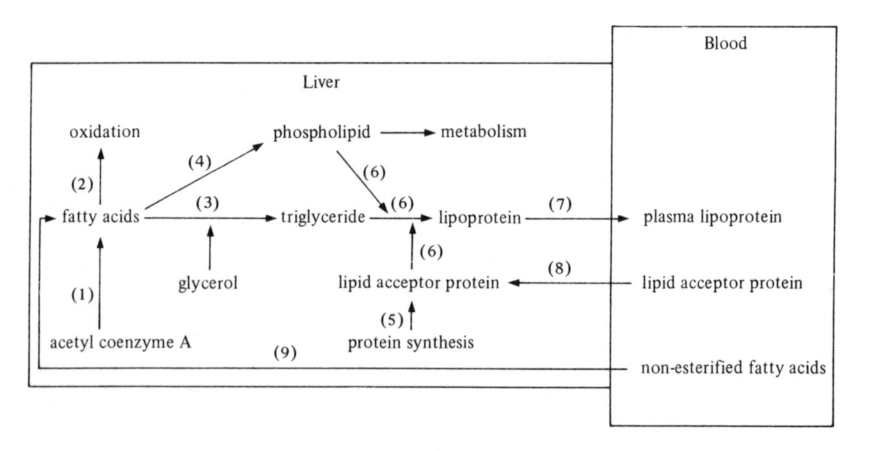

Figure 9.2. Pathways of fatty acid metabolism in liver.
Routes: (1) fatty acid biosynthesis; (2) fatty acid oxidation (β and ω); (3) triglyceride (TG) synthesis; (4) phospholipid synthesis; (5) synthesis of the lipid-carrier protein (LPC); (6) combination of phospholipid, triglyceride and lipid-carrier protein to form lipoprotein (LP); (7) secretion of lipoprotein out of the liver; (8) re-entry of lipid-carrier protein; (9) non-esterified fatty acids (NEFA) from depot fat utilised by the liver.

The increase in liver triglyceride does not appear to result from a failure in fatty acid β-oxidation to any significant extent. β-Oxidation of fatty acids is largely confined to the mitochondrial fraction of rat liver and no disturbances in mitochondrial fatty acid oxidation have been observed that would be compatible with the time-scale of triglyceride accumulation. For example, Reynolds *et al.* (1962) found no change in the oxidation of octanoate, succinate, β-hydroxybutyrate and glutamate in liver mitochondria 10 h after dosing with carbon tetrachloride; these oxidation rates were strongly decreased, however, 20 h after dosing.

Electron-microscopic examination of liver sections taken at successive intervals after administering carbon tetrachloride also failed to show significant alterations in mitochondrial structure during the first few hours of poisoning, in contrast to the marked swelling of the membranes of the endoplasmic reticulum that can be observed within 60 min of dosing (Oberling and Rouiller, 1956; Smuckler *et al.* 1962). Other evidence that major properties of the mitochondrial membrane are not affected early on in carbon tetrachloride intoxication is that citrate is retained within the mitochondrion and that fluoroacetate administration after dosing with carbon tetrachloride results in a normal accumulation of fluorocitrate in the mitochondria. Moreover, several hours after giving carbon tetrachloride, and when liver triglyceride has increased substantially, the mitochondria still re-accumulate potassium after a period of potassium depletion (Shore and Recknagel, 1959). Some relatively subtle changes do occur, however, in the mitochondrial fraction very early in the intoxication; these are connected with nucleotide shifts. Slater and Delaney (1970), using a rapid microfiltration procedure, found that the percentages of ATP and of NADPH in the mitochondrial fraction change in opposing directions 1–2 h after giving carbon tetrachloride. The data are shown in Table 9.5. These

Table 9.5. Effects of oral dosing with carbon tetrachloride (1·25 ml/kg body wt.) on the intracellular distributions of ATP and of $NADP^+$ +NADPH. The cell sap fraction (S) was isolated by a microfiltration technique (see Delaney and Slater, 1970). Results are expressed as a percentage of the total amount of component in the whole homogenate. Data are from Slater and Delaney (1970) and unpublished results of T. F. Slater and B. C. Sawyer. Mean values ±S.E.M. are given with the number of determinations in parentheses.

	Time (h)	ATP (% in fraction S)	Time (h)	$NADP^+$ + NADPH (% in fraction S)
Control	2	47 ± 1 (4)	1	37 ± 4 (6)
CCl_4	2	41 ± 1 (4)[a]	1	57 ± 6 (6)[b]

[a] $P < 0{\cdot}001$. [b] $P < 0{\cdot}01$.

nucleotide shifts may be of relevance to the necrogenic action of the compound and are further discussed in the following Chapter[(3)].

Although β-oxidation of fatty acids in the mitochondrial fraction appears relatively normal in the early phase of carbon tetrachloride intoxication there are no data available on the effects of carbon tetrachloride on the ω-oxidation of fatty acids in the endoplasmic reticulum (see Lu *et al.*, 1969). Since ω-oxidation is associated with the NADPH-cytochrome P_{450} electron-transport chain that undergoes early changes after administration of carbon tetrachloride (see Section 10.11) it is possible that ω-oxidation is inhibited early on in the injury process. The significance of such an inhibition, if it occurs, to the overall liver injury is not clear.

Another possible mechanism that would contribute to the increase in liver triglyceride in carbon tetrachloride poisoning is an increase in the new synthesis of triglycerides from fatty acids. This, however, has not been observed experimentally: when isotopically-labelled fatty acid was fed to rats poisoned with carbon tetrachloride and the specific activity of extracted lipid determined at various time-intervals later, Rees and Shotlander (1963) found a *decreased* incorporation rate in the poisoned rats (Table 9.6).

No changes in plasma non-esterified fatty acids have been found that are consistent with the mechanism for an increased liver triglyceride; the

Table 9.6. Effects of carbon tetrachloride on the incorporation *in vitro* of acetate, pyruvate or palmitate into rat liver lipids. In each case the carbon tetrachloride was given 3 h before the rat was killed and slices of liver were incubated for incorporation studies. Data are from Rees and Shotlander (1963).

Substrate	Incorporation (c.p.m. at infinite thickness)	
	Control	CCl_4
[1-^{14}C]Acetate	10·254	3·841
[3-^{14}C]Pyruvate	5·591	1·469
[1-^{14}C]Palmitate	31·456	17·631

[(3)] Another very early change reported to affect the mitochondrial fraction is the rapid influx of calcium ions into the mitochondria (Thiers *et al.*, 1960), particularly of the *mid-zonal* cells (Reynolds, 1963). This phenomenon has been re-examined by Cohn *et al.* (1968), who concluded that the major part of the calcium ion changes occur during the isolation of the mitochondria and may therefore be considered artefactual. In relation to the effects of carbon tetrachloride on membrane structures, Wands *et al.* (1970) have reported it to cause a rapid depolarisation of the transmembrane potential of rat liver cells. The control value of −37 mV changing to −11 mV by 24 h after dosing.

The relatively low variation in the values obtained from case to case led the authors to conclude that carbon tetrachloride produced a uniform alteration in transmembrane potential in all cells of the liver.

changes in non-esterified fatty acids are either too little or too late to be responsible for the early and massive increase in liver fat. Similarly, several studies have shown that there is no gross disturbance in phospholipid synthesis during the early stage of carbon tetrachloride intoxication.

It can be seen therefore that the increased triglyceride content after administration of carbon tetrachloride does not appear to result from an increased triglyceride synthesis, a decreased triglyceride catabolism via β-oxidation, an increased flux of non-esterified fatty acids to the liver or from a failure in phospholipid synthesis required for lipoprotein secretion. This leaves the question of lipoprotein assembly and its subsequent secretion to be considered.

Lipoproteins are complex proteins in which the proportions of fat, phospholipid and protein vary rather widely according to the source of the material. It is important to note in connection with the liver injury produced by carbon tetrachloride that the lipoproteins contain a phospholipid component, often to a considerable extent. A review of the properties of lipoproteins is given in Putnam (1960).

In man about 1% by wt. of serum is lipid; this means that 15–30 g of lipid is contained in the total blood compartments. Virtually all of the serum lipid is in a lipoprotein complex having a density of 0·93–1·16. Lombardi and Ugazio (1965) found that in plasma from starved rats 65% of the plasma triglyceride was bound in the form of a very low-density (less than 1·019) lipoprotein: it is as low-density lipoprotein that triglyceride is secreted from the liver.

In several species so far examined, the triglyceride component of plasma lipoprotein turns over much faster than the protein component, which, in the rat, constitutes only approximately 10% of the total mass (Lombardi and Ugazio, 1965). The half-life for triglyceride is 20–100 fold shorter than that known for the protein component. This suggests that the protein component can be utilised several times over in the formation of low-density lipoprotein in the liver, each time carrying further triglyceride from the liver to the peripheral depots. Evidence bearing on this has been obtained by Roheim *et al.* (1965b) with the isolated perfused liver preparation. When the perfusion medium used was Krebs–Ringer bicarbonate plus washed erythrocytes, the liver secreted triglyceride, cholesterol and phospholipid as lipoprotein at a very slow rate. When a lipid-acceptor protein fraction was added to the medium, however, the secretion rate of lipoprotein into the perfusate was markedly increased. The active material was found to be a glycoprotein[(4)] normally present in the globulin fraction. It seems likely that triglyceride secretion from the liver is dependent on the formation, in the endoplasmic reticulum, of very low-density lipoprotein, of which a necessary component is the lipid-acceptor protein. This has a relatively long half-life in comparison with

[(4)] For background references to glycoproteins and role in secretion and membrane specific properties see Eylar (1965) and Spiro (1969).

triglyceride, and although it is synthesised continuously by the liver, the major part of the triglyceride secretion is dependent on the recirculation of pre-existing lipid-acceptor protein.

If a failure in lipoprotein secretion is responsible for the triglyceride accumulation that is a feature of carbon tetrachloride intoxication, then the disturbance could involve (a) an inhibition of synthesis of the lipid-acceptor protein necessary for lipoprotein formation, (b) a failure in the combination of triglyceride, phospholipid and lipid-acceptor protein to produce lipoprotein or (c) a failure in the secretion of normally synthesised lipoprotein from the liver cell. These possibilities will now be discussed in turn.

9.5.1 Protein synthesis

It has long been known that carbon tetrachloride produces an early disturbance of cytoplasmic nucleic acid (i.e. RNA). For example, Rosin and Doljanski (1946) found that the numbers of pyronin-staining granules in liver parenchymal cells decrease shortly after rats are given carbon tetrachloride. Leduc and Wilson (1958) reported that centrilobular basophilia became less pronounced 15–30 min after giving carbon tetrachloride to mice. Quantitative study of such qualitative observations began in the early 1960s with Smuckler and Benditt's work in Seattle. It was shown that carbon tetrachloride caused a rapid loss of ribosomal particles from the membranes of the endoplasmic reticulum. This dislocation was accompanied by dissociation of the larger polysomes into smaller units (see Figure 9.3; Smuckler and Benditt, 1963). It was also found that carbon tetrachloride produces a very early decrease in the rate

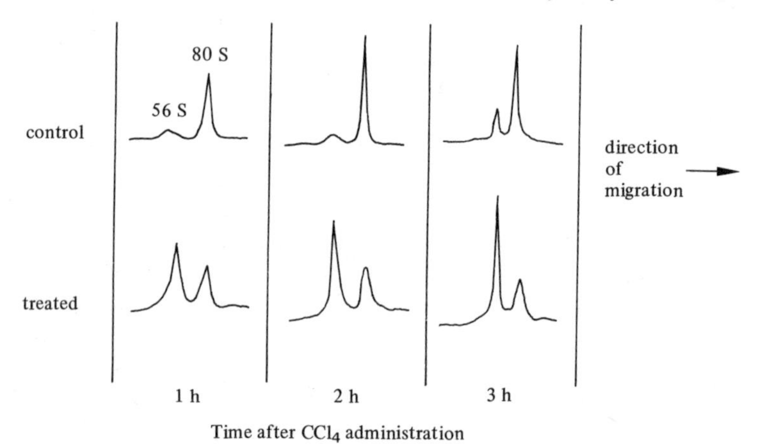

Figure 9.3. Effect of administering carbon tetrachloride (2·5 ml/kg body wt.) to rats on the ultracentrifugal sedimentation pattern of liver ribosomes. In control animals the major peak is at 80 S with a smaller peak at 54 S. In the treated rats there is a progressive loss of the 84 S peak with a corresponding increase in the 54 S peak. Data are from Smuckler and Benditt (1965).

of incorporation of labelled amino acids into plasma protein and especially into microsomal protein[5]. Table 9.7 quotes some of the early results of Smuckler and Benditt (1965).

Table 9.7. Effect of various oral doses of carbon tetrachloride on the incorporation *in vitro* of leucine into microsomal protein. The microsomal fraction was isolated 2 h after dosing with carbon tetrachloride. Data are from Smuckler and Benditt (1965).

Dose of CCl_4 (ml/kg body wt.)	% of incorporation into protein of control microsomes
0·00	100
0·7	47
1·3	27
2·5	28
5·0	29

After the discovery that carbon tetrachloride decreased protein synthesis in the liver the suggestion was made that the cause of fatty liver in intoxication with the compound was an inhibition of protein synthesis, with a subsequent failure of lipoprotein formation and secretion (see Rees and Shotlander, 1963). Indirect evidence favouring this view was put forward by Seakins and Robinson (1963), who gave **puromycin** to rats and obtained fatty livers in the treated animals. Puromycin, an antibiotic with a structure (Figure 9.4) resembling the terminal portion of transfer RNA, is known to produce a powerful inhibition of protein synthesis in the rat (Villa-Trevino *et al.*, 1964a).

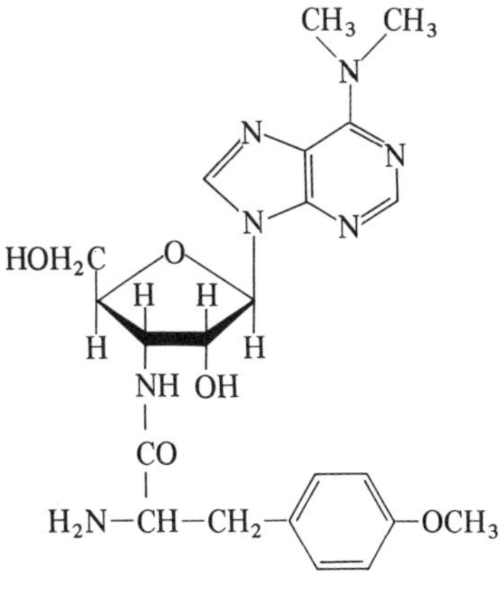

Figure 9.4. Structure of puromycin.

[5] Radioautographic studies (Monlux and Smuckler, 1969) have shown that the depression in protein synthesis is localised around the central veins and extends into the mid-zone; a similar localisation of the inhibition of protein synthesis occurs in the mouse after the administration of carbon tetrachloride (Gerhard *et al.*, 1970).

The discovery that carbon tetrachloride drastically inhibits microsomal protein synthesis led to suggestions by various authors that this was possibly a major feature not only in fatty degeneration but in the pathogenesis of necrosis as well. The relevance to necrosis, however, was sharply questioned after work, chiefly by Farber's group in Pittsburgh (see Farber *et al.*, 1964), on the fatty liver produced by **ethionine**. Ethionine is an analogue of the essential amino acid methionine and competes with it for ATP in the reaction shown in Figure 9.5.

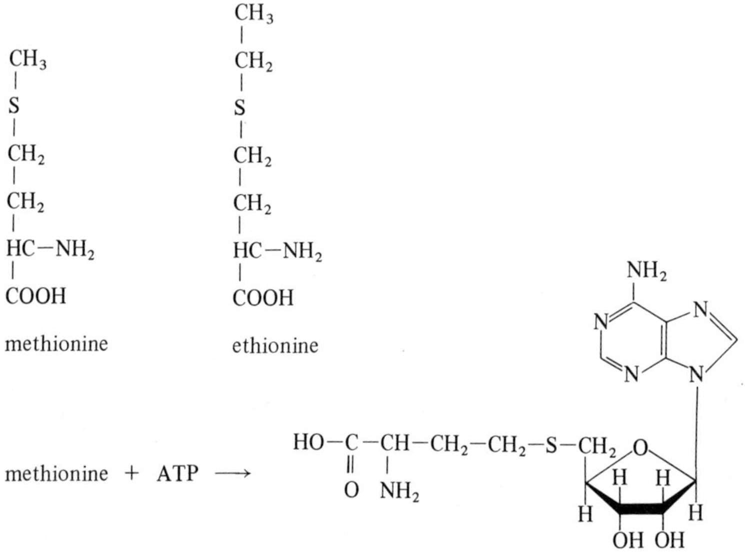

$$CH_3-S-CH_2-CH_2-CH(NH_2)-COOH$$

methionine

$$CH_3-CH_2-S-CH_2-CH_2-CH(NH_2)-COOH$$

ethionine

methionine + ATP ⟶ HO–C(=O)–CH(NH₂)–CH₂–CH₂–S–CH₂–(ribose, OH OH)–adenine

S-adenosylmethionine

Figure 9.5. Structures of methionine and ethionine, and the conversion of methionine into *S*-adenosylmethionine.

S-Adenosylmethionine is metabolically active and can donate C_1 units to a variety of acceptors; *S*-adenosylethionine, on the other hand, is relatively unreactive in such methyl transfers and thereby acts as a drain on cellular ATP. This was clearly shown first by Schull (1962) and confirmed shortly after by Villa-Trevino *et al.* (1964b), who also found that there was a rapid decrease in protein synthesis (see Figure 9.6).

Although the administration of ethionine produces a rapid loss of liver ATP and a rapid decrease in protein synthesis, it does *not* result in liver necrosis. A related observation was made by McLean *et al.* (1965), who showed that actinomycin D administered to rats causes severe inhibition of protein synthesis and eventual death, but liver necrosis is not a conspicuous feature of the injury. Such observations suggest that the decreased protein synthesis found after administration of carbon tetrachloride may not be responsible *per se* for the ensuing necrosis.

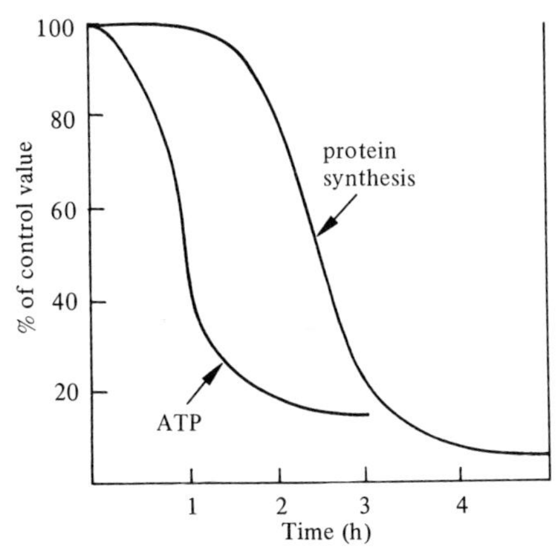

Figure 9.6. Effects of administering DL-ethionine (100 mg/100 g body wt.) to female rats on the liver's content of ATP and on the rate of incorporation of [^{14}C]leucine into liver ribosomal protein *in vitro*. Data are from Villa-Trevino *et al.* (1964b). The values are shown as percentages of the corresponding control values.

This still leaves unanswered the question whether the decreased protein synthesis is responsible for the fatty liver. Recent work with orotic acid helps to answer this difficult question. The administration to rats of a large dose of orotic acid (Figure 9.7) results in the slow accumulation of triglyceride in the liver (Standerfer and Handler, 1955; see Table 9.8), and a severe decrease in liver ATP (Rubin and Pendleton, 1964) and other nucleotides. The change in liver ATP occurs less rapidly than in ethionine intoxication; moreover orotic acid did not cause a significant decrease in liver protein synthesis, in contrast to the situation with ethionine. It seems unlikely therefore that the accumulation of triglyceride is due to a failure to synthesise sufficient quantities of lipid-acceptor protein, and this further suggests that the triglyceride disturbance in orotic acid intoxication [(6)]

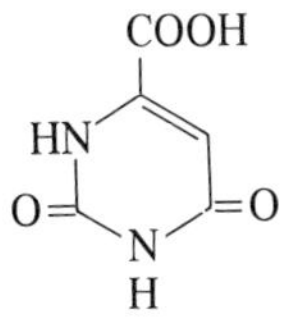

Figure 9.7. Structure of orotic acid.

[(6)] In relation to the fatty liver that is produced by feeding with orotic acid it is noteworthy that Kinsella (1967) has reported a large increase in lipid peroxidation in the treated livers. Rats were placed on a diet containing 1·5% orotic acid and liver samples taken after 4 and 8 days of treatment. Malonaldehyde estimations on lipid extracts showed that the control value of 7 nmol/liver was increased to 42 and 30 nmol/liver respectively after 4 and 8 days of treatment. It will be of interest to see the effect of concomitant treatment with vitamin E or *NN*′-diphenyl-*p*-phenylenediamine on these increased amounts of malonaldehyde.

involves the *coupling* of triglyceride with lipid-acceptor protein, an essential process before triglyceride secretion. This speculation seems justified by experiments with orotic acid and the isolated perfused liver technique (Roheim *et al.*, 1965a, 1965b), where a failure to incorporate the apoprotein of plasma lipoprotein into low-density lipoprotein was observed after orotic acid administration.

Table 9.8. Effects of orotic acid on rat liver lipid content and nucleotide concentrations. Orotic acid and adenine sulphate were included in the diet where required at 1% and 0·25% respectively. Data in section (a) are from Marchetti *et al.* (1964); rats were kept on the appropriate diets for 4 weeks before analysis. The nucleotide values are in μmol/100 g of defatted liver; the lipid content is in mg/g wet wt. Data in section (b) are from Von Euler *et al.* (1963); rats were kept on diets for 5 days and nucleotide values are given as μmol/100 g wet wt. of liver.

Section	Component	Diet		
		Control	Orotic acid	Orotic acid + adenine sulphate
(a)	ATP	59	24	30
	ADP	105	38	54
	AMP	153	124	152
	Σ Adenine nucleotides	317	186	236
	GTP	9	trace	3
	UMP	36	115	83
	Total lipid	55	153	56
(b)	NAD^+ + NADH	76	65	
	$NADP^+$ + NADPH	33	19	
	Σ Adenine nucleotides	100%	50%	86%

The results found with orotic acid, together with other evidence discussed by Recknagel (1967), led that author to postulate that the fatty liver observed in carbon tetrachloride intoxication does not result primarily from a failure in protein synthesis but is the result of a coupling failure between triglyceride and the lipid-acceptor protein. The half-life of apoprotein is probably of the order of hours, whereas triglyceride accumulation can be detected less than 1 h after dosing with carbon tetrachloride. This suggests that there is a more or less immediate disruption of the coupling process leading to a rapid increase in liver triglyceride. Of course, the inhibition in protein synthesis plays an increasingly important role in later stages of the injury, for as the apoprotein is metabolised it is not replaced. Chloroform intoxication is also accompanied by a decreased incorporation of amino acids into liver protein, whereas similar exposures to halothane are without effect (Table 9.9). As the results show, the depression produced by chloroform is most obvious in the centrilobular zone.

Table 9.9. Effects of chloroform (1%) and halothane (1·5%) anaesthesia on protein synthesis (leucine incorporation) in rat liver. Data are from Scholler *et al.* (1968). The incorporation of leucine into cells at the centre or periphery of liver lobules was assessed by a radioautographic procedure. The exposure times to the anaesthetics are indicated. Results are expressed as the number of silver grains per cell using autoradiography with leucine-T_3 labelling.

Cell location	Control	Halothane (4 h)	Chloroform	
			(1 h)	(4 h)
Centrilobular	31	31	12	2
Periportal	29	32	29	20

9.6 Effects of antioxidants on fatty degeneration

The detailed mechanism that controls the coupling of triglyceride, phospholipid and lipid-acceptor protein to produce lipoprotein is not known. The coupling process does, however, involve the endoplasmic reticulum, and it seems probable for reasons outlined below that the production of free radicals at that site is responsible for the disturbance in lipoprotein secretion during carbon tetrachloride intoxication.

Many agents have been tested *in vivo* for a protective action against centrilobular necrosis and fatty degeneration of the liver resulting from the administration of carbon tetrachloride. The mechanisms of protection are discussed in detail in the next Chapter with respect to necrosis, but it is relevant to note here the effects of a variety of agents on the accumulation of liver triglyceride in carbon tetrachloride intoxication. Most of the examples to be discussed are antioxidants and their use was largely prompted by a current hypothesis that acute liver injury induced by carbon tetrachloride proceeds via a free-radical-motivated reaction that may involve lipid peroxidation as an important part of the overall process (see Recknagel, 1967). The antioxidants that have been tested include: (1) *NN′*-diphenyl-*p*-phenylenediamine (DPPD); (2) vitamin E; (3) promethazine; (4) reduced ubiquinone-4; (5) ethoxyquin; (6) butylated hydroxytoluene (BHT); (7) propyl gallate (Figure 9.8). The results found with each of these agents will be discussed in turn, but first it is important to stress the varying conditions used by different investigators. Variables in experimentation have included the species, sex and strain of animal; the route, dosage and timing of the administration of the protective agent; the dose and route of administration of carbon tetrachloride; the dietary status of the animal; and the parameters measured at the time of killing.

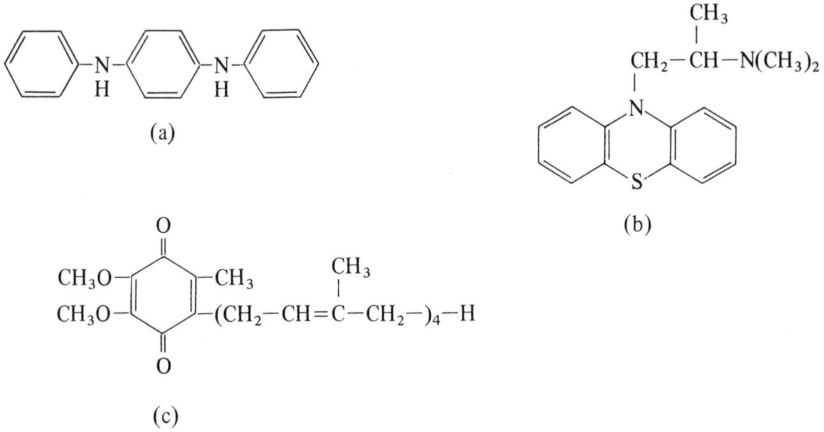

Figure 9.8. Structures of (a) *NN'*-diphenyl-*p*-phenylenediamine, (b) promethazine (Phenergan) and (c) ubiquinone-4.

9.6.1 *NN'*-Diphenyl-*p*-phenylenediamine

Table 9.10 summarises the reports in the literature for this compound. It can be said in general that although prior dosage with DPPD can attenuate the increase in liver triglyceride due to carbon tetrachloride, it has little effect on the decreased activities of aminopyrine or ethylmorphine demethylases, on the decrease in cytochrome P_{450} or on the decrease in protein synthesis. As will be discussed more fully in the next Chapter, prior dosage with DPPD is partially effective in decreasing the extent of liver necrosis that is produced by carbon tetrachloride.

9.6.2 Vitamin E

The data on the protective action of vitamin E on liver injury due to carbon tetrachloride in the rat are given in abbreviated form in Table 9.11. As with DPPD the administration of vitamin E has no effect in protecting against the decreased content of cytochrome P_{450}, the decrease in aminopyrine demethylation[7] or the decrease in protein synthesis. It has, however, a protective effect against the decrease in extinction at 260 nm in acid-soluble extracts of liver (Gallagher, 1962) and some activity against the increase in liver triglyceride and the extent of necrosis. For example, in animals deficient in vitamin E, administration of carbon tetrachloride produces a 300% increase in liver triglyceride; if, however, vitamin E had been previously administered to the vitamin E deficient rats prior to

[7] It is not legitimate to equate mortality rates with the extent of liver damage in such situations. The causes of death in carbon tetrachloride and chloroform intoxications are not clearly defined but probably involve cardiovascular collapse and widespread central nervous system disturbances. A protective action on such shock-effects may involve totally different mechanisms from protective action against liver necrosis.

Table 9.10. Protective action of *NN'*-diphenyl-*p*-phenylenediamine (DPPD) on the livers of rats dosed with carbon tetrachloride.

Reference	Sex	Carbon tetrachloride		Time killed (h)	DPPD			Liver				
		Route	Dose		Route	Times of dosing (h)	Dose	Necrosis	TG	P_{450}	AP	IPS
Gallagher (1962)	F	st	4·0	+24	ip	−48, −24 0	100*	P		–	–	–
Di Luzio & Costales (1965)	F	st	4·0	+24	ip	−48, −24 −2	699	P	P	–	–	–
Di Luzio (1967)	F	st	2·0	+24	iv	0		–	P			
						−2, 0	7·5	–	P			
						−24, 0		–	P			
Castro *et al.* (1968)	M	st	2·5	+24 +3	ip	−48, −24 −10 min	600	P	–	NP	NP	–
Dingell & Heimberg (1968)	M	st	1·25	+24	ip	−40, −20 0	100*			–	NP	–
Alpers *et al.* (1968)	?	st	1·25	+1	ip	−48, −24	600	–	–	–	–	NP
				+4				–	P			
				+24				–	P			
Cawthorne *et al.* (1970)	M	st	2·0	+24	ip	−48, −24 0	500		P			
	F	st	2·0	+24	st	−72, −48 −24	500	P	NP			

Abbreviations:

M Male.
F Female.
st Oral.
ip Intraperitoneal.
iv Intravenous.
P Protective effect.
NP No protective effect observed.
AP Demethylation of aminopyrine.
IPS Inhibition of protein synthesis.
P_{450} Depression of cytochrome P_{450} concentration.
TG Increase in liver triglyceride followed by histological or chemical estimations.

Doses of carbon tetrachloride and DPPD are given as ml or mg/kg body wt. except where marked with an asterisk (*), where the dose is mg/rat.

Table 9.11. Effects of vitamin E on liver disturbances resulting from administration of carbon tetrachloride to rats. Where vitamin E was found to have a protective action (usually only partial) against disturbances induced by carbon tetrachloride this is indicated by 'P'; 'NP' signifies that no protective action was reported.

Reference	Sex	Carbon tetrachloride			Vitamin E			Liver							Comments
		Route	Dose	Killed (h)	Route	Times (h)	Dose	Necrosis	TG	P_{450}	DM	PS	E_{260}	PO	
Hove and Hardin (1951)	M	sc	1·0	24	d	5 weeks	–		NP						E-deficient
Krone (1952)	F	st							P						
Tamura *et al.* (1959)	M	sc	0·6	24	sc	−72, −48, −24	665		P					P	E-deficient
Gallagher (1962)	F	st	4·0	20	ip[d]	−72, −48, −24, −6	115						P		
Eliseo and Pesaresi (1963)	–	ip	2·5	–	im[d]	−48	600		P						
Di Luzio (1967)	F	st	4·0	24	iv[d]	−24, 0	500		P						starved
McLean (1967)	M	st	2·5	24	st[c]	−24	430	NP	P						starved, E-deficient
Meldolesi (1968)	M	st	2·5	24	ip[e]	−72, −48, −26	125	P	P						starved
Dingell and Heimberg (1968)	M	st	1·0	24	ip[b]	−40, −20, 0	365				NP				
Castro *et al.* (1968)	M	st	2·5	3	ip[a]	−40	575			NP	NP				starved
Alpers *et al.* (1968)	–	st	1·25	1	ip[d]	−48, −24	100					NP			starved
Green *et al.* (1969)	F	st	2·0	6, 12	[c]	daily (8 days)	585	NP	NP						E-deficient
Cawthorne *et al.* (1970)	M/F	st	2·0	24	varied	varied	varied	NP	NP						

[a] DL-α-Tocopherol acetate. [b] α-Tocopherol. [c] D-α-Tocopherol acetate. [d] α-Tocopherol acetate. [e] α-Tocopherol succinate.

Abbreviations for liver disturbances: TG increased liver triglyceride; P_{450} decrease in cytochrome P_{450}; DM decrease in drug metabolism; PS decrease in protein synthesis; E_{260} decrease in extinction at 260 nm of acid-soluble extracts of liver homogenates; PO decrease in oxidative phosphorylation P/O ratios.

Other abbreviations: sc subcutaneous; st stomach intubation; d dietary intake; ip intraperitoneal; iv intravenous; im intramuscular; M male; F female.

Carbon tetrachloride and vitamin E doses are in ml/kg body wt. and mg/kg body wt. respectively; the values for vitamin E have where necessary been calculated from the reported dose/rat and average body wt. Where rats were starved before dosing with carbon tetrachloride or were on a vitamin E-deficient diet this is shown under Comments.

dosing with carbon tetrachloride, then the latter agent produced a rise of 188% in liver triglyceride. Compared to the effect of carbon tetrachloride on vitamin E deficient rats, the prior administration of vitamin E had effectively inhibited triglyceride accumulation by 37% (see McLean, 1967). Green *et al.* (1969) and Cawthorne *et al.* (1970), however, in comprehensive experiments could find no protective effect of α-tocopherol or α-tocopheryl acetate against either the accumulation of liver triglyceride or the extent of liver necrosis but some protective effect on mortality was observed.

It appears likely from the reports cited in Table 9.8 that vitamin E is not very effective either in preventing fatty degeneration or in attenuating necrosis; the latter action of vitamin E is considered later in more detail (Section 10.9).

9.6.3 Promethazine

A number of investigators (Rees and Spector, 1961; Rees *et al.*, 1961; Slater *et al.*, 1966; Wigglesworth, 1964; Fox *et al.*, 1962; Serratoni *et al.*, 1969) have found that promethazine protects against centrilobular necrosis in rats and rabbits but has little effect on the accumulation of fat.

Table 9.12. Effects of treatment with promethazine on various liver disturbances produced by the oral administration of carbon tetrachloride to rats at a dose of 1·25 ml/kg body wt.; promethazine was given intraperitoneally at a dose of 25 mg/kg body wt. at the time of feeding with carbon tetrachloride (time 0); a second dose (12·5 mg/kg body wt.) was given at +5 h to rats maintained for periods longer than 3 h. Data are from Rees and Shotlander (1963), Slater *et al.* (1964), Slater and Sawyer (1969). Values are given as percentage changes with respect to control values. Where the promethazine treatment gave values that were not significantly different from the control group this is shown by (NS).

Component studied	Time of analysis (h)	CCl_4	CCl_4 + promethazine
Triglyceride	22	+740	+408
Protein synthesis *in vitro*	3	−70	−58
NADPH	1	−27	(NS)
RNA P ratio[a]	1	−28	(NS)
P_{450}	1	−22	−32
	2	−28	
NADH-neotetrazolium reductase	1	+39	(NS)
Glucose 6-phosphatase	1	−24	−14
Inosine[b]	2	−17	(NS)
Necrosis		++++	+

[a] Ratio of RNA P in microsomes/cell sap.
[b] CCl_4 dose 2·5 ml/kg body wt.

In the experiments of Rees *et al.* (1961) liver triglyceride was increased approximately sevenfold by the administration of carbon tetrachloride, and concomitant dosing with promethazine gave a 30% protection. Promethazine has no effect in preventing the decreased synthesis of protein that results from administration of carbon tetrachloride, or the decreased cytochrome P_{450} (Slater and Sawyer, 1969), but does partly stop the decreased activity of glucose 6-phosphatase and the loss of RNA from the endoplasmic reticulum.

9.6.4 Ubiquinone-4

This agent has been studied by Di Luzio (1967), who found that when given intravenously it was active against the increased triglyceride level resulting from the administration of carbon tetrachloride.

9.6.5 Ethoxyquin and butylated hydroxytoluene

These two antioxidants have been investigated by Cawthorne *et al.* (1970), who found that both agents were partly effective against liver necrosis and the accumulation of fat. The effects were noticeable even when the antioxidants were given 48 h before the carbon tetrachloride. The concentration of antioxidant in the liver at the time of killing was consequently extremely low.

9.6.6 Propyl gallate

Ugazio and Torrielli (1969) have shown that this water-soluble antioxidant is very effective against the increased triglyceride in rat liver at least over the first 8 h after dosing. Its effect later is not significant, nor has propyl gallate any pronounced effect on centrilobular necrosis 24 h after carbon tetrachloride is given. This lack of effect of propyl gallate at late stages of the investigation may be related to its rapid excretion from the animal.

In general it can be concluded that the antioxidants studied have some protective effect against the increased liver triglyceride that results from the administration of carbon tetrachloride, but that the extent of protection is variable from agent to agent and depends on the route and timing of administration, and on the dose of carbon tetrachloride used. In several studies mentioned the dose of carbon tetrachloride used has been very high and secondary effects due to the compound's lipophilic nature may be of importance.

Where the dose of carbon tetrachloride used has been relatively low, however, it appears that the antioxidants that are comparatively water-soluble (e.g. propyl gallate) are more effective than non-polar antioxidants (e.g. vitamin E). This suggests that the interaction site in the endoplasmic reticulum that is damaged by carbon tetrachloride to result in a lipoprotein-coupling failure involves a *relatively* polar region. Before the site can be discussed more thoroughly it is necessary to consider the structure of the NADPH-cytochrome P_{450} chain in some detail. This is dealt with in the following Chapter.

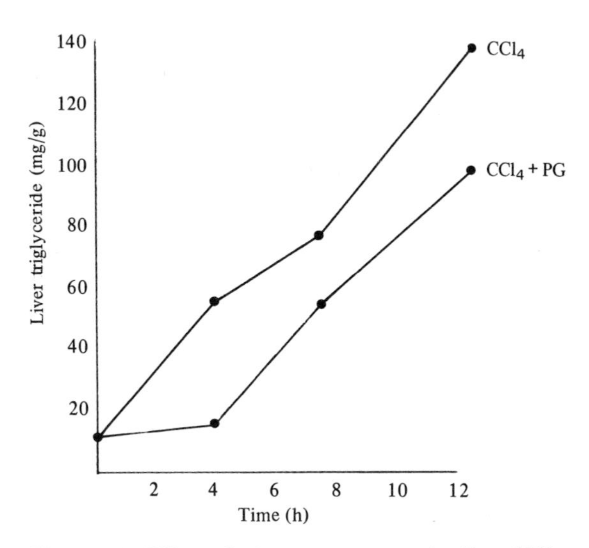

Figure 9.9. Effect of administering propyl gallate (PG) on the increase in liver triglycerides after intoxication with carbon tetrachloride. Data are from Ugazio and Torrielli (1969). The propyl gallate (30 mg/100 g body wt.) was given 1 h before carbon tetrachloride (2·5 ml/kg body wt.).

References

Alexander, N. M., Scheig, R., Klatskin, G., 1967, *Biochem. Pharmac.*, **16**, 1091.

Alpers, D. H., Solin, M., Isselbacher, K. J., 1968, *Mol. Pharmac.*, **4**, 566.

Biswas, M., Ghoshal, S., 1966, *Chem. Ind.*, 1717.

Brody, T. M., 1963, "Hepatotoxicity of therapeutic agents", *Ann. N.Y. Acad.Sci.*, **104**, 1065.

Browning, E., 1965, *Toxicity and Metabolism of Industrial Solvents* (Elsevier, Amsterdam).

Cameron, G. R., Karunaratne, W. A. E., 1936, *J. Path. Bact.*, **42**, 1.

Castro, J. A., Sesame, H. A., Sussman, H., Gillette, J. R., 1968, *Life Sci.*, **7**, 129.

Cawthorne, M. A., Bunyan, J., Sennitt, M. V., Green, J., Grasso, P., 1970, *Br. J. Nutr.*, **24**, 357.

Cohn, D. V., Bawdon, R., Newman, R. R., Hamilton, J. W., 1968, *J. biol. Chem.*, **243**, 1089.

Cottrell, T. L., 1958, *The Strengths of Chemical Bonds* (Butterworths, London).

Delaney, V. B., Slater, T. F., 1970, *Biochem. J.*, **116**, 299.

Di Luzio, N. R., 1967, *Prog. Biochem. Pharmac.*, **3**, 325.

Di Luzio, N. R., Costales, F., 1965, *Exp. molec. Path.*, **4**, 141.

Dingell, J. V., Heimberg, M., 1968, *Biochem. Pharmac.*, **17**, 1269.

Dodson, V. N., Friberg, R., Ketchum, D., 1965, *Proc. Soc. exp. Biol. Med.*, **120**, 355.

Duncan, W. A. M., Raventós, J., 1959, *Brit. J. Anaesth.*, **31**, 302.

Duncum, B. M., 1947, *The Development of Inhalation Anaesthesia* (Oxford University Press, Oxford).

Eliseo, V., Pesaresi, C., 1963, *Folia med., Napoli*, **46**, 641.

Espinosa, E., Insunza, I., 1962, *Proc. soc. exp. Biol. Med.*, **111**, 174.

Euler, L. H. von, Rubin, R. J., Handschumacher, R. E., 1963, *J. biol. Chem.*, **238**, 2464.

Eylar, E. H., 1965, *J. theor. Biol.*, **10**, 89.

Farber, E., Schull, K. H., Villa-Trevino, S., Lombardi, B., Thomas, M., 1964, *Nature, Lond.*, **203**, 34.
Fox, C. F., Dinman, B. D., Frajola, W. J., 1962, *Proc. Soc. exp. Biol. Med.*, **111**, 731.
Gallagher, C. H., 1962, *Aust. J. exp. Biol. med. Sci.*, **40**, 241.
Gerhard, H., Schultze, B., Maurer, W., 1970, *Virchows Arch. Abt. B. Zellpath.*, **6**, 38.
Goodman, L. S., Gilman, A., 1955, *The Pharmacological Basis of Therapeutics,* second edition (Macmillan, New York), p.64.
Graham, E. A., 1915, *J. exp. Med.*, **22**, 48.
Green, J., Bunyan, J., Cawthorne, M. A., Diplock, A. T., 1969, *Br. J. Nutr.*, **23**, 297.
Gregory, N. L., 1966, *Nature, Lond.*, **212**, 1460.
Hirschfelder, J. O., 1941, *J. chem. Phys.*, **9**, 645.
Hove, E. L., Hardin, J. O., 1951, *Proc. Soc. exp. Biol. Med.*, **78**, 858.
Kinsella, J. E., 1967, *Biochim. biophys. Acta,* **137**, 205.
Klatskin, G., Kimberg, D. V., 1969, *New Engl. J. Med.*, **280**, 515.
Krone, H. A., 1952, *Int. Z. Vitam. Forsch.*, **24**, 12.
Leduc, E. H., Wilson, J. W., 1958, *Archs. Path.*, **65**, 147.
Lombardi, B., Ugazio, G., 1965, *J. Lipid Res.*, **6**, 498.
Lu, A. Y. H., Junk, K. W., Coon, M. J., 1969, *J. biol. Chem.*, **244**, 3714.
Marchetti, M., Puddu, P., Caldarera, C. M., 1964, *Biochem. J.*, **92**, 46.
McLean, A. E. M., 1967, *Br. J. exp. Path.*, **48**, 632.
McLean, A. E. M., McLean, E. K., 1966, *Biochem. J.*, **100**, 564.
McLean, A. E. M., McLean, E. K., Judah, J. D., 1965, *Int. Rev. exp. Path.*, **4**, 127.
Meldolesi, J., 1968, *Expl. molec. Path.*, **9**, 141.
Monlux, G., Smuckler, E. A., 1969, *Am. J. Path.*, **54**, 73.
National Halothane Study, 1966, *J. Am. med. Ass.*, **197**, 775.
Oberling, C., Rouiller, C., 1956, *Annls. Anat. path.*, **1**, 401.
Oettingen, W. F. van, 1955, *The Halogenated Hydrocarbons, Toxicity and Potential Dangers* (US Department of Health, Education and Welfare, Public Health Service Publication number 414, Washington DC).
Plaa, G. L., Evans, E. A., Hine, C. H., 1958, *J. Pharmac. exp. Ther.*, **123**, 224.
Putnam, F. W. (Ed.), 1960, *The Plasma Proteins,* volume 2 (Academic Press, New York).
Recknagel, R. O., 1967, *Pharmac. Rev.*, **19**, 145.
Rees, K. R., Shotlander, V. L., 1963, *Proc. Roy. Soc. B.*, **157**, 517.
Rees, K. R., Sinha, K. P., 1960, *J. Path. Bact.*, **80**, 297.
Rees, K. R., Sinha, K. P., Spector, W. G., 1961, *J. Path. Bact.*, **81**, 107.
Rees, K. R., Spector, W. G., 1961, *Nature, Lond.*, **190**, 821.
Reynolds, E. S., 1963, *J. Cell. Biol.*, **19**, 139.
Reynolds, E. S., Lee, A. G., 1967, *Lab. Invest.*, **16**, 591.
Reynolds, E. S., Thiers, R. E., Vallee, B. L., 1962, *J. biol. Chem.*, **237**, 3546.
Roheim, P. S., Miller, L., Eder, H. A., 1965a, *J. biol. Chem.*, **240**, 2994.
Roheim, P. S., Switzer, S., Girard, A., Eder, H. A., 1965b, *Biochem. biophys. Res. Commun.*, **20**, 416.
Rosin, A., Doljanski, L., 1946, *Proc. Soc. exp. Biol. Med.*, **62**, 62.
Rubin, R. J., Pendleton, R. G., 1964, *Fedn Proc. Fedn Am. Socs exp. Biol.*, **23**, 126.
Scholler, K. L., Müller, E., Plehwe, U., 1968, *Anaesthesist,* **17**, 87.
Schotz, M. C., Recknagel, R. O., 1960, *Biochim. biophys. Acta,* **41**, 151.
Schull, K. H., 1962, *J. biol. Chem.*, **237**, PC1734.
Seakins, A., Robinson, D. S., 1963, *Biochem. J.*, **86**, 401.
Serratoni, F. T., Schnitzer, B., Smith, E. B., 1969, *Archs Path.*, **87**, 46.
Shore, L., Recknagel, R. O., 1959, *Am. J. Physiol.*, **197**, 121.
Slater, T. F., 1971, *Anaesthesiologie,* in the press.

Slater, T. F., Delaney, V. B., 1970, *Biochem. J.,* **116**, 303.
Slater, T. F., Greenbaum, A. L., 1965, *Biochem. J.,* **96**, 484.
Slater, T. F., Sawyer, B. C., 1969, *Biochem. J.,* **111**, 317.
Slater, T. F., Sawyer, B. C., 1971, *Biochem. J.,* **123**, 805, 815, 823.
Slater, T. F., Sawyer, B. C., Strauli, U. D., 1966, *Biochem. Pharmac.,* **15**, 1273.
Slater, T. F., Sträuli, U. D., Sawyer, B. C., 1964, *Biochem. J.,* **93**, 260.
Smuckler, E. A., Benditt, E. P., 1963, *Science,* **140**, 308.
Smuckler, E. A., Benditt, E. P., 1965, *Biochemistry, Easton,* **4**, 671.
Smuckler, E. A., Iseri, O. A., Benditt, E. P., 1962, *J. exp. Med.,* **116**, 55.
Sosnowsky, G., 1964, *Free Radical Reactions in Preparative Organic Chemistry* (Macmillan, New York).
Spiro, R. G., 1969, *New Engl. J. Med.,* **281**, 991.
Standerfer, S. B., Handler, P., 1955, *Proc. Soc. exp. Biol. Med.,* **90**, 270.
Tamura, U., Tsuchiya, T., Harada, T., Kuroiwa, H., Kitani, T., 1959, *Med. J. Osaka Univ.,* **10**, 91.
Thiers, R. E., Reynolds, E. S., Vallee, B. L., 1960, *J. biol. Chem.,* **235**, 2130.
Ugazio, G., Torrielli, M. V., 1969, *Biochem. Pharmac.,* **18**, 2271.
Urey, W. H., Eiszner, J. R., 1952, *J. Am. chem. Soc.,* **74**, 5822.
Villa-Trevino, S., Farber, E., Staeheli, T., Wettstein, F. O., Noll, H., 1964a, *J. biol. Chem.,* **239**, 3826.
Villa-Trevino, S., Schull, K. H., Farber, E., 1964b, *J. biol. Chem.,* **238**, 1757.
Walling, C., 1957, *Free Radicals in Solution* (John Wiley, New York).
Wands, J. R., Smuckler, E. A., Woodbury, W. J., 1970, *Amer. J. Pathol.,* **58**, 499.
Wells, H. G., 1908, *Archs. intern. Med.,* **1**, 589.
Wigglesworth, J. S., 1964, *J. Path. Bact.,* **87**, 333.

10

Hepatotoxicity of carbon tetrachloride: necrosis

10.1 Introduction

The necrosis that develops in the liver after *oral* dosing with carbon tetrachloride is centred mainly around the hepatic veins where the sinusoidal blood collects before leaving the liver. When the concentration of carbon tetrachloride that reaches the liver is increased, as happens for example when it is injected *intra-portally* (Cameron and Karunaratne, 1936), a different histological picture results. Under these conditions the toxic agent reaches a concentration that is lethal for parenchymal cells throughout the lobule and a massive toxic necrosis develops. The implication is that after oral dosing the concentration of carbon tetrachloride reaching the liver via the portal system is sufficient to kill the cells around the central veins but is not high enough to kill the cells around the portal tracts. In other words, the centrilobular cells are *more sensitive* to the effects of carbon tetrachloride than cells in the periportal region.

Histochemical investigations show that orally administered carbon tetrachloride causes severe disturbances in liver cell metabolism, and these are often localised in the centrilobular region where necrosis later develops. [Hashimoto *et al.* (1968) have shown that carbon tetrachloride produces changes in *human* liver similar to those discussed here for the rat.] For example, shortly after giving a very low dose of carbon tetrachloride orally to rats there is a marked central suppression of glucose 6-phosphatase activity (Reynolds, 1967). This lobular distribution of metabolic disturbance, clearly shown by histochemical procedures (see Table 10.1), has received confirmation by the technically difficult experiments of Morrison and colleagues (Morrison *et al.*, 1965) in which intralobular microdissection of thin liver sections was performed, and ultramicro-enzyme assays were carried out on the surgically separated *intra*lobular fragments.

Table 10.1. Relative localisation of enzymes in either the centrilobular or periportal zones of the liver is marked by +. Only a few examples are quoted; extensive discussion of distributions of intralobular enzymes is given by Rappaport (1963) and Pette and Brandau (1966).

Enzyme	Centrilobular	Periportal	Reference
1 Alcohol dehydrogenase		+	Greenberger *et al.* (1965)
2 Drug-metabolising enzymes	+		Wattenberg and Leong (1962)
3 Succinate dehydrogenase		+	
4 Cytochrome oxidase		+	Pette and Brandau (1966)
5 Glutamate dehydrogenase	+		
6 Glucose 6-phosphatase		+	

10.2 Liver necrosis and carbon tetrachloride: general comments

The toxic effects of carbon tetrachloride on liver parenchyma that have been outlined above raise the questions: (a) what chemical mechanism(s) are primarily responsible for the disturbances in the cell's metabolism that ultimately result in its death? (b) Why is necrosis restricted mainly to the liver[(1)] and, within the liver, why is it concentrated around the central veins? The major question (a) may be divided into several important ancillary questions: (1) is the necrogenic action of carbon tetrachloride the result of the agent itself or does carbon tetrachloride have to undergo metabolism to a more toxic form? (2) If carbon tetrachloride undergoes metabolism, what is the route of metabolism and what components of the liver parenchyma are involved? (3) What are the metabolites of carbon tetrachloride: are they toxic or non-toxic? (4) What reactions would the metabolites undergo with neighbouring cell constituents that are likely to be deleterious to the continued well-being of the cell? The remainder of this Chapter will be concerned with attempts to answer these questions.

Many studies have been concerned with situations *in vitro* in which, for example, the metabolism of carbon tetrachloride has been studied under a variety of experimental conditions. The information collected in this way relates to the toxic action of the compound on liver components, intracellular fractions, enzyme activities etc. *in vitro* and is now very extensive. With such information it is possible to predict what effects should occur *in vivo*, the time-sequence of the major changes and the modifying actions of protective agents. Direct analysis of the situation *in vivo* then permits a test of the predictions that were based on findings *in vitro*. In this way an hypothesis can be developed to account for the necrogenic action of carbon tetrachloride.

First it will be convenient to outline some of the early schemes that have been proposed to explain the toxic action of carbon tetrachloride on the liver[(2)]. This discussion leads to the view that the compound undergoes metabolism as an essential step in the pathways to necrosis.

10.3 Early theories of the necrogenic action of carbon tetrachloride

Attempts to explain how carbon tetrachloride (and chloroform) cause liver necrosis have been numerous and may be traced back for more than 50 years. Wells (1908), for example, in discussing the hepatotoxic action of

(1) Some damage, of course, occurs in other tissues, principally the kidney (see Striker *et al.*, 1968), the intestine (Alpers and Isselbacher, 1968) and also the adrenals (Higgins and Cragg, 1937; Sesame *et al.*, 1968).

(2) The discussion will be restricted mainly to studies with rats. A considerable amount of work on the hepatotoxicity of carbon tetrachloride has been performed on other species, however. In general, the results obtained have been similar to those that will be discussed for the rat. Particularly extensive work on the mouse has been carried out by Frunder's group (see, for example, Frunder, 1968; Frunder *et al.*, 1961; Thielmann *et al.*, 1963).

chloroform, suggested that the lipophilic solvent greatly decreased oxidative enzymes concerned with energy production and synthesis leaving autolytic enzymes uninhibited and free to initiate cell digestion. It is noteworthy, in view of Wells' speculation, that Slater and Greenbaum (1965) found no early changes in liver lysosomal enzymes after oral dosing of rats with carbon tetrachloride[(3)].

The majority of the mechanisms proposed to explain the necrogenic action of carbon tetrachloride are: (a) intracellular acid production; (b) central ischaemic anoxia; (c) mitochondrial damage with loss of intramitochondrial nucleotides and failure of oxidative reactions. These will be discussed briefly in turn.

10.3.1 Toxic metabolites

In an interesting paper, Graham (1915) suggested that the halogenomethanes cause necrosis by producing excess of **acid** within the liver cells. Chloroform is well known to hydrolyse (e.g. photochemically) to hydrochloric acid via the intermediate formation of phosgene and this hydrolyses spontaneously [Reactions (10.1)–(10.2)].

$$CHCl_3 + O \longrightarrow COCl_2 + HCl \qquad (10.1)$$

$$COCl_2 + H_2O \longrightarrow CO_2 + 2HCl \qquad (10.2)$$

Graham considered that evidence for his hypothesis was the relative order of necrogenic activities: carbon tetrachloride > chloroform > dichloromethane, which is the same order as their *potential* capabilities to produce hydrochloric acid.

Phosgene was suggested by Müller (1911) as the major factor in the production of liver necrosis by chloroform and its possible role in intoxication by carbon tetrachloride has been more recently stressed by McLean (1967b). The possibility that phosgene itself produced cell death was dismissed by Graham in view of its rapid hydrolysis.

The possible routes of metabolism of carbon tetrachloride to phosgene and other known degradative products will be considered in a later Section. However, it can be seen that even in the early studies some authors were already thinking in terms of the metabolism of carbon tetrachloride and chloroform being a necessary factor in the production of necrosis.

10.3.2 Ischaemic anoxia

In 1913 Fischler noted that centrilobular necrosis was often observed in dogs after an Eck fistula had been established; the centrilobular lesion was particularly marked when the oxygen supply via the hepatic artery was low. This led Graham (1915) and other workers to suggest that the centrilobular distribution of the injury was due to a relative **anoxia** existing around the central veins in comparison with the portal region.

[(3)] The participation of the lysosomal fraction in the early events that occur during intoxication with carbon tetrachloride has been reviewed (Slater, 1969).

This role of a central hypoxia in establishing the site of necrosis was taken up later by Himsworth (1950), who stressed the importance of central ischaemic anoxia in the development of the lesion. However, direct experimental evidence for liver ischaemia *in vivo* after the administration of carbon tetrachloride has not been obtained. Stoner (1956) could find no indication of early changes in liver blood flow after rats were given carbon tetrachloride. In an elegant experiment Brauer (1965) reversed the blood flow through an isolated perfused liver preparation so that the perfusate *entered* the lobule via the central veins and left through the portal region; in such preparations chloroform still caused the earliest demonstrable effects around the central vein[(4)]. From the results of such experiments it seems unlikely that central anoxia is a major early feature in the aetiology of centrilobular necrosis although it is probably of considerable importance in later stages of the injury.

10.3.3 Mitochondrial damage

During the rapid expansion of knowledge concerning the intracellular distribution of enzymes and coenzymes that occurred in the period 1945–1955, several groups of workers laid emphasis on the role of mitochondrial damage in the aetiology of liver necrosis induced by carbon tetrachloride. Christie and Judah (1954) and Dianzani (1954) regarded disturbance in the mitochondrial production and maintenance of ATP to be an important factor after dosing with this agent. However, later experiments failed to provide experimental evidence of gross mitochondrial damage within the first 8–12 h of oral dosing with carbon tetrachloride. Although Gallagher (1962) found that liver mitochondria isolated from rats poisoned with carbon tetrachloride lost intramitochondrial nucleotides more rapidly than did control mitochondria, this was regarded by Recknagel and Anthony (1959) as an artefact resulting from the dispersion of mitochondria in the presence of carbon tetrachloride during the process of homogenising the liver. The addition of carbon tetrachloride to normal mitochondria *in vitro* produced a similar disturbance in the mitochondrial membrane. No evidence could be obtained for mitochondrial membrane changes *in vivo*. Recent work with a rapid microfiltration method, however, has provided evidence that carbon tetrachloride *in vivo* produces alterations in the nucleotide content of the mitochondrial fraction (see Table 9.5) so that the previous views on the occurrence of an early mitochondrial disturbance in intoxication with this compound have to some extent been justified. Such mitochondrial shifts appear, however, to be consequent upon even earlier changes in the endoplasmic reticulum; this aspect of the liver disturbance is considered later.

[(4)] A similar result has been obtained more recently by Sigel *et al.* (1967) using transplants of liver.

10.4 Relationship between the metabolism of carbon tetrachloride in the endoplasmic reticulum and liver necrosis

A detailed discussion of metabolic pathways will be given in the next Section. It is pertinent to consider first whether the metabolism of carbon tetrachloride is essential to the necrogenic activity of this toxic agent, or whether it is a secondary phenomenon connected perhaps with the detoxication of the compound.

First, it can be noted that high concentrations of carbon tetrachloride can be detected shortly after oral administration in such tissues as bone marrow, brain and adipose tissue; lower but significant concentrations are found in liver, kidney and muscle (van Oettingen, 1955). In the adult rat, however, *extensive* necrosis occurs only in the liver. If the toxic properties of carbon tetrachloride were due solely to its lipophilic nature and, as a consequence, to unspecific damage to lipid-rich structures, it is strange that those tissues that accumulate it most readily do not suffer the most severe damage. Clearly the indications are that some more specific type of interaction is necessary for necrosis to develop, and the simplest explanation is that the compound is converted into a more toxic metabolite only in those tissues that afterwards become extensively necrotic. A similar conclusion can be reached by considering the relative activity of carbon tetrachloride on the adult rat, the newborn rat and the chicken. Although oral administration of the compound produces similar concentrations of the toxic agent in the liver in all three instances, the newborn rat and the chicken do not develop significant centrilobular necrosis of the liver (see Dawkins, 1963; Bhattacharyya, 1965).

Secondly, experiments concerning the toxicity of trichlorofluoromethane may be discussed. This substance has physical properties very similar to carbon tetrachloride (e.g. lipid solubility, density, molecular weight etc.); it can be predicted, however, to have an increased resistance to homolytic degradation compared to carbon tetrachloride. Experiments with trichlorofluoromethane are complicated by its high volatility. A single dose of trichlorofluoromethane is thereby rapidly eliminated from the body. For studying the activity of trichlorofluoromethane in terms of necrosis it is preferable to expose the animals to the agent by continuous inhalation. Jenkins *et al.* (1970) have performed such experiments with rats, guinea pigs, monkeys and dogs and have shown that trichlorofluoromethane is virtually non-toxic to the liver. Similarly, Slater (1966a) reported that oral dosing of rats with trichlorofluoromethane did not result in liver necrosis. It may be concluded that carbon tetrachloride undergoes some fairly specific change in the adult rat liver that initiates the necrogenic chain of events. This change does not take place in other tissues of the adult rat to a comparable extent, nor in the liver of newborn rats or chickens.

Where in the liver does the metabolism of carbon tetrachloride occur that is essential for necrosis? The earliest morphological and biochemical changes that have been detected in the liver cell fractions have been in the endoplasmic reticulum. As a working hypothesis it can be suggested that the metabolism of the compound occurs in this fraction. Compelling evidence that this hypothesis is correct has been provided by McLean and McLean (1966), who subjected rats to various treatments that altered the activities of enzyme systems in the liver endoplasmic reticulum. Treatment with sodium phenobarbital or DDT increases the amount of endoplasmic reticulum in the whole liver in general, and the activity of the NADPH-cytochrome P_{450} sequence in particular[(5)]. Subjecting the rat to a low protein diet was found to have the reverse effect; that is there was a decrease in both the endoplasmic reticulum in the whole liver and in the activity of the NADPH-cytochrome P_{450} sequence (Table 10.2).

Table 10.2. Effects of 'protein-free diet' on liver components after dosing with carbon tetrachloride (data from McLean and McLean, 1966). Figures in parentheses show the number of days on the protein-free diet.

Parameter	Control diet		Protein-free diet	
	carbon tetrachloride		carbon tetrachloride	
	−	+	−	+
1 Pyramidone demethylation (μmol/g wet wt. per h)	0·68		0·3 (2) 0·06 (4)	
2 Benzpyrene hydroxylation ('quinine units'/g wet wt.)	46		30 (2) 9 (4)	
3 Plasma isocitrate dehydrogenase log (nmol/ml per min)	0·1	2·1		2·2 (2) 1·5 (4)
4 Plasma bilirubin (mg/100 ml)	0·2	0·6		0·9 (2) 0·3 (4)
5 Liver fat (g/kg fat-free dry wt.)	71	371	213	509 (8–14)
6 Liver glycogen (g of glucose/g wet wt.)	7·4	0·8	10·4	5·3 (8–14)
7 LD_{50} to carbon tetrachloride (ml/kg body wt.)		6·4		14·7 (8–22)
8 NADPH-cytochrome *c* reductase[a]	100%		59% (4)	

[a] Unpublished data of M. N. Eakins and T. F. Slater.

(5) Treatment with sodium phenobarbital and similar 'inducers' also produces increases in some mitochondrial and soluble components of the liver (see Platt and Cockrill, 1967, 1969). The most marked changes in amount of new material synthesised, however, occur in the liver endoplasmic reticulum. An interesting role for peroxidised fatty acids in the induction process has been reported by McLean and Marshall (1971). It was observed that rats fed a purified diet containing 20% casein did not respond to the inductive action of phenobarbital as well as rats on a normal commercially produced diet. Addition of peroxidised fats to the purified diet, however, allowed full induction of cytochrome P_{450} by phenobarbital.

When such treated rats were intoxicated with carbon tetrachloride, it was found that rats with more endoplasmic reticulum were more affected by the compound, whereas rats with less endoplasmic reticulum were much less affected by it. Such experiments strongly suggest that the necrogenic action[6] of carbon tetrachloride is limited to the activity of some enzyme or enzyme system localised in the endoplasmic reticulum. Since the NADPH-cytochrome P_{450} satisfies this criterion, and is also the sequence concerned with the metabolism of a wide variety of other foreign materials, it is natural to consider it as the most likely site of metabolism of carbon tetrachloride. The choice of this system as the site of metabolism of carbon tetrachloride would, moreover, explain certain facets of the toxicity of carbon tetrachloride that have been outlined above: the NADPH-cytochrome P_{450} sequence is not present in tissues other than liver and kidney[7] to any significant extent (Knecht, 1966), nor is it very active in newborn rats (Dallner *et al.*, 1965) or in chickens (Slater, 1968). These points are summarised in Tables 10.3 and 10.4.

It is important to note that treatment such as sodium phenobarbital administration or low protein dietary regimes alter the activity of the NADPH-cytochrome P_{450} sequence as a whole and do not affect only the amount of cytochrome P_{450} (see, for example, Orrenius and Ernster, 1964). In particular, the activity of the initial flavoprotein dehydrogenase, NADPH-cytochrome *c* reductase, usually changes in unison with

Table 10.3. Relative activities of various rat tissues in demethylating dimethylnitrosamine (data from Knecht, 1966). The organ weights are from Long (1961).

Organ	Demethylating activity (%)	Organ weight (g/kg body wt.)	Relative demethylating activity per organ (%)
Liver	100	33·5	100
Kidney	79	10·9	26
Spleen	12	2·9	1
Lung	15	7·9	4
Brain	13	12·2	5
Muscle	10		

(6) The low-protein-diet treatment alleviated necrosis but had little effect on the accumulation of liver triglyceride.

(7) The relatively high activity of drug-metabolising enzymes in the kidney is of interest with respect to the damaging action of carbon tetrachloride on that organ, especially in man (see van Oettingen, 1955). In the rat, however, administration of carbon tetrachloride was found to have no effect on the cytochrome P_{450} concentration in kidney microsomes (Sesame *et al.*, 1968), whereas a considerable decrease occurs in liver microsomes. Slater and Sawyer (unpublished data) have found that carbon tetrachloride stimulates malonaldehyde production in a suspension of kidney microsomes in buffer-NADPH mixture, although the stimulation is less than that observed with liver microsomes under identical conditions.

cytochrome P_{450}.[8] The significance of this will be more apparent after the discussion in Section 10.5. Having established the likelihood that metabolism of carbon tetrachloride is essential for the production of necrosis, and that the metabolism most probably occurs in liver endoplasmic reticulum through an interaction with the NADPH–cytochrome P_{450} sequence, we should now consider what form the metabolism takes.

Table 10.4. Changes in aminopyrine demethylase, cytochome P_{450} and the NADPH–ADP–Fe^{2+} lipid peroxidation system during the neonatal period in rat liver microsomes and (for the last-mentioned parameter) in adult chicken liver microsomes. Values are given as percentages of the corresponding value obtained with adult rat liver microsomes (data from Dallner *et al.*, 1965, and Slater, 1968).

Days after birth	Aminopyrine demethylation (%)	Cytochrome P_{450} (%)	NADPH–ADP–Fe^{2+} lipid peroxidation (uptake of O_2) (%)
1	17	18	
2			5
4	25	50	
8	36	81	
Chicken			11

10.5 Metabolism of carbon tetrachloride

There are a number of reports showing that the halogenomethanes are metabolised by tissue suspensions and from these it appears probable that several quite distinct routes of metabolism are operative.

Heppel and Porterfield (1948) showed that rat liver slices, homogenates and extracts dehalogenated a number of bromo- and chloro-alkanes that were studied. The products were halide ion and formaldehyde [Reaction (10.3)].

$$CH_2BrCl + H_2O \longrightarrow Br^- + Cl^- + CH_2O + 2H^+ \qquad (10.3)$$

The system was activated by cyanide, glutathione or cysteine, and dehalogenation was more active under nitrogen than in the presence of oxygen. The process was decreased by heating the tissue sample for 5 min at 60°C and liver activity was increased by repeated exposure of the rat to bromochloromethane; kidney slices were more active than liver cells. This particular enzymic system appears to be non-oxidative in its mechanism, the properties indicating that it is *unrelated* to the metabolic degradation of carbon tetrachloride that is important for the development of necrosis.

[8] An exception occurs after feeding rats with methylcholanthrene (Conney *et al.*, 1959); this material also produces a shift in the absorption maximum of the microsomal cytochrome from 450 and 448 nm (see Glaumann *et al.*, 1969).

Bray *et al.* (1952) studied dehalogenation in rabbit liver supernatant fractions. Chloride ion was liberated slowly from chloroform but more rapidly from benzyl chloride and benzotrichloride. There was a decrease in activity after dialysis but some activity was always retained and no induction of the system was apparent after continued treatment of rabbits with chloral hydrate. Further, there was no great decrease in activity after boiling the tissue extract under reflux for 1 h. Since the liberation of halide ions was increased by the addition of thiol compounds Bray *et al.* suggested that the reaction is largely a non-enzymic dehalogenation involving reaction with thiol groups.

In 1961 Butler examined the metabolism of carbon tetrachloride and chloroform both in intact dogs and *in vitro* with mouse tissue homogenates. He detected the *conversion* of carbon tetrachloride into chloroform by using gas chromatography of the expired air *in vivo*, but no further reduction to dichloromethane was detected. A non-enzymic conversion of carbon tetrachloride into chloroform was found in the presence of glutathione, cysteine or ascorbic acid. No appreciable tissue specificity was found in the same reaction and much of the tissue reaction detected could be non-enzymic in character. In an interesting discussion, Butler proposed that carbon tetrachloride and chloroform are degraded by reaction with tissue thiol compounds and that these reactions are **homolytic** in nature (see also Wirtschafter and Cronyn, 1964). A clear suggestion was made by Butler (1961) that metabolism of carbon tetrachloride and chloroform is essential for the hepatotoxic action of these materials but no direct evidence for this point was presented.

Paul and Rubenstein (1963) carried out experiments *in vivo* similar to those described above. After intraduodenal dosing of rats with [^{14}C]carbon tetrachloride or [^{14}C]chloroform more than 75% of the isotope was recovered in the expired air in 18 h. Less than 5% of the dose was recovered as carbon dioxide and the production of this gas was considerably greater *in vivo* with chloroform than with carbon tetrachloride. Slices of liver were approximately four times as active as kidney slices in degrading carbon tetrachloride and chloroform to carbon dioxide; with equivalent volumes of the two compounds in the side arms of Warburg flasks there was approximately twice as much conversion of chloroform into carbon dioxide as found with carbon tetrachloride. However, from the relative solubilities of the two compounds in tissue suspensions (given in Table 9.1) the data of Paul and Rubenstein (1963) indicate that on a molar basis carbon tetrachloride is converted *more* rapidly into carbon dioxide than is chloroform under their *in vitro* conditions. The conversion of both compounds into carbon dioxide was strongly inhibited by 15 mM-cyanide, but not by glutathione. Only a very slow conversion of carbon tetrachloride into carbon dioxide occurred in heat *denatured* slices and this was stimulated by glutathione or cysteine.

Seawright and McLean (1967) have examined the conversion of carbon tetrachloride into carbon dioxide by rat liver tissue suspensions in some detail. They showed that the process occurs in the endoplasmic reticulum and is largely dependent on NADPH and oxygen. The metabolism of carbon tetrachloride was partially inhibited by 0·1 mM-SKF 525A[9] and 0·1 mM-promethazine: 45% and 46% inhibition respectively; there was a more extensive inhibition (88%) by 5 mM-aniline. It may be concluded that the conversion of carbon tetrachloride into carbon dioxide occurs mainly through an interaction with the enzyme system in the liver endoplasmic reticulum that is concerned with the oxidative metabolism of foreign materials.

The above discussion shows that there are several distinct routes by which carbon tetrachloride and chloroform can be degraded by liver tissue:

1. A non-enzymic dehalogenation that is stimulated by thiol compounds.
2. Enzyme-dependent mechanisms that include oxidative and non-oxidative routes. Of these processes the oxidative degradation to carbon dioxide appears the most important so far identified in quantitative terms, although we shall see below that the route to carbon dioxide is not necessarily of major importance in the necrogenic activity of carbon tetrachloride.

The detailed mechanism by which carbon tetrachloride is converted into carbon dioxide by interaction with the drug-metabolising system in liver endoplasmic reticulum is not clear. As mentioned, carbon tetrachloride appears relatively more reactive in the pathway producing carbon dioxide than is chloroform. Some indication of possible routes of metabolism can be obtained from the work of Hine (1956). In the alkaline **hydrolysis** (or alcoholysis) of the halogenomethanes in the presence of a strongly basic *nucleophilic* reagent (OH^-), the order of hydrolysis that was found was chloromethane $>$ chloroform $\gg$ carbon tetrachloride. The reaction proceeds as shown in Reactions (10.4)–(10.6).

$$CHCl_3 \longrightarrow CCl_3^- + H_2O \tag{10.4}$$

$$CCl_3^- \longrightarrow :CCl_2 + Cl^- \tag{10.5}$$

$$:CCl_2 \longrightarrow CO + \text{formate} \tag{10.6}$$

Carbon tetrachloride gives the same products as chloroform on hydrolysis or alcoholysis and the dissociation to the trichloromethide anion is rate-limiting.

$$CCl_4 \longrightarrow CCl_3^- + Cl^+ \tag{10.7}$$

The ionic mechanism shown above proceeds very much faster with chloroform than with carbon tetrachloride and gives a product (carbon monoxide) that is not known to be produced during the metabolism of the two compounds *in vivo*. Moreover, in a highly *non-polar* environment,

[9] SKF 525A: β-diethylaminoethyl-3,3′-diphenylpropyl acetate.

as exists in the endoplasmic reticulum in the region of the drug-metabolising sequence[(10)], ionic reactions would not be favoured. This suggests that the metabolism involves *homolysis* rather than *heterolysis* [Reactions (10.8) and (10.9) where $R^{\bullet}$ is an endogenous radical present in the endoplasmic reticulum].

$$CCl_4 \longrightarrow CCl_3^- + Cl^+ \qquad \text{heterolysis} \qquad (10.8)$$

$$CCl_4 + R^{\bullet} \longrightarrow CCl_3^{\bullet} + RCl \qquad \text{homolysis} \qquad (10.9)$$

Homolytic reactions involving carbon tetrachloride and chloroform to yield carbon dioxide could proceed as shown in Figure 10.1. The scheme involves two interesting intermediates, **phosgene** and **chlorine monoxide.** Phosgene is shown arising from trichloromethanol in Figure 10.1 (7), and also by the combination of oxygen with trichloromethyl [Figure 10.1 (4)]; the latter reaction has been described by Steacie (1954). Phosgene spontaneously hydrolyses to carbon dioxide and hydrochloric acid but no data are available to indicate whether the *kinetics* of such a scheme as shown in Figure 10.1 are compatible with the known production rate of

(1) $CCl_4 + OH^{\bullet} \longrightarrow CCl_3^{\bullet} + HO{-}Cl$

(2) $CHCl_3 + OH^{\bullet} \longrightarrow CCl_3^{\bullet} + H{-}OH$

(3) $CCl_3^{\bullet} + OH^{\bullet} \longrightarrow C(OH)Cl_3$

(4) $C(OH)Cl_3 \longrightarrow COCl_2 + HCl$

(5) $CCl_3^{\bullet} + O_2 \longrightarrow COCl_2 + Cl{-}O$

(6) $COCl_2 + H_2O \longrightarrow CO_2 + 2HCl$

(7) $COCl_2 + 2NH_3 \longrightarrow CO(NH_2)_2 + 2HCl$

Figure 10.1. A possible route for the homolytic breakdown of carbon tetrachloride (and chloroform) by interaction with hydroxyl radicals in the endoplasmic reticulum. The interaction between trichloromethyl radicals and oxygen has been described by Steacie (1954).

$$\begin{array}{l} \quad\ \ \delta-\ \ \delta+ \\ Cl_3{-}C{-}Cl \\ \quad\ \ |\quad\ | \\ \quad\ \ O{-}O^{2-} \end{array} \longrightarrow Cl_3{-}C{-}O^- + OCl^- \xrightarrow{+H^+} CCl_3OH \longrightarrow COCl_2 + HCl \xrightarrow{+H_2O} CO_2 + 2HCl$$

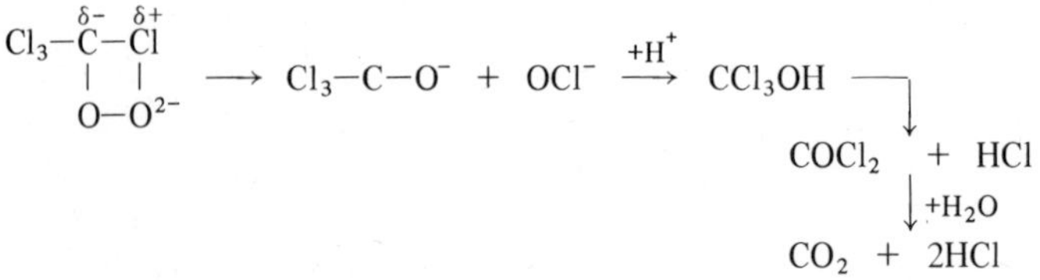

Figure 10.2. A route for the breakdown of carbon tetrachloride involving the doubly-charged oxygen anion that has been postulated as an intermediate in drug hydroxylations mediated by cytochrome P_{450} (see Hill *et al.*, 1969; Staudinger *et al.*, 1965). The transition state is shown as splitting into two anionic products, trichloromethoxylate and hypochlorite.

[(10)] Drug metabolism has been shown to require the presence of a lipid component (see May and McCay, 1968a, 1968b; Lu *et al.*, 1969; Strobel *et al.*, 1970).

carbon dioxide *in vivo* during intoxication with carbon tetrachloride. The potential formation of chlorine monoxide is significant, for this material is extremely toxic to living systems. An alternative scheme to that shown in Figure 10.1 can be written in which homolytic degradation involves interaction with O_2^{2-} rather than $OH^{\bullet}$, but at present no experimental data are available to favour either scheme (Figure 10.2), both of which should be regarded as highly speculative.

The introduction of *homolytic* considerations into the discussion of the metabolism of carbon tetrachloride means that products of its degradation other than carbon dioxide have to be considered. The trichloromethyl radical, for example, would interact with neighbouring molecules such as nucleotide bases, thiol groups, unsaturated fatty acids etc. In consequence we may expect that the metabolism of carbon tetrachloride to carbon dioxide via a homolytic pathway may represent only part of the homolytic reactions involving carbon tetrachloride. In relation to the binding of $CCl_3^{\bullet}$ to neighbouring molecules, Reynolds and Lee (1967, 1968) have found that shortly after administration of radioactive carbon tetrachloride and chloroform to rats radioactivity can be detected bound covalently to protein and lipid. The extent of this covalent bound material has been estimated by Reynolds (1963) to be approximately 1 μmol/g of liver by 2 h after dosing, which is about 40% of the conversion of carbon tetrachloride into carbon dioxide during the first 2 h of intoxication (Garner and McLean, 1969). The rapidity of the binding between metabolites of carbon tetrachloride and rat liver lipids and proteins has been emphasised by Rao and Recknagel (1969). After giving C^{14}-labelled carbon tetrachloride to rats, microsomal lipids were maximally labelled within five minutes; the labelling of mitochondrial lipids was about one-quarter to one-seventh that of the microsomal lipids. The rapidity of the microsomal lipid binding is similar to the very rapid stimulation of lipid peroxidation (as measured by diene conjugation, Rao and Recknagel, 1968) and is the earliest known disturbance to the liver consequent to the administration of carbon tetrachloride. The authors conclude that the data are presumptive evidence that homolytic cleavage of the carbon–chlorine bond with resulting lipid peroxidation are important features of the hepatotoxic action of carbon tetrachloride.

Gordis (1969) has produced further evidence that homolysis of carbon tetrachloride occurs *in vivo* by studying the composition of the lipid material that contains radioactivity shortly after administering [^{14}C]carbon tetrachloride to rats. The radioactive metabolites were found to comprise a heterogeneous group of branched long-chain chlorinated fatty acids, probably containing a $CCl_3^{\bullet}$ side-chain. A simple chemical system that generates $CCl_3^{\bullet}$ gave the same products on reaction with lipid as those isolated from the situation *in vivo*.

The discussion so far has indicated that carbon tetrachloride is metabolised in the rat mainly by the liver endoplasmic reticulum involving the NADPH–cytochrome P_{450} enzyme sequence. The suggestion has been made that such metabolism is probably homolytic in character and is dependent on an interaction of carbon tetrachloride with an endogenous free radical in the NADPH–cytochrome P_{450} complex. The composition of the complex electron-transport chain is shown in Figure 10.3.

The flavoprotein (FP) is identical with the NADPH–cytochrome *c* reductase flavoprotein described by Horecker (1950). There is a thiol-dependent site distal to the flavoprotein and an unknown component X_1 that has been proposed by Dallner *et al.* (1965) to account for the changes in drug metabolism that occur shortly after birth in the livers of newborn rats; X_1 is believed to be the rate-limiting component for the whole sequence. Cytochrome P_{450} is a cytochrome of the *b*-type that, when reduced and exposed to carbon monoxide, gives a characteristic peak at 450 nm. The interaction between a drug to be metabolised, cytochrome P_{450} and oxygen is believed to involve an activated form of oxygen; mechanisms involving $OH^\bullet$, O_2^{2-}, $O_2H^\bullet$ and the perferryl ion have been proposed.

The drug-metabolising sequence of liver endoplasmic reticulum contains at least two endogenous radical-reactive sites. Electron-spin-resonance studies on endoplasmic reticulum demonstrate this point clearly (see Section 6.3). There is the flavoprotein site that oscillates between the fully reduced and half-reduced semiquinone radical and the microsomal Fe_x component that has been correlated with the cytochrome P_{450} content. Both of these radical sites appear to be in a lipid environment: the flavoprotein requires a lipid component for activity (May and McCay, 1968a, 1968b) and the interaction of drugs with cytochrome P_{450} is known to be increased as the lipophilic character of the drug is increased.

If carbon tetrachloride reacts with one or both of the endogenous radical sites in the endoplasmic reticulum then it may be expected that trichloromethyl radicals will be produced. This radical is chemically very

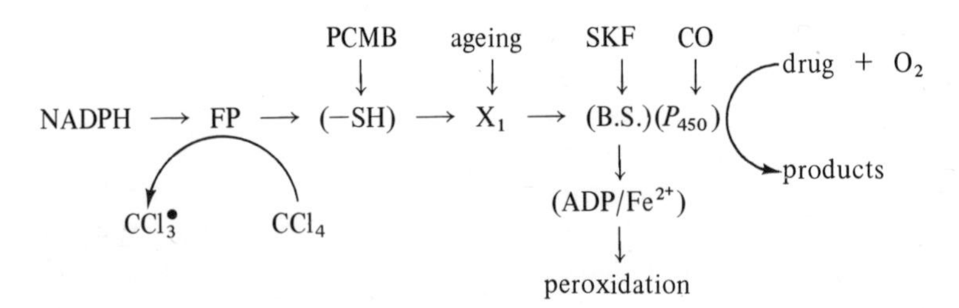

Figure 10.3. Components of the NADPH electron-transport chain in rat liver endoplasmic reticulum concerned with the metabolism of drugs.
Abbreviations: FP, flavoprotein; –SH, thiol-dependent region; X_1, rate-limiting component; (B.S.), binding site of drug; *p*-CMB, *p*-chloromercuribenzoate.
The points of interaction of metabolic inhibitors (*p*-CMB, SKF 525A and carbon monoxide) are indicated by broken arrows. Ageing microsomes at 0°C appears to decrease component X_1 (Slater and Sawyer, 1971). The lipid peroxidation route stimulated by ADP and Fe^{2+} is indicated (Hochstein and Ernster, 1963).

reactive, and so far it has only been possible to infer its production *in vivo* by indirect means[11]. For example, experiments *in vitro* indicate the type of product that is formed by the interactions of $CCl_3^\bullet$ with unsaturated lipid, nucleotides and thiol groups; the detection of such products in liver samples after administering carbon tetrachloride *in vivo* can be regarded as indirect evidence for the formation of $CCl_3^\bullet$ *in vivo*.

More direct evidence of the formation of $CCl_3^\bullet$ can be obtained by experiments with liver endoplasmic reticulum preparations *in vitro*. If the reasoning given above for the reaction of carbon tetrachloride with the NADPH–cytochrome P_{450} chain is correct, then it is possible to devise experiments to demonstrate the reaction not only by indirect identification of the reaction products but by kinetic analysis of the interaction and, ideally, by electron spin resonance identification of the $CCl_3^\bullet$ radical. Moreover, the use of metabolic inhibitors of the NADPH–cytochrome P_{450} electron transport chain should allow a decision to be made as to the actual site or sites of interaction of carbon tetrachloride with the NADPH–cytochrome P_{450} chain. The latter approach can be made by an analysis of the products that result from the interaction in the endoplasmic reticulum between $CCl_3^\bullet$ and either unsaturated lipid (to yield lipid peroxides) or neighbouring nucleotides and thiol groups. For practical reasons, however, almost all of the work so far performed in this approach has been in relation to lipid peroxidation. In the following Section the stimulatory action of carbon tetrachloride on lipid peroxidation in rat liver microsomes will be discussed. This will be followed by a discussion of some kinetic results and the effects of metabolic inhibitors on the metabolism of carbon tetrachloride.

10.6 Lipid peroxidation

A pro-oxidant action of carbon tetrachloride has been known for a long time; Hove (1953), for example, found that the compound increased the rate of deterioration of carotene *in vitro*. More extensive and recent work on the action of carbon tetrachloride on lipid autoxidation has come from studies on lipid peroxidation in liver suspensions.

Ghoshal and Recknagel (1965a) and Comporti *et al.* (1965) independently reported that low concentrations of carbon tetrachloride stimulated the production by suspensions of rat liver (microsomes + cell sap) of material that gave a positive reaction in the thiobarbituric acid reaction. Slater (1966b) reported that a similar (although considerably smaller) stimulation of malonaldehyde production could be obtained in washed microsomal suspensions alone, provided that a source of NADPH was supplied. These experiments show that carbon tetrachloride will stimulate free-radical

[11] The formation of $CCl_3^\bullet$ *in vivo* can be expected to lead to the subsequent production of hexachloroethane by dimerisation. Fowler (1969) has identified hexachloroethane in liver extracts after feeding carbon tetrachloride to rabbits and so has obtained further indirect evidence for the formation of $CCl_3^\bullet$ *in vivo*.

processes in liver endoplasmic reticulum and as such are of considerable importance in explaining the toxic effects of the compound on the liver. The present situation will be described in relatively extensive terms.

When a small quantity [12] of carbon tetrachloride is placed in the side arm of a Warburg flask and allowed to *diffuse* into the central compartment containing a suspension of microsomes, or (microsome + cell sap) together with NADPH, there is an increased production of malonaldehyde during incubation at 37°C. The results of numerous experiments performed in the author's laboratory are summarised in Table 10.5. The stimulation was dependent on a source of NADPH (usually $NADP^+$ + glucose 6-phosphate + glucose 6-phosphate dehydrogenase was added to give a

Table 10.5. Stimulation of production of malonaldehyde by carbon tetrachloride *in vitro* in suspensions of rat liver microsomes or (microsomes plus supernatant) in a stock buffer mixture[a] containing a source of NADPH. Malonaldehyde was determined by the thiobarbituric acid reaction at the end of the incubation. Mean values ± S.E.M. are shown as percentages of the production of malonaldehyde in the absence of carbon tetrachloride. The carbon tetrachloride (2 μl of 1 : 1 solution in liquid paraffin) was placed in the side arms of Warburg flasks and diffused into the stock suspension (2·5 ml) contained in the central compartment. Incubations were carried out in the dark with shaking at 60 cyc./min. Data are from Slater and Sawyer (1971).

	Incubation time (min)	Number of estimations	Production of malonaldehyde (%)
Microsome suspension			
Control	10–15	30	100
+ CCl_4	10–15		119 ± 2
Microsomes + supernatant suspension			
Control	60	35	100
+ CCl_4	60		166 ± 23

[a] The composition of the stock buffer mixture used in experiments concerned with the stimulation of malonaldehyde production by carbon tetrachloride was as follows: 83·5 mM KCl; 37·2 mM tris-HCl buffer, pH 8·0; 5·5 mM sodium glucose 6-phosphate; 0·245 mM sodium $NADP^+$; 10 mM acetamide; 8·4 international units glucose 6-phosphate dehydrogenase; total volume 32·4 ml. The stock buffer was mixed with either 4 ml of a suspension of microsomes (1 ml suspension contained microsomes from 1 g original wet wt. liver) or 6 ml of a microsome plus supernatant suspension (1 ml equivalent to 0·2 g wet wt. liver). The microsomes plus supernatant fraction were isolated by centrifuging a liver homogenate in 0·25 M sucrose at 11 700 g_{av} for two periods of 10 min; the microsomal pellet was prepared by a further centrifuging at 157000 g_{av} for 40 min and was resuspended in 0·15 M KCl solution.

[12] High concentrations of carbon tetrachloride not only fail to stimulate lipid peroxidation in liver microsomes but also depress *endogenous* lipid peroxidation, as do other lipophilic solvents. This concentration effect explains the failure of some investigators to obtain a stimulation of lipid peroxidation with carbon tetrachloride (for example see Wills, 1969).

continuous production of NADPH); NADH was not found suitable as a replacement. Table 10.6 gives the relevant data obtained with microsomal suspensions. A particularly important result was that the stimulation in the production of malonaldehyde by carbon tetrachloride was proportional to the *square root* of the concentration of the carbon tetrachloride in the incubation mixture (see Figures 10.4 and 10.5). This shows that the compound is stimulating the process of lipid peroxidation by acting as an *initiator* of the process, presumably after being converted first into $CCl_3^{\bullet}$. It is well known from studies on polymer kinetics (see Section 3.3.2) that the rate of polymerisation is proportional to the square root of the initiator concentration. This result for carbon tetrachloride shows that the stimulation of lipid peroxidation is not simply due to the lipophilic solvent having an indirect effect (for example via a conformational change) on the lipid-rich structures which facilitate oxidative degradation, but is entering the reaction in a direct manner as an initiator of the free-radical chain process.

Further evidence for the involvement of homolytic bond cleavage of carbon tetrachloride in the stimulation of lipid peroxidation comes from a comparison of the relative efficiencies of a series of halogenomethanes in stimulating lipid peroxidation in suspensions of rat liver (microsomes + cell sap). On an equimolar basis the order of relative activities obtained

Table 10.6. Stimulatory effect of carbon tetrachloride on the production of malonaldehyde in suspensions of rat liver microsomes in buffer mixtures supplemented where necessary with NADPH or NADH[a]. Results of two separate experiments are shown (unpublished data of T. F. Slater and B. C. Sawyer).

	Malonaldehyde production (nmol ml^{-1} of suspension)			
	Experiment I		Experiment II	
	−NADPH (a)	+NADPH (b)	+NADH (a)	+NADPH (b)
Control	4·2	7·4	3·2	7·4
+ CCl_4	3·9	9·3	3·2	8·3
Difference	−0·3†	+1·9¶	0·0†	+0·9§

† Difference not significant.
¶ $P < 0.001$ for the difference between control and CCl_4 values.
§ $P < 0.01$ for the difference between control and CCl_4 values.

[a] Standard buffer mixture for experiment (I); in experiment (IIb) a modified standard buffer was used with 5 mM nicotinamide replacing acetamide. In experiment (IIa) glucose 6-phosphate, glucose 6-phosphate dehydrogenase, $NADP^+$ and acetamide were omitted and replaced by 10 mM ethanol, 720 units alcohol dehydrogenase, 0·25 mM NAD^+ and 5 mM nicotinamide.

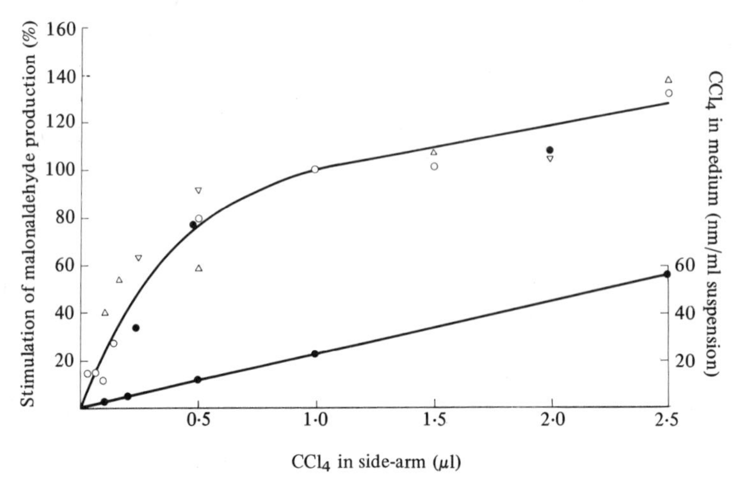

Figure 10.4. Stimulation of malonaldehyde production by various concentrations of carbon tetrachloride *in vitro* in suspensions of rat liver microsomes plus supernatant in stock buffer. The data of four separate experiments are shown with distinguishing symbols. The lower straight line shows the concentration of carbon tetrachloride in the incubation mixture contained in the central compartment of Warburg flasks as a function of the amount of carbon tetrachloride added to the side arm of the flask. Incubations were for 60 min at 37°C. The values have been normalised to a 100% value for malonaldehyde production with 1 μl of carbon tetrachloride in the side arms. For further details see Slater and Sawyer (1971).

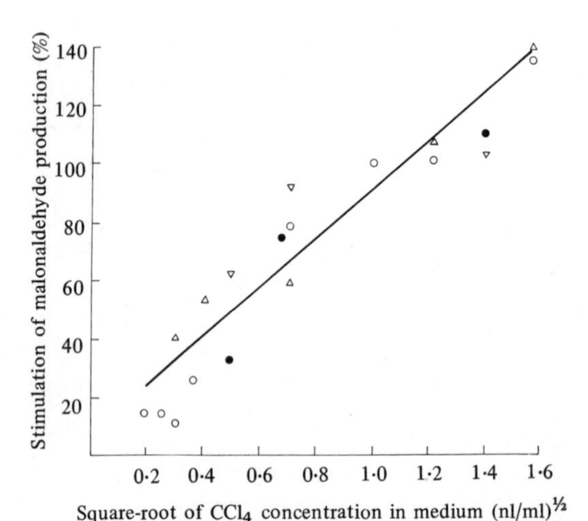

Figure 10.5. The upper curve of Figure 10.4 has been replotted to show the relationship between the stimulation of malonaldehyde production in rat liver microsomes plus supernatant suspensions and the square root of the concentration of carbon tetrachloride in the incubation medium. The regression line is drawn ($r = 0{\cdot}82$; $P \ll 0{\cdot}001$). Data of Slater and Sawyer (1971).

was bromotrichloromethane > carbon tetrachloride > fluorotrichloromethane > chloroform (Table 10.7). This relative order of activities is that to be expected if a homolytic bond fission of the halogenomethane was a prerequisite of the stimulated lipid peroxidation.

The increased lipid peroxide produced by low concentrations of carbon tetrachloride is accompanied by a *decreased* activity of glucose 6-phosphatase, an important enzyme in the regulation of carbohydrate metabolism. The relative efficiencies of bromotrichloromethane, carbon tetrachloride, chloroform and fluorotrichloromethane in decreasing glucose 6-phosphatase activity is the same as that described above (Table 10.7) in connection with the stimulation of malonaldehyde production. In fact, as will be seen below, the same inhibitors that block the *increased* production of malonaldehyde also block the *decrease* in glucose 6-phosphatase activity so that the two events appear closely related.

Table 10.7. Relative activities of halogenomethanes and halothane in stimulating production of malonaldehyde in suspensions of rat liver microsomes plus supernatant in stock buffer mixtures containing a source of NADPH. Incubation was for 60 min at 37°C with 2 μl of halogenomethane (or halothane)-liquid paraffin (1 : 1, *v*/*v*) in the side arms of Warburg flasks where necessary. Results are expressed as relative *molar* activities with respect to the action of carbon tetrachloride (unpublished data of T. F. Slater and B. C. Sawyer). The number of experiments with each material is in parenthesis.

Agent	Relative molar activity in stimulating production of malonaldehyde
CCl_4	100
$CHCl_3$	7 (7)
$CFCl_3$	34[a] (2)
$CBrCl_3$	~3000 (3)
Halothane	< 5 (5)

[a] This value may be too high due to difficulties in measuring the concentration of this very volatile material.

10.7 Interaction site with the P_{450} chain

We have seen above that carbon tetrachloride and related halogenomethanes react with the liver endoplasmic reticulum in the presence of NADPH to stimulate the production of malonaldehyde and to decrease the activity of glucose 6-phosphatase. For reasons already outlined the reaction most likely occurs with the NADPH–cytochrome P_{450} chain that is concerned with the metabolism of foreign compounds. The question is: where along this chain is the interaction site or sites with carbon tetrachloride?

One obvious interaction is with cytochrome P_{450} itself. McLean (1967a) has shown that carbon tetrachloride interacts with cytochrome P_{450} to produce a **difference spectrum** resembling that produced with other drugs, such as hexobarbital. Since the cytochrome P_{450} region is associated with endogenous radicals (microsomal Fe_x and oxygen radicals involved in drug metabolism) it is possible that the cytochrome P_{450} site is where carbon tetrachloride is homolytically cleaved to $CCl_3^{\bullet}$ with subsequent stimulation of lipid peroxidation. Considerable evidence, however, is against this speculation. First, other solvents such as chloroform and trichlorofluoromethane interact as strongly with cytochrome P_{450} as does carbon tetrachloride to give difference spectra (Figure 10.6) and yet do not stimulate lipid peroxidation or decrease glucose 6-phosphatase activity to the same extent as carbon tetrachloride. Secondly, experiments with a number of inhibitors that block the endoplasmic reticulum electron-transport chain from NADPH to cytochrome P_{450} suggest that the interaction site of carbon tetrachloride with the chain is in the region of the NADPH–flavoprotein rather than with cytochrome P_{450}. Some relevant data are given in Table 10.8. It can be seen that carrying out the incubation in a partial atmosphere of *carbon monoxide,* which binds to reduced cytochrome P_{450}, actually increases the stimulation in malonaldehyde production normally seen with carbon tetrachloride; similar results were found with *SFK 525A,*

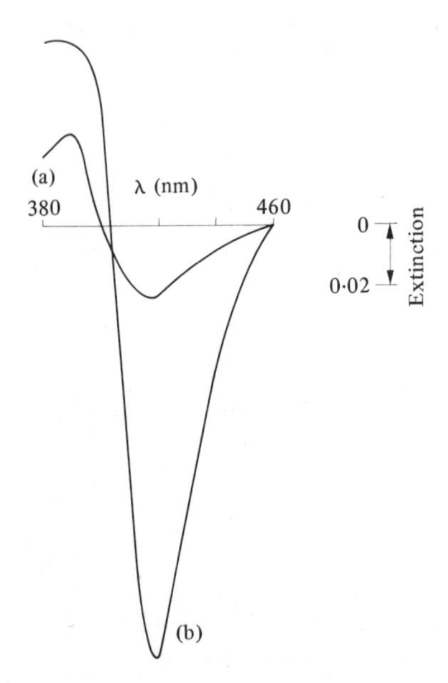

Figure 10.6. Difference spectra obtained with (a) fluorotrichloromethane (18 mM) and (b) SKF 525A (0·18 mM) and rat liver microsome suspensions in 0·05 M-tris buffer, pH 7·4 (see Schenkman *et al.*, 1967, for experimental procedure). Data of Slater and Sawyer (1971).

which binds strongly with cytochrome P_{450}, and with *p-chloromercuribenzoate* (*p*-CMB), which inhibits the electron-transport chain between the flavoprotein and cytochrome P_{450}. *Cytochrome c*, however, strongly inhibits the increased production of malonaldehyde due to carbon tetrachloride (see also Glende and Recknagel, 1969). This evidence suggests that carbon tetrachloride interacts with the endoplasmic reticulum in the vicinity of the NADPH–flavoprotein. The product of the interaction, probably $CCl_3^{\bullet}$, then stimulates lipid peroxidation in the neighbouring membrane and this is associated with a decrease in glucose 6-phosphatase activity.

In connection with the above experiments it is relevant to note that when using inhibitors to selectively block particular regions of an electron transport chain (e.g. the NADPH–cytochrome P_{450} chain) it is important to use the minimum concentration of inhibitor that will produce its selective effect. For example, with the cytochrome P_{450} chain, a

Table 10.8. Effects of metabolic inhibitors and cytochrome *c* on the stimulation of production of malonaldehyde due to carbon tetrachloride in suspensions of rat liver microsomes in buffer mixture. Incubation was at 37°C for the periods shown, with 2 μl of carbon tetrachloride–liquid paraffin (1 : 1, *v*/*v*) in the side arms of Warburg flasks. Other details were as described in the legend to Table 10.5. Stimulation in production due to carbon tetrachloride under standard conditions in the absence of added inhibitor is described as 'control stimulation'; in the presence of inhibitor the observed stimulation due to carbon tetrachloride is described as 'inhibitor stimulation'. Unpublished data of T. F. Slater and B. C. Sawyer.
Abbreviations: cyt. *c*, cytochrome *c*; CO, carbon monoxide; SKF, β-diethylaminoethyl-3,3′-diphenyl propyl acetate; *p*-CMB, sodium *p*-chloromercuribenzoate.

Exp.	Stimulation	Concentration of drug (μM)	Incubation time (min)	Stimulation of production of malonaldehyde (nmol/ml of suspension
a	Control		15	+0·75
	SKF	100	15	+1·00†
b	Control		10	+0·73
	p-CMB	94	10	+1·35‡
c	Control		15	+4·75
	CO	–[a]	15	+7·35§
d	Control		10	+0·95
	cyt. *c*	50[b]	10	−0·28¶

[a] 5 min gassing of suspension with CO or O_2 (control) before incubation ± CCl_4.
[b] mg 100^{-1} ml of suspension.

For differences between the control stimulation and the inhibitor stimulation:
† not significant. ‡ $P < 0{\cdot}05$. § $P < 0{\cdot}01$. ¶ $P < 0{\cdot}001$.

concentration of SKF 525A of 100 μM causes substantial inhibition of drug metabolism by virtue of its interaction in the neighbourhood of P_{450} itself. A much higher concentration, however, has an unspecific effect on the NADPH–cytochrome P_{450} chain as a whole due to its surface active action on the membrane (see Lee *et al.*, 1968). Similarly, *p*-CMB inhibits electron transport between the flavoprotein and cytochrome P_{450} when present in 100 μM; higher concentrations affect the activity of the flavoprotein itself (Kamin *et al.*, 1965). The principle of the minimum effective inhibitor concentration is an important one and is of particular value when attempting to elucidate interaction sites such as those between carbon tetrachloride and the P_{450} chain.

An important finding in the course of these experiments was that the stimulation in malonaldehyde production by carbon tetrachloride was inversely proportional to the already existing level of peroxidation of the lipids in the microsome or (microsome + cell sap) fractions. If the

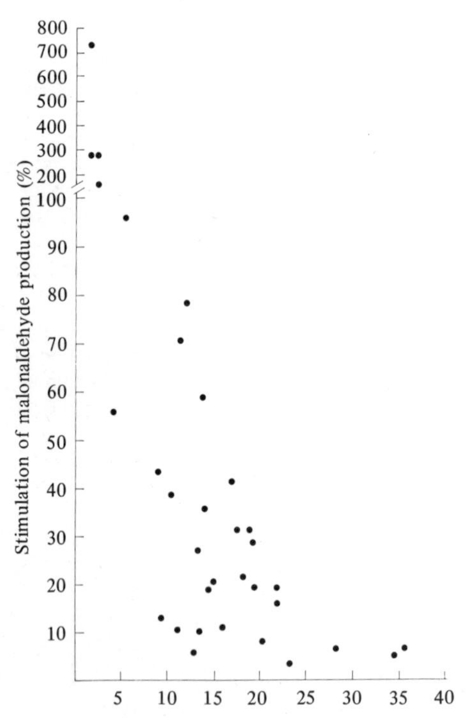

Figure 10.7. Relationship between the stimulation of production of malonaldehyde in rat liver microsomes plus supernatant suspensions (2·5 ml volumes in the central compartments of Warburg flasks) by carbon tetrachloride and the endogenous malonaldehyde production in its absence. Incubation time 60 min at 37°C; 2 μl of carbon tetrachloride–liquid paraffin (1 : 1, *v*/*v*) was placed in the side arms of Warburg flasks. Data of Slater and Sawyer (1971).

endogenous malonaldehyde content was high then carbon tetrachloride produced little if any stimulation in further malonaldehyde production. This was most clearly found with suspensions of (microsomes + cell sap) (Figure 10.7). The implication of the finding is that carbon tetrachloride can only stimulate the peroxidation of a *limited* amount of lipid that is associated with, or surrounds, the active site where homolytic cleavage of the compound occurs. Once this is peroxidised no further stimulation of peroxidation by carbon tetrachloride is possible. This is very unlike the situation where lipid peroxidation is initiated in the endoplasmic reticulum by (ferrous ions + ADP) and which proceeds to an almost complete peroxidation of the total lipid with solubilisation of the microsomal membrane. If the above conclusion is warranted it indicates that some fairly active chain-breaking process operates in the region of the NADPH–flavoprotein to prevent the lipid peroxides produced by $CCl_3^{\bullet}$ from involving lipid-rich areas away from the initiation site. There is, however, another possible explanation of the data shown in Figure 10.7. A high endogenous peroxidation will result in the rapid utilisation of NADPH but may also result in an increased destruction of this coenzyme which could then become limiting. If so then the stimulatory action of

Table 10.9. Effect of promethazine on the stimulation of production of malonaldehyde due to carbon tetrachloride in suspensions of rat liver microsomes in buffer mixture. Incubation was for 30 min at 37°C in the dark with 2 μl of a carbon tetrachloride-liquid paraffin solution (1 : 1, *v*/*v*) in the side arms of Warburg flasks. Other details were as described in the legend to Table 10.5. The stimulation in production due to carbon tetrachloride under standard conditions in the absence of added drug is described as the 'control stimulation'; in the presence of promethazine the observed stimulation due to carbon tetrachloride is described as 'promethazine stimulation'. Unpublished data of T. F. Slater, B. C. Sawyer and P. J. Jose. In experiment c the rats had been starved of food for 42 h before sacrifice, but had free access to water.

Exp.	Stimulation	Concentration of promethazine (μM)	Stimulation in production of malonaldehyde (nmol/ml of suspension)
a	Control		+1·1
	Promethazine	0·1	+0·5¶
b	Control		+0·9
	Promethazine	1·0	+0·4¶
c	Control		+3·8
	Promethazine	10	+0·4¶

For differences between control stimulations and the stimulation in the presence of promethazine: ¶ $P < 0{\cdot}001$.

carbon tetrachloride would be far less obvious, since it exhibits a need for a higher concentration of NADPH than does endogenous peroxidation (see Table 10.6).

Obviously more experimental data are required in order to evaluate the relative contributions of the two suggested mechanisms as explanations of the effect shown in Figure 10.7.

The increased lipid peroxidation in microsomal suspensions due to low concentrations of carbon tetrachloride is sensitive to very low concentrations of free-radical scavengers. For example, Table 10.9 shows that the phenothiazine drug promethazine in concentrations of 10^{-5}–10^{-7} M

Table 10.10. Effects of various drugs on the stimulation of production of malonaldehyde due to carbon tetrachloride in suspensions of rat liver microsomes plus supernatant in buffer mixture. Incubation was for 60 min at 37°C with 2 μl of carbon tetrachloride-liquid paraffin (1:1, *v*/*v*) in the side arms of Warburg flasks. Other details were as described in the legend to Table 10.5. Stimulation in production due to carbon tetrachloride under standard conditions in the absence of added drug is described as 'control stimulation'; in the presence of drug the observed stimulation due to carbon tetrachloride is described as 'drug stimulation'. Unpublished data of T. F. Slater and B. C. Sawyer.
Abbreviations: DPPD, *NN'*-diphenyl-*p*-phenylenediamine; vitamin E, α-tocopherol polyethylene glycol 1000 succinate.

Exp.	Stimulation	Concentration of drug (μM)	Stimulation of production of malonaldehyde (nmol/ml of suspension)
a	Control		+4·0
	Propyl gallate	2	+1·2§
b	Control		+4·1
	Inosine	10	+1·0‡
c	Control		+1·6
	EDTA	20	+0·4¶
d	Control		+4·3
	DPPD	0·1	+1·0¶
e	Control		+2·5
	Vitamin E	5·5	+0·2¶
f	Control		+2·4
	Sodium phenobarbital	50	+2·6†
g	Control		+4·3
	Sodium phenobarbital	500	+2·8§

For differences between the control stimulation and that in the presence of the drug:
† not significant. ‡ $P < 0{\cdot}05$. § $P < 0{\cdot}01$. ¶ $P < 0{\cdot}001$.

prevents the stimulation in malonaldehyde production due to CCl_4 but has little effect on *endogenous* peroxidation at these concentrations[13]. Similar effects have been found with a number of other free-radical scavengers and their effects on endogenous peroxidation and on the stimulation in malonaldehyde production due to carbon tetrachloride are shown in Table 10.10.

In this Section it has been shown that carbon tetrachloride interacts with the endoplasmic reticulum probably at the flavoprotein site to initiate a localised lipid peroxidation. This is associated with a decreased activity of glucose 6-phosphatase and both events are inhibited by low concentrations of free-radical scavengers. Since these same scavengers also retard the development of centrilobular necrosis *in vivo* (see Section 10.9) it is plausible that the metabolism of carbon tetrachloride required for the development of necrosis is that associated under conditions *in vitro* with lipid peroxidation[14]. If this is accepted then changes should occur *in vivo* similar to those found *in vitro* to be the result of lipid peroxidation in the membranes of the endoplasmic reticulum.

Of course these remarks also apply to reactions of the $CCl_3^\bullet$ radical other than lipid peroxidation; for example, reaction with nucleotides or with thiol groups. In the following Section the changes in such components observed *in vivo* during intoxication with carbon tetrachloride will be compared with the known reactions of $CCl_3^\bullet$ *in vitro*.

10.8 Relevance of the data *in vitro* to the injury *in situ*

If there is an interaction *in vivo* between carbon tetrachloride and the NADPH–flavoprotein resulting in the production of the chemically reactive and readily diffusible trichloromethyl radical, as already described for the situation *in vitro*, then certain consequences can be expected to follow.

1. The trichloromethyl radicals would react with neighbouring unsaturated fatty acid components of the endoplasmic reticulum resulting in lipid peroxidation. This process may or may not be *localised* in position and extent and may or may not be *significant* to the production of necrosis.
2. The formation of $CCl_3^\bullet$ will be associated with a decreased activity of glucose 6-phosphatase in the endoplasmic reticulum although this may not be a direct result of lipid peroxidation (see Section 10.12.1).

[13] The response to promethazine is dependent to some extent on the previous nutritional history of the rats used; for example, prior starvation for 24 h increases the stimulatory action of carbon tetrachloride and makes the system more sensitive to promethazine.

[14] Results with free-radical scavengers as protective agents *in vivo*, together with observations on their inhibitory action on lipid peroxidation *in vitro*, do not necessarily imply that lipid peroxidation is an *essential* process for the production of necrosis. Lipid peroxidation is, after all, only one of several free-radical pathways that can be initiated by $CCl_3^\bullet$. It is a relatively convenient one to demonstrate whereas other routes may be biologically more important.

3. Any considerable production of extraneous free radicals in the endoplasmic reticulum can be predicted to lead to an increased rate of breakdown of endogenous antioxidants near to the site of free-radical initiation, and also to some degree of radical annihilation (see footnote 11).
4. The trichloromethyl radical can be expected to react with neighbouring thiol groups, nucleic acids and proteins giving rise to altered biological activities.
5. Protection against the above effects can be obtained by concomitant administration of free-radical scavengers that penetrate to the initiation site.

Let us examine the evidence for each of these predictions in turn.

10.8.1 Lipid peroxidation *in vivo*

It has proved difficult to demonstrate that carbon tetrachloride stimulates lipid peroxidation in the liver *in vivo*. There are several major reasons for this difficulty. First, as seen above from the studies *in vitro*, it is probable that the actual amount of lipid peroxide formed owing to the carbon tetrachloride is small; moreover, production of lipid peroxide will be limited to a particular *region* of the endoplasmic reticulum (see Section 10.6). If the production of lipid peroxide by carbon tetrachloride is an important event in the onset of necrosis then we can imagine a further restriction: the stimulation of lipid peroxidation will be restricted to the centrilobular cells. Studies on whole liver samples would thus dilute out the stimulation of lipid peroxidation by carbon tetrachloride by including a large amount of extraneous non-reactive material. Secondly, it is difficult to determine very low concentrations of lipid peroxide in tissue suspensions without a considerable risk of artefactual peroxidation occurring during manipulation of the lipid-rich material during extraction and estimation. To a certain extent this is overcome by measuring the malonaldehyde content of tissue extracts and assuming that this bears a direct relation to the concentration of lipid peroxide, an assumption that has been borne out wherever this point has been checked (see Chapter 4). However, this procedure is itself complicated by the findings of Holtcamp and Hill (1951), Placer *et al.* (1965), Recknagel and Ghoshal (1966c) and Horton and Packer (1970), that malonaldehyde is rapidly *metabolised* by liver mitochondria. Thus attempts to subfractionate liver suspensions to measure malonaldehyde in the endoplasmic reticulum are complicated by the continuous disappearance of material under test. This means that determinations of malonaldehyde should be carried out on whole samples of liver as rapidly as possible after removing the sample from the liver *in situ*. Thirdly, even with the precautions noted above, it may still be very difficult to demonstrate that carbon tetrachloride increases malonaldehyde *in vivo*, since the mechanism of the decomposition of lipid peroxide to malonaldehyde is complex (see Chapter 4) and the kinetics relating to the utilisation of malonaldehyde by the mitochondria in response to a stimulated production of lipid peroxide in the endoplasmic reticulum are totally unknown. Thus it is quite conceivable that the steady-state concentration of malonaldehyde is normally very low owing to the rate of metabolism by the mitochondria exceeding the rate of formation in the cell; in such circumstances a relatively small stimulation in the rate of production would have no significant effect on the steady-state concentration.

Early attempts by Priest *et al.* (1962) and Recknagel and Ghoshal (1965, 1966c) to find an increased concentration of malonaldehyde after dosing with carbon tetrachloride *in vivo* were unsuccessful. In more recent work Slater and Sawyer (1971) have taken liver samples from rats poisoned 1–2 h previously with carbon tetrachloride and immediately extracted the samples with cold trichloroacetic acid. The trichloroacetic acid-soluble fraction was then heated with thiobarbituric acid reagent to detect malonaldehyde. Liver, however, contains a number of compounds other than malonaldehyde that give colours with thiobarbituric acid and these give powerful absorption over the wavelength range where the malonaldehyde product absorbs. Thus a small increase in malonaldehyde is hidden by a large background reading. To overcome this difficulty the boiled solutions may be extracted with butan-1-ol, which removes the malonaldehyde colour and leaves most of the residual blank material in the aqueous layer. Absorption spectra of such butanol extracts clearly show that carbon tetrachloride *in vivo* increases the amount of malonaldehyde in whole liver samples (Figure 10.8, Table 10.11). The absolute amounts of malonaldehyde shown in Table 10.11 are obviously very small but if allowance is made for a possible restriction of the observed increase in production to the centrilobular zone, and to the endoplasmic reticulum in

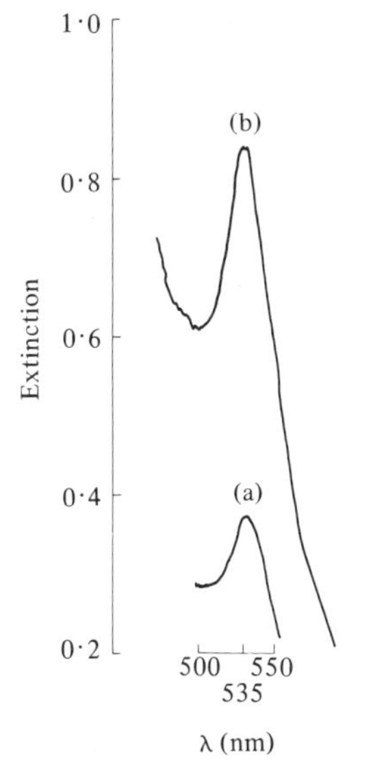

Figure 10.8. 'Malonaldehyde' content of liver samples 2 h after orally dosing rats with (a) liquid paraffin or (b) 2·5 ml of carbon tetrachloride/kg body wt. as a suspension (1 : 1, *v*/*v*) in liquid paraffin. The liver samples were frozen, weighed and extracted with cold 10% trichloroacetic acid. The extracts were then heated with thiobarbituric acid; the malonaldehyde colour was extracted with butan-1-ol and the spectrum measured. Malonaldehyde gives a peak at 531 nm. Data of Slater and Sawyer (1971).

the cells of that region, then it may be that in such locations the increase in malonaldehyde concentration approaches that observed *in vitro* in the presence of equivalent concentrations of the toxic agent. Clearly the data of Table 10.11 are evidence in favour of a stimulation of lipid peroxidation *in vivo* from intoxication with carbon tetrachloride.

The problem of detecting the presence of increased lipid peroxidation due to intoxication with carbon tetrachloride *in vivo* has been tackled in another way by Recknagel and Ghoshal (1966a). Free-radical attack by trichloromethyl radicals on the methylene bridges of polyunsaturated fatty acids can be expected to occur in the vicinity of the homolytic cleavage of carbon tetrachloride to $CCl_3^\bullet$. One consequence is a rearrangement of the bonding electrons leading to intense diene absorption at 234 nm (see Section 3.3.2). Recknagel and Ghoshal (1966a) were able to detect an increased diene absorption in the lipid fraction isolated from the liver endoplasmic reticulum within 1–2 h of dosing with carbon tetrachloride[15].

The time-course of the increase in diene conjugates during intoxication with carbon tetrachloride has been studied in male rats by Klaassen and Plaa (1969). A simplified version of their data is shown in Figure 10.9 where it can be seen that the maximum increase occurs about 1 h after dosing and is followed by a quite rapid fall-off. This decrease in the diene conjugate concentration suggests that the liver can metabolise these materials to an appreciable extent, and that it is only in the early stage of the injury that the rate of formation is greater than the rate at which the dienes can be removed[16].

Table 10.11. Effect of oral dosing with carbon tetrachloride (2·5 ml/kg body wt.) on the malonaldehyde content of rat liver. Mean values ±S.E.M. are given as pmol of malonaldehyde/g wet wt. of liver. Rats were killed 2 h after dosing with carbon tetrachloride and a piece of liver was immediately extracted with trichloroacetic acid. The acid-soluble extract was heated with thiobarbituric acid and the resultant pink colour extracted with butan-1-ol. Results are the means of 12 experiments (unpublished data of T. F. Slater and B. C. Sawyer).

	Malonaldehyde
Control	370 ± 60
Carbon tetrachloride	568 ± 102

[15] Recknagel's group have recently extended their data on this point (Rao *et al.*, 1970). The increase in diene conjugation produced by carbon tetrachloride *in vivo* was found to be accentuated by prior treatment of the rats with phenobarbital, and to be suppressed by prior treatment with SKF 525A.

[16] Although Green *et al.* (1969) observed a diene spectrum in microsomal lipids isolated from male rats that had received carbon tetrachloride, they could find no such evidence for lipid peroxidation in microsomal lipids prepared from female rat livers when analysed 6 h after dosing with the compound. In view of the results of Klaassen and Plaa (1969) it is obviously important to check for the production of a diene spectrum at several times during the early period of intoxication.

Further indirect evidence of free-radical attack on unsaturated lipid material *in vivo* is given from a study of the relative proportion of unsaturated to saturated fatty acids in the lipid fraction of liver endoplasmic reticulum. May *et al.* (1966) and Recknagel and Ghoshal (1966b) have shown that, *in vitro*, peroxidation of microsomal lipid markedly decreases the contents of arachidonic acid and docosahexanoic acid. Similar changes were observed *in vivo* after the administration of carbon tetrachloride: within 90 min after dosing the arachidonic acid concentration was decreased 20% and by 24 h after dosing the decrease was 33% (Horning *et al.*, 1962; Table 10.12).

Although each of the experimental procedures outlined above is open to criticism (see, for example, Green *et al.*, 1969, for an alternative explanation of the diene shift), when considered together they constitute compelling evidence that carbon tetrachloride stimulates free-radical processes in the lipid fraction of liver endoplasmic reticulum *in vivo*.

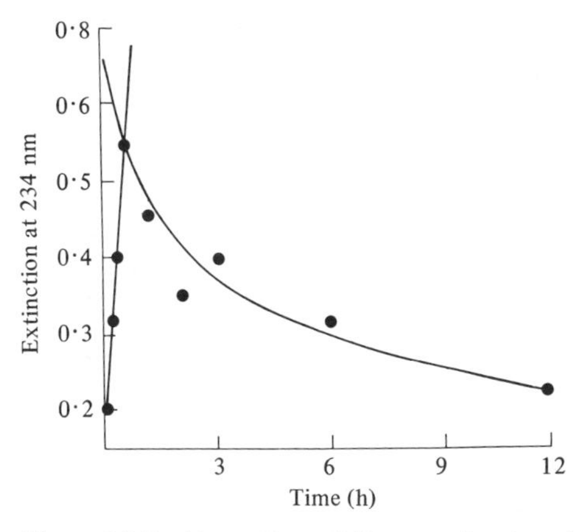

Figure 10.9. Absorption of diene conjugates at 234 nm in liver microsomal lipids as a function of time after administration of carbon tetrachloride (1·42 ml/kg body wt.) to rats. Data of Klaassen and Plaa (1969).

Table 10.12. Effect of dosing with carbon tetrachloride (2·5 ml/kg body wt.) on the fatty acid composition of liver phospholipids. Rats were killed 24 h after dosing (data of Horning *et al.*, 1962). Values are given as a percentage of the total phospholipid fatty acid fraction.

Fatty acid	Control	CCl_4
Palmitic (C_{16})	21	21
Stearic (C_{28})	34	34
Oleic (C_{18-1})	5	11
Linoleic (C_{18-2})	13	16
Arachidonic (C_{20-4})	26	17

10.8.2 Changes in the activity of glucose 6-phosphatase *in vivo*

One of the most sensitive indices of the effects of carbon tetrachloride on the liver is the decrease in the activity of glucose 6-phosphatase in the centrilobular region. Reynolds and Lee (1967) have shown that the effect may be detected after a dose as low as 9 μl of carbon tetrachloride/100 g body wt. The change occurs very rapidly after dosing and is only partially prevented by concomitant administration of promethazine (see Table 9.12). Clearly the glucose 6-phosphatase is remarkably sensitive to the presence of low concentrations of carbon tetrachloride *in vivo*. The addition of the agent *in vitro* to microsomal suspensions (with no added source of NADPH) prepared from normal rats had no effect on enzyme activity until very high concentrations of it were present (Ghoshal and Recknagel, 1965b). Evidently it is a product of the metabolism of carbon tetrachloride that is responsible for the decreased activity of microsomal glucose 6-phosphatase.

It is known that unspecific lipid peroxidation in liver microsomes (for example, that produced by the addition of ferrous ions plus ADP) is accompanied by a decreased activity in glucose 6-phosphatase (Hochstein and Ernster, 1963; Nordenbrand *et al.*, 1964; Ghoshal and Recknagel, 1965b). If the microsomal fraction is protected against lipid peroxidation by the addition of for example, vitamin E or EDTA, then only minimal changes in glucose 6-phosphatase activity are observed[17]. It has already been seen that carbon tetrachloride *in vitro* not only stimulates lipid peroxidation in isolated microsomes but decreases glucose 6-phosphatase activity, and that both processes are prevented by low concentrations of free-radical scavengers. Evidently a similar situation occurs *in vivo*, for as discussed in Section 10.8.1, carbon tetrachloride stimulates lipid peroxidation *in vivo* and, as just mentioned, there is an early suppression of glucose 6-phosphatase activity. The two events appear to be closely related to one another, and to reflect an early stimulation of free-radical reactions in the liver endoplasmic reticulum.

Glucose 6-phosphatase is an important enzyme in carbohydrate metabolism. It stands not only at the cross-roads of glycolysis, the pentose shunt and *gluconeogenesis*, but has recently been shown to have a multifunctional character (Nordlie and Arion, 1965). The enzyme catalyses the hydrolysis of (1) glucose 6-phosphate or (2) pyrophosphate and also acts as (3) a glucophosphotransferase with pyrophosphate or carbamyl phosphate.

$$\text{(1) Glucose 6-phosphatase} \longrightarrow \text{glucose} + P_i \qquad (10.10)$$

$$\text{(2) Pyrophosphate} \longrightarrow 2P_i \qquad (10.11)$$

[17] It should be remembered that the site of the pathway concerned in lipid peroxidation in the presence of ADP plus Fe^{2+} is different from that which operates in the presence of carbon tetrachloride (see Figure 10.3).

(3) Glucose + PP_i ⟶ glucose 6-phosphate + P_i (10.12)

(4) Glucose + carbamyl phosphate ⟶ glucose 6-phosphate + carbamate (10.13)

A large decrease in the activity of this enzyme can be expected to have severe consequences on the organised metabolism of the normal liver cell, which has a key role to play in maintaining blood sugar by gluconeogenesis. It is of no surprise therefore to observe that acute intoxication with carbon tetrachloride is accompanied by marked *hypo*glycaemia.

A further indication of the serious effects that a large decrease in glucose 6-phosphatase activity would have, can be obtained by references to the glycogen storage disease first discovered by von Giercke. In this condition there is a congenital deficiency of liver and kidney glucose 6-phosphatase; these tissues become packed with glycogen and death usually occurs early in infancy.

10.8.3 Changes in liver antioxidants

If carbon tetrachloride is metabolised through an interaction with the endoplasmic reticulum to produce free-radical products then some decrease in the concentration of endogenous antioxidant material should occur and be detectable. Although this appears a feasible assessment at first sight there are several points that complicate the picture.

First, the various endogenous antioxidants in the microsomes may not be affected equally by the localised production of $CCl_3^\bullet$. In any reaction between a fixed radical generated in a lipid-rich membrane and a variety of antioxidants, the rate of interaction will vary from one to another antioxidant depending on such factors as the accessibility of the initiation site, steric factors, activation energy of the interaction, concentration of the antioxidant at the reaction site[18] etc. An example of such specificity in antioxidant behaviour is that vitamin E, which is often used as a standard for biological antioxidant activity, has been reported to be inactive in retarding the oxidation of carotene by sugar beet enzyme whereas vitamin C and quercetin were active; in the conversion of ubiquinol-6 into ubiquinone-6 by the haem-catalysed decomposition of cumene hyperperoxide, α-tocopherol was not effective as an inhibitor and did not compete with ubiquinol for free radicals (Mellors and Tappell, 1966).

Secondly, the free-radical products of homolysis of carbon tetrachloride may react more vigorously with neighbouring thiol and nucleotide molecules (which have good scavenging properties; Section 5.2.1) with little involvement of other antioxidant material such as vitamin E. Although, as we have seen, carbon tetrachloride stimulates lipid peroxidation *in vivo*, the overall amount of peroxidation in the liver is small and borders on the limits of detection. As a consequence significant changes in the amounts

[18] Low concentrations of some antioxidants may act as pro-oxidants, as discussed in Section 5.1.

of antioxidants involved in interaction with lipid peroxidation chains may be expected to be difficult to detect.

The major antioxidants present in endoplasmic reticulum are vitamin E, inosine and ubiquinol (recent work by Nyquist *et al.*, 1970, suggests that ubiquinone is localised in the Golgi apparatus rather than the endoplasmic reticulum, and it will be most interesting to see whether α-tocopherol shows a similar distribution pattern); there is also a relatively high concentration of glutathione in the cell sap that bathes the membranes of the endoplasmic reticulum. Of these, vitamin E has been studied most extensively, particularly by Green *et al.* (1969) with [^{14}C]tocopherol. These workers failed to find any evidence that significant oxidation of α-tocopherol to the corresponding quinone occurred during the course of several types of experimentally produced liver necrosis. The experimental design of these experiments has recently been criticised by Witting (1969), however, who carried out kinetic studies on vitamin E oxidation and concludes that even if vitamin E had been coupled to lipid peroxidation in Green's experiments, the oxidation rate of labelled tocopherol *in vivo* would have been too low for detection. Csallany *et al.* (1970) have shown that the products of α-tocopherol oxidation formed during the autoxidation of methyl linoleate are analogous to those produced *in vivo*. They conclude that α-tocopherol is metabolised *in vivo* by way of reaction with lipid free radicals or peroxides.

Of the other antioxidants previously listed there are no data in the literature on the concentrations of ubiquinol in endoplasmic reticulum after dosing with carbon tetrachloride; protein-bound and non-protein thiol-group concentrations are reported to be unchanged (Icén and Huovinen, 1959) in early stages of intoxication with carbon tetrachloride. The concentration of inosine in normal rat liver endoplasmic reticulum is very high (Siekevitz, 1955). Inosine is closely involved in the metabolism of the purine nucleosides (Figures 5.5 and 10.10) but the significance of the high concentration in liver microsomal suspensions is not clear. Like many heterocyclic nitrogen ring systems inosine can function as an

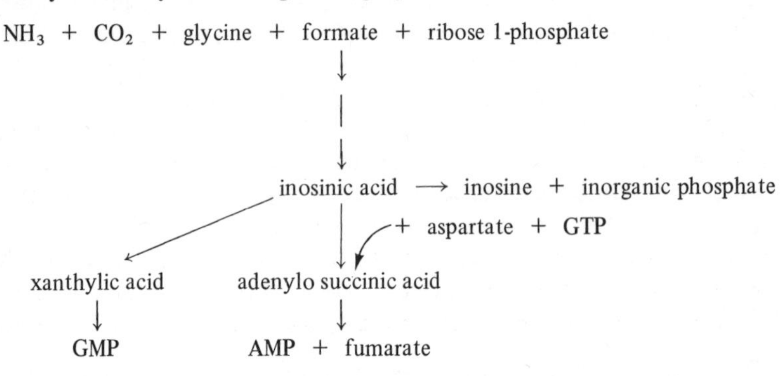

Figure 10.10. Metabolic routes involving inosinic acid.

antioxidant. Table 10.10 shows that low concentrations of inosine inhibit the stimulation of production of malonaldehyde due to carbon tetrachloride in microsomal suspensions *in vitro*, as do many other antioxidants. The concentrations of inosine in liver microsomes *in vivo* after dosing rats with carbon tetrachloride have been measured by Slater and Sawyer (1971) and their results are shown in Table 10.13. It can be seen that the toxic compound produces approximately a 20% decrease in the content of inosine in the microsomal fraction obtained from the *whole* liver and that this decrease is prevented by concomitant dosing of the rats with promethazine. Thus carbon tetrachloride *in vivo* produces a *decrease* in at least one microsomal component known to function as an antioxidant.

The lack of change in vitamin E in liver endoplasmic reticulum after dosing with carbon tetrachloride has been mentioned in connection with Witting's findings. There is one further important point that must be made in this connection. It has been claimed (McLean, 1967b) that the lack of effect with added vitamin E given to rats previously maintained on an E-deficient diet and dosed with carbon tetrachloride indicates that free-radical processes (as would be required to produce a stimulation of lipid peroxidation) are not of much significance in connection with liver necrosis. This view, however, presupposes several properties for vitamin E that may not be warranted. It assumes that the added vitamin E can gain access to the site of free-radical initiation in the endoplasmic reticulum. Although studies with labelled vitamin E show that it is quickly absorbed from the gut and can be detected in the endoplasmic reticulum, it is a *particular zone* of the endoplasmic reticulum that we are concerned with here: the zone surrounding the NADPH–cytochrome *c* flavoprotein. It has been seen from the studies *in vitro* that lipid peroxidation stimulated by carbon tetrachloride is possibly confined to a particular region of the lipid-rich membrane. Secondly, the experimental approach implicitly admits that vitamin E depletion would be conducive to increased homolytic damage due to carbon tetrachloride. This would only be the case if the homolysis of the compound were rate-limited by the concentration of vitamin E

Table 10.13. Effect of oral dosing with carbon tetrachloride (2·5 ml/kg body wt.) on the concentration of inosine in rat liver microsomes. Inosine was determined by the method of Siekevitz (1955). In some instances the rats were dosed with promethazine (25 mg/kg body wt.) at the time of feeding with the carbon tetrachloride; rats were killed 2 h after dosing with carbon tetrachloride (unpublished data of T. F. Slater and B. C. Sawyer).

	No. of animals	Inosine (%)
Control		100
+ CCl_4	9	83 ± 3
Promethazine		100
+ CCl_4	3	96

normally present. There is no experimental evidence on this point and if the assumption is not valid then depletion of vitamin E would obviously have little effect on the subsequent injury produced by carbon tetrachloride. There is a further complication that may arise from such studies in that depletion of vitamin E for long periods (as in the experiments of McLean, 1967b, or Green *et al.*, 1969) may result in depletion of other antioxidants and damage to other tissues [e.g. adrenals (Kitabchi and Williams, 1968)] that may secondarily affect the liver. In such a complex situation it is difficult to elucidate and separate those effects due simply to vitamin E deficiency in the liver from those due to depletion of other antioxidants or to damage to other tissues. Further, the addition of vitamin E to such deficient animals may in the short term re-establish amounts of vitamin E without ensuring that the other depleted antioxidants are also restored. If one of these is rate-limiting for the carbon tetrachloride injury then vitamin E supplementation will have no effect in the short term. It may be concluded that experiments involving animals reared on vitamin E deficient diets for protracted periods are somewhat difficult to interpret simply with present knowledge of antioxidant interactions *in vivo*.

10.8.4 Reaction with neighbouring metabolites

Whilst the localised production of trichloromethyl radicals around the NADPH–cytochrome *c* reductase flavoprotein (or at the cytochrome P_{450} site) can probably not be expected to affect greatly the whole-liver amounts of free thiol groups (as has been found for glutathione) it may well be that thiol groups essential for the function of various enzymes around these particular sites are irreversibly affected. There is, however, no information on this point.

One important aspect of the generation of trichloromethyl radicals by the reaction of carbon tetrachloride with the NADPH–flavoprotein is that the important coenzyme NADPH approaches the binding site of the flavoprotein near where the chemically very reactive trichloromethyl radicals are being produced. Thus a situation exists analogous to the irradiation experiments *in vitro* (see Section 7.4.3 and Table 10.14). It may confidently be expected that some destruction of NADPH will occur owing to such a reaction with trichloromethyl radicals. What would be the consequences of this reaction?

Initially, there would be a decrease in the *total* liver content of $NADP^+$ + NADPH as has been reported by Slater *et al.* (1964) and as illustrated in Figure 10.11. This decrease was found to be prevented by dosing the rats with promethazine, which, as already mentioned, inhibits the changes observed in inosine content, glucose 6-phosphatase activity and malonaldehyde production. NADPH is produced in liver by the cytoplasmic enzyme NAD kinase:

$$NAD^+ + ATP \xrightarrow{\text{kinase}} NADP^+ + ADP \qquad (10.14)$$

No changes in the total liver concentration of NAD, ATP[(19)] or NAD kinase were found in the early hours after administration of carbon tetrachloride (Table 10.15), indicating that the decreased NADPH concentration in the liver was not due to a failure in synthesis but is probably due either to an increased rate of catabolism or to a loss of coenzyme from the liver by leakage into the blood, lymph or bile. No indication of leakage of NADPH into plasma or bile, however, was found.

Table 10.14. Destruction of NADPH by bromotrichloromethane during irradiation with ultraviolet light for 5 min. The irradiation medium contained tris buffer, pH 8·0, acetone, bromotrichloromethane (120 mM) and NADPH. Other additions are shown in the Table (unpublished data of T. F. Slater, P. J. Jose and B. C. Sawyer).

Additions	NADPH[a] (nmol ml^{-1})
Control	480
CCl_3Br	215
CCl_3Br + nicotinamide (38 mM)	227
CCl_3Br + AMP (2·4 mM)	227
CCl_3Br + propyl gallate (4·4 mM)	395

[a] NADPH remaining at the end of the irradiation period; NADPH at the start of incubation was 490 nmol ml^{-1}.

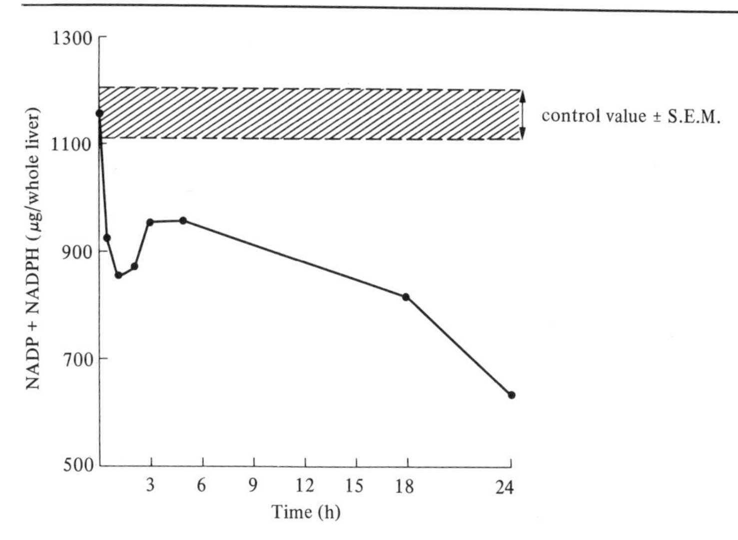

Figure 10.11. Effect of oral dosing with carbon tetrachloride (1·25 ml/kg body wt.) on the $NADP^+$ + NADPH content of rat liver. Data from Slater *et al.* (1964).

[(19)] Striking decreases in ATP concentration in rat liver 3 h after dosing with carbon tetrachloride in rats subjected to oxygen enrichment during the period of tissue sampling have been reported by Smuckler *et al.* (1968) but the mechanism of this effect is not known.

Table 10.15. Effect of oral dosing with carbon tetrachloride (1·25 ml/kg body wt.) on liver components (means ± S.E.M.) associated with the synthesis and destruction of $NADP^+$. Unpublished data of T. F. Slater and B. C. Sawyer; Slater *et al.* (1964) and Slater and Delaney (1970).

Component	Time of assay (h)	Control	CCl_4
1 NAD-kinase[a]	1	752 ± 90	699 ± 115
2 NADP-glycohydrolase[a]	1	322 ± 29	485 ± 64
3 Cell sap ATP (μmol/liver per 100 g body wt.)	2	4·82 ± 0·33	4·86 ± 0·37
4 NAD^+ + NADH (μg/liver per 100 g body wt.)	2	1671 ± 209	1938 ± 143

[a] In arbitrary extinction units.

The possible reaction between trichloromethyl and NADPH to yield unreactive coenzyme products will have a more subtle effect on the functioning of NADPH-linked enzymes than is immediately apparent.

$NADP^+$ and NADPH are localised normally mainly in the mitochondrial fraction of liver. Results obtained by a rapid microfiltration procedure, which enables intracellular distribution studies to be obtained have been given in Table 9.5. After administration of carbon tetrachloride there is an intracellular redistribution of $NADP^+$ + NADPH so that relatively more is in the supernatant fraction, although it must be remembered that this is superimposed on the overall net *decrease* of $NADP^+$ + NADPH in the whole liver. At the same time there is a movement of ATP in the opposite direction into the mitochondria. It can be argued that the destruction of NADPH on the membranes of the endoplasmic reticulum through an interaction with trichloromethyl drains more NADPH out of the mitochondria, thus producing the altered distribution observed. The movement of ATP in the reverse direction may be related to this loss of mitochondrial NADPH and, if so, is presumably accompanied by subtle changes in mitochondrial membrane permeabilities[20].

In these experiments the conclusions reached are similar to those proposed by Gallagher (1960, 1962) some 10 years ago, but the mechanisms suggested are different. In the sense that carbon tetrachloride appears to give a depletion of mitochondrial NADPH (Gallagher's experiments did not distinguish between a loss of NADH or NADPH) liver injury induced by the toxic compound can be viewed as a specific anomaly of mitochondrial nucleotide metabolism, an anomaly that is reflected by decreased synthetic and oxidative reactions, manifested at a later stage by mitochondrial abnormality and later still by necrosis.

[20] For another indication of an early change in mitochondrial membrane properties see footnote 3 in Section 13.2.

10.9 Protectors

A large number of agents have been reported to protect the liver against liver necrosis induced by carbon tetrachloride. A representative list is given in Table 10.16. The agents listed were assessed by histological examination of the liver, usually 24 h after dosing with carbon tetrachloride; a few of the agents have also been examined for their protective action during the early stages of intoxication on various parameters such as glucose 6-phosphatase activity, NADPH concentration, serum enzyme release, plasma bilirubin, protein synthesis and drug-metabolising activity.

'Protection', of course, may be against one or more of the various aspects of liver disturbance produced by carbon tetrachloride. For example, against the very complex sequence of events leading to necrosis or against a disturbance produced in a *single* component. Protection may be very variable in the time that the protective effects are demonstrable. For example, promethazine given at time 0 and +5 h after dosing with carbon tetrachloride retards the appearance of necrosis normally produced at 24 h; after 48 h, however, necrosis is very noticeable. On the other hand, phenobarbital or cysteamine given at time 0 and +5 h have no significant effect on the extent of necrosis assessed at 24 h after giving carbon tetrachloride but prevent the early effects of the agent on the liver's content of $NADP^+$ + NADPH (Slater *et al.*, 1964). Presumably, the protective actions of these compounds *in vivo* are partially determined by their excretion rates in the urine, which are known to be very fast; as a consequence they are protective against early effects of carbon tetrachloride but any retardation of the overall process of centrilobular necrosis is probably too short, say approximately 3 h, for any significant difference to be noticeable by subjective examination of the tissue sections some 20 h later. Another important factor that has to be considered in this discussion is the powerful effect that some of these 'protective' agents have on **body temperature**. Larson and Plaa (1965) have demonstrated how the decreased body temperature resulting from the combination of spinal cord sectioning with carbon tetrachloride intoxication modifies the course of the liver damage in the rat; similar studies concerning the effect of environmental temperature on the toxicity have been reported recently by Adam and Thorp (1970) using mice. Some chemical 'protective' agents also produce large decreases in body temperature. For example, cysteamine (150 mg/kg body wt.) was found to produce a 12°C drop in the body temperature of rats (Liebecq-Hutter and Bacq, 1958). The effects on body temperature of a number of other agents previously examined for their 'protective' action on liver necrosis induced by carbon tetrachloride are shown in Table 10.17. None of the agents reported in Table 10.17 has such a striking action on body temperature as that mentioned above for cysteamine.

The histological appearance of necrosis represents the culmination of a complex network of metabolic disturbances and perturbations of normal

Table 10.16. Agents that have been reported to modify the liver disturbances produced by carbon tetrachloride *in vivo* in rats. Effectiveness is listed as follows: P, strongly protective effect; p, significant though weaker protective effect; NP, lack of protective effect; V, very variable results. In necrosis and often also for fat accumulation the assessment has involved a subjective microscopic decision usually 24 h after dosing with carbon tetrachloride. The protective effects on early biochemical changes [decrease in NADPH at 1 h post-dosing, decreased protein synthesis (PrSy) or drug metabolism at 1–3 h post-dosing] that are listed here are in some instances representative of numerous measurements reported in the literature.
Abbreviations: DPPD, *NN'*-diphenyl-*p*-phenylenediamine; BHT, butylated hydroxytoluene; SKF 525A, β-diethylaminoethyl-3,3'-diphenylpropyl acetate.
Note. The experiments with reserpine were performed with mice and not rats.

Agent	References	Necrosis	Fat	NADPH	PrSy	Drug metabolism
Promethazine	1	P	p	P	NP	NP
Propyl gallate	2		P (4 h)	P		P?
DPPD	3	P	P			
Vitamin E	4	V	p		NP	NP
Ethoxyquin	5	P	P			
BHT	6	P	P			
Ubiquinone-4	7		P			
Cysteamine	8	NP		P		
Triperidol	9			P		
SKF 525A	10	NP	NP	NP	NP	NP
Nicotinamide	11	P	P	P		
Nupercaine	12	P		P		
Cetab	13	P		P		
Anthisan	14	p		p		
Benadryl	15	p		p		
Phenobarbital	16	NP		P		
Asparagine	17	NP	P			
Dodecyl sulphate	18	NP		NP		
Aminoacetonitrile	19				P	
Trypan blue	20	P	P			
Reserpine	21	P	P			
3,5-Dimethylpyrazole	22		P			
Cysteine	23		P			

References: 1, Rees and Spector (1961); Fox *et al.* (1962); Wigglesworth (1964); Slater *et al.* (1966); Serratoni *et al.* (1969). 2, Ugazio and Torrielli (1969); T. F. Slater (unpublished work). 3, see Table 9.7. 4, see Table 9.8. 5, Cawthorne *et al.* (1970). 6, Cawthorne *et al.* (1970). 7, Di Luzio (1966). 8, Slater *et al.* (1966). 9, T. F. Slater (unpublished work). 10, Slater *et al.* (1966); Smuckler and Hultin (1966); Castro *et al.* (1968). 11, Gibb and Brody (1967).
12–16, 18, Bangham *et al.* (1962); Slater *et al.* (1966). 17, Alexander *et al.* (1967). 19, Mager *et al.* (1965). 20, Petrelli and Stenger (1969). 21, Clower *et al.* (1968). 22, Bizzi *et al.* (1966). 23, Agostini and Comi (1968).

Table 10.17. Effect of various treatments on the rectal temperature of rats. Mean values are given in °C at various times after dosing (unpublished data of T. F. Slater and B. C. Sawyer).

Treatment[a]	Number of rats	Rectal temperature (°C)					
		Time (h)					
		0	1	2	3	4	6
Control	8	37·9 ± 0·1					
$CHCl_3$	2	37·0	34·8	33·8	32·6	32·9	
CCl_4	2	36·6	36·2	36·7	36·6		
CCl_4[b]	4		37·4				
CCl_4 + promethazine	4	37·6	33·7	34·1	34·6		
CCl_4 + promethazine[b]	4		36·9				
CCl_4 + Cetab	4	37·6	34·3			34·8	36·7
CCl_4 + dodecyl sulphate	2		36·2				
CCl_4 + phenobarbital	2	36·6	36·7	36·0	36·1		

[a] Doses: CCl_4, 1·25 ml/kg body wt.; $CHCl_3$, 1·0 ml/kg body wt.; promethazine, 25 mg/kg body wt.; Cetab, 5 mg/kg body wt.; dodecyl sulphate, 25 mg/kg body wt.; phenobarbitone, 24 mg/kg body wt.
[b] Rats kept in incubator with air temperature 36°C.
Rats were lightly anaesthetised with ether to facilitate administration of CCl_4 and $CHCl_3$; zero time values in these cases are therefore with anaesthetised animals that were then allowed to recover spontaneously. Control rats were not anaesthetised.

steady-state conditions. As such we may expect that certain routes through the network are preferentially favoured so that blockage of these routes, leading to the opening up of alternative pathways, will result in a slowing down of the process as a whole. Thus it can be imagined that many sites of 'protection' are available with respect to the development of centrilobular necrosis, some acting near to the primary site of activation of carbon tetrachloride on the endoplasmic reticulum and others acting somewhat later along the network. As a consequence a *multiplicity* of mechanisms must be postulated to explain the action of the known protective agents.

Many of the protectors that are active against centrilobular necrosis are also active in preventing the decrease in liver NADPH. In fact there appears to be a correlation between the decrease in liver NADPH at 1 h and the extent of centrilobular necrosis shown histologically 24 h after dosing with carbon tetrachloride. Figure 10.12 gives the results obtained. Thus if the decrease in liver NADPH is due to a reaction between the trichloromethyl free radical and NADPH at the flavoprotein site it is probable that many of the protectors are acting near to the primary site of cellular damage produced by carbon tetrachloride. (For cysteamine the effect on body temperature must also be considered.) We may also infer from Figure 10.12 that results are further evidence for the importance of the decrease in liver NADPH in relation to the overall process of necrosis.

The point is often raised in connection with the mechanism by which several antioxidants act as protectors against necrosis induced by carbon tetrachloride that the agents also react with the cytochrome P_{450} site and thereby may interfere with the *metabolism* of carbon tetrachloride at that site. In such a situation the protectors are acting not as scavengers of trichloromethyl radicals, but are acting as competitive inhibitors of the metabolism of carbon tetrachloride to toxic products of an unknown nature. Although this is conceivable it is unlikely since promethazine, which is very active in retarding necrosis and many early biochemical changes in the liver, is present in the liver 1 h after dosing in very low concentration; carbon tetrachloride, on the other hand, is present in concentrations of approximately 1 mM. Further, the drug SKF 525A, which binds very strongly to cytochrome P_{450}, has a weaker protective action against necrosis induced by carbon tetrachloride than has promethazine (Smuckler *et al.*, 1967; Castro *et al.*, 1968), although Rao *et al.* (1970) have reported that SKF 525A (100 mg/kg body wt.) given 30 min before carbon tetrachloride (0·25 ml/kg body wt.) had a protective action against the normally observed increase in diene conjugation in rat liver microsomal lipids. Further, whereas SKF 525A, which binds strongly to cytochrome

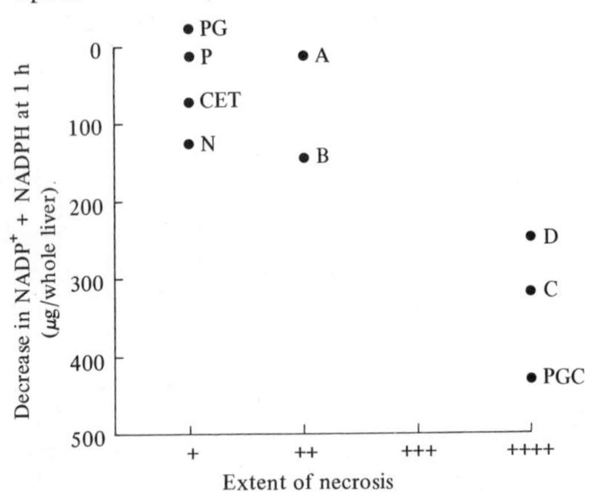

Figure 10.12. Effect of drug administration on the decrease in liver $NADP^+$ + NADPH content and on liver necrosis produced by oral dosing with carbon tetrachloride (1·25 ml/kg body wt.). Extensive centrilobular necrosis is graded (+–++++) and was assessed 24 h after dosing. The decrease in nucleotide content was measured 1 h after dosing. Abbreviations, with doses as mg/100 g body wt., and times of administration relative to the time of giving carbon tetrachloride: P, promethazine (2·5 and 1·25; time 0 h and +5 h respectively); CET, Cetab (0·5; time 0 h); PG, propyl gallate (20, time 0 h); N. Nupercaine (1·44; −15 min); A, Anthisan (1·5; −15 min); B, Benadryl (2·5; −15 min); D, sodium dodecyl sulphate (2·5; time 0 h); PGC, propyl gallate plus ascorbic acid (30 + 80; time 0 h); C, carbon tetrachloride treatment alone. Data of Slater *et al.* (1966) and T. F. Slater (unpublished work).

P_{450}, has little effect on the liver damage produced by carbon tetrachloride *in vivo*, the free-radical scavenger propyl gallate is extremely effective in counteracting a number of early effects of the toxic agent on the liver such as the decreased content of NADPH and the increased triglyceride content. Despite such activity, however, propyl gallate produces no significant difference spectrum with liver microsomes. Torrielli and Slater (1970) have found that this substance strongly *inhibits* NADPH–cytochrome *c* reductase and so interferes with the primary activation stage of carbon tetrachloride to $CCl_3^{\bullet}$. Its effect is relatively transient, however, as it is rapidly excreted in the bile. The inhibitory action of the antioxidant propyl gallate on NADPH–cytochrome *c* reductase illustrates the complexities of the reactions of such drugs with tissue or tissue suspensions: reactions other than those to be expected on the basis of known chemical behaviour may often be of considerable interest and even of prime importance. Recent work by Marchand *et al.* (1971) illustrates this point again, this time for SKF 525A.

This drug is often used as a powerful inhibitor of the NADPH–cytochrome P_{450} system. As a consequence its effects on the liver are generally judged to result from inhibitory action at the cytochrome P_{450} site. Marchand *et al.* (1971), however, find that SKF 525A has a marked action in diminishing the uptake of carbon tetrachloride from the blood into the liver following oral administration of the toxic halogenomethane. As a consequence, any modifying actions of SKF 525A on the hepatotoxicity of carbon tetrachloride *in vivo* must be considered as resulting from a decreased liver concentration of carbon tetrachloride. For example, Rao *et al.* (1970) reported that the administration of SKF 525A thirty minutes before dosing with carbon tetrachloride attenuated the rise in diene conjugation in microsomal lipids. The authors interpreted this finding as resulting from an inhibition by SKF 525A of the cytochrome P_{450} chain. Although this may indeed be the case it does appear as a consequence of the report of Marchand *et al.* (1971) that further data are required before an unequivocal conclusion can be reached.

Finally in this Section a quite different mechanism of protection can be mentioned. Gibb and Brody (1967) have shown that pre-treatment with nicotinamide prevents the early decrease in liver NADPH and attenuates the extent of necrosis. This agent is probably effective through its action (Kaplan, 1968) in increasing liver NAD^+ and NADPH: the liver's content of NAD^+ + NADH may be increased some tenfold, and the content of $NADP^+$ + NADPH may double after nicotinamide injection (Figure 10.13). As a consequence, although carbon tetrachloride still produces a decrease in NADPH after treatment with nicotinamide the final concentration is still higher than in normal untreated rats. These results suggest that the depletion of NADPH in the cells of the centrilobular zone is a critical process and is one that is incompatible with the continued integrity of the cells in that region. Strangely enough there is no significant change in the

NAD^+ + NADH content after administration of 1·25 ml of carbon tetrachloride/kg body wt., suggesting either that the activity of the NAD-kinase reaction is not responsive to the decrease in NADPH (for otherwise more NAD^+ would be channelled into $NADP^+$ synthesis) or that the activity of the NAD-synthetase system *is* closely linked to NAD^+ concentrations and can maintain the normal concentration despite an increased drain of product into $NADP^+$ synthesis.

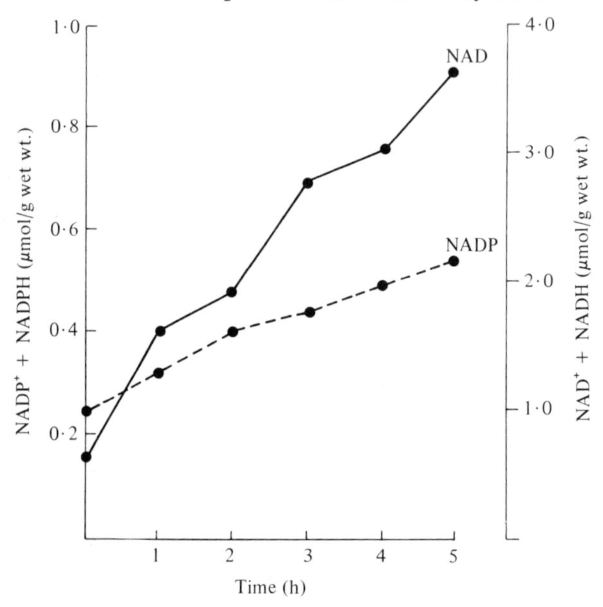

Figure 10.13. Effect of administration of nicotinamide (500 mg/kg body wt.) on the liver concentrations of NAD^+ + NADH and of $NADP^+$ + NADPH. Data from Clark and Pinder (1969).

10.10 Summary of evidence from studies *in vivo*

At this point it will be worthwhile to summarise the discussion of the last few Sections. It has been found that a metabolic conversion of carbon tetrachloride into some more toxic product or products is necessary for the full development of necrosis. This metabolism involves the NADPH–cytochrome P_{450} sequence in the liver endoplasmic reticulum. Of the possible routes of metabolism that have been demonstrated for carbon tetrachloride in tissue suspensions it is likely that a homolytic cleavage to $CCl_3^{\bullet}$ and $Cl^{\bullet}$ is the major route in the lipophilic environment of the endoplasmic reticulum. Evidence has been presented that $CCl_3^{\bullet}$ (or primary products resulting from its formation) can react readily with nucleotide bases, thiol groups and unsaturated lipids. The last-named reaction can be easily demonstrated *in vitro* by the peroxidative mechanism that results in the production of malonaldehyde. The stimulation of lipid peroxidation in liver endoplasmic reticulum has been shown to involve an interaction of carbon tetrachloride with the NADPH–flavoprotein (or a component in its immediate vicinity); this interaction is inhibited by very low concentrations of free-radical scavengers. Associated with the increased lipid peroxidation *in vitro* is a decrease in the activity of glucose 6-phosphatase.

The studies *in vivo* have produced analogous results to those obtained *in vitro*; the metabolism is dependent on the NADPH–cytochrome P_{450} sequence, the development of necrosis is not blocked by SKF 525A but is blocked by promethazine and other scavengers, there is an associated decrease in glucose 6-phosphatase and a decreased concentration of the purine riboside inosine. Other changes to be expected from the formation of $CCl_3^{\bullet}$ (i.e. decreased nicotinamide–adenine dinucleotides) are also blocked by concomitant dosing with scavengers. It seems reasonable to accept therefore that the metabolism of carbon tetrachloride *in vivo* that is essential for necrosis results from an interaction with the NADPH–flavoprotein. In the sense that this interaction is the earliest that has yet been demonstrably linked with necrosis (namely the blocking effect of scavengers on the necrosis), it can be regarded at present as the primary event in the development of necrosis resulting from the administration of carbon tetrachloride. While it is difficult to obtain direct proof for many of these submissions the indirect evidence is substantial. Further, there is no doubt that carbon tetrachloride induces a severe disturbance of homolytic reactions in the liver soon after dosing. Figure 10.14 shows *electron spin resonance spectra of microsomal fractions*

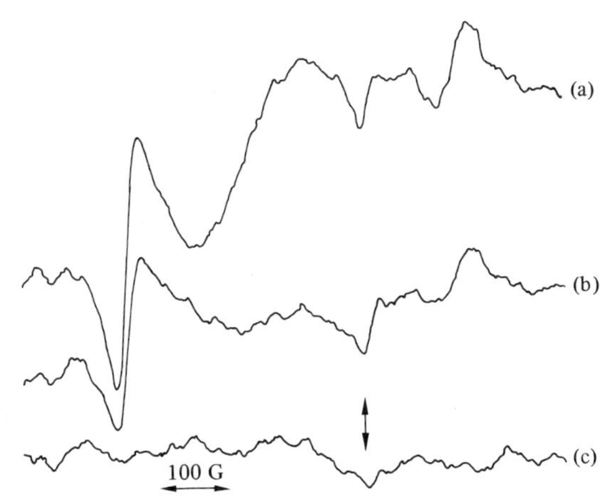

Figure 10.14. Electron spin resonance spectra of liver microsome fractions studied at −180°C (unpublished data of T. F. Slater and J. W. R. Cook). Microsomes were suspended in tris buffer, pH 7·4, approximately 70 mg of protein ml^{-1}. (a) Rat given 0·5 ml of liquid paraffin/100 g body wt. by stomach intubation 2 h previously; (b) rat given carbon tetrachloride–liquid paraffin (1 : 3, *v*/*v*; 0·5 ml/100 g body wt.) by stomach intubation 2 h previously; (c) cavity background in the absence of microsomal samples. The vertical arrow indicates the position of the free electron spin; the horizontal scale indicates 100 gauss.

obtained from normal and carbon tetrachloride-treated rats 2 h after dosing. There are obviously very large qualitative differences between the spectra. Approximate summation of the signal area at $g = 2 \cdot 002$ in three preparations of control microsomes and in three microsomal suspensions obtained from rats dosed 2 h previously with carbon tetrachloride indicated that the signal is more powerful after carbon tetrachloride dosing. Claridge and Willard (1965) have shown that the trichloromethyl radical gives a simple absorption line at $g = 2 \cdot 002$; it is difficult therefore to identify this species with certainty with microsomal samples since other natural components of liver give a signal in that region (21). Experiments with [^{13}C]carbon tetrachloride should be helpful in this respect as the ^{13}C isotope gives a triplet signal. Finally, in this context, Calligaro *et al.* (1970) have reported that administration of carbon tetrachloride or trichlorobromomethane to rats results in an accentuation of the electron spin resonance signal in the $g = 2$ region, although quantitative evaluation of their work is not presented.

There is one other facet of the action of carbon tetrachloride on the liver that requires discussion. Its interaction with cytochrome P_{450} has already been briefly mentioned and this may well have an important role in the overall liver disturbance. In the sense that these changes are not blocked by promethazine or other scavengers it can be assumed that such events are not essential for necrosis. Nevertheless they are potentially of considerable interest and importance.

10.11 Interaction of carbon tetrachloride with the P_{450} site

The oxidation of carbon tetrachloride to carbon dioxide has been discussed in Section 10.2, where a reaction scheme was proposed involving homolytic mechanisms. Since the oxidation is blocked by drugs known to undergo metabolism at the cytochrome P_{450} site (Seawright and McLean, 1967), it is probable that the breakdown of carbon tetrachloride also occurs at that site. If so, and if the breakdown follows a course or courses similar to those given in Figures 10.1 and 10.2, then there is an obvious risk of damage to the P_{450} end of the electron-transport chain by the $CCl_3^{\bullet}$, phosgene and OCl^- intermediates. This may be the reason underlying the rapid decrease in cytochrome P_{450} content in intoxication with carbon tetrachloride and which is associated with an extensive decrease in general drug metabolism (22). These events have been

(21) With samples of whole liver it is far more difficult to obtain unequivocal evidence through electron spin resonance. In whole liver the main contribution to the signal in the $g = 2$ region is from mitochondrial components (see Wyard, 1969). Thus changes in the minor contribution from endoplasmic reticulum would be completely hidden.

(22) Although most drugs are metabolised more slowly in the presence of carbon tetrachloride this is not the case with tetralin (Lin and Chen, 1969). The hydroxylation of tetralin appears to involve a radical-initiated formation of the intermediate hydroperoxide. The overall process is *stimulated* by low concentrations of carbon tetrachloride and inhibited by radical scavengers.

recorded by several groups. (1) Concerning drug metabolism, see Neubert and Maibauer, 1959; Clifford and Rees, 1966; Smuckler *et al.*, 1967; Castro *et al.*, 1968; Sesame *et al.*, 1968; Dingell and Heimberg, 1968; Klinger *et al.*, 1968. The last two groups have shown that the process *in vivo* is very rapid, there being a 40% decrease in the metabolism of phenazine and aminophenazone within 20 min of dosing with carbon tetrachloride. Aminopyrine metabolism also decreases rapidly whereas hexobarbital and *p*-nitrobenzoic acid metabolism decreased at a slower rate. (2) For information concerning cytochrome P_{450}, see Smuckler *et al.*, 1967; Castro *et al.*, 1968; Sesame *et al.*, 1968; Dingell and Heimberg, 1968; Slater and Sawyer, 1969. A summary is provided in Table 10.18.

As mentioned in Section 9.6, these changes in drug metabolism and cytochrome P_{450} (and in protein synthesis) are not prevented *in vivo* by dosing with free-radical scavengers. This suggests that if free radicals are involved as intermediates in the oxidation of carbon tetrachloride to carbon dioxide they must have very transient lifetimes (or are formed perhaps in a zone that is not easily accessible to the exogenous antioxidants), probably reacting in a 'bound form' with oxygen to yield

Table 10.18. Effect of carbon tetrachloride on cytochrome P_{450} content of rat liver microsomes—values in nmol/mg protein except for reference 4. Doses of 2·5 ml of carbon tetrachloride/kg body wt. were used except for reference 3, where 1·25 ml/kg body wt. was used.
Abbreviations: DPPD, *NN*′-diphenyl-*p*-phenylenediamine; SKF, β-dimethylamino-ethyl-3,3′-diphenylpropyl acetate.

Ref.	Time after dosing (h)	Treatment	Cytochrome P_{450}		
			Liver	Kidney	Adrenal
1	3	Control	0·99	0·15	0·53
		CCl_4	0·53	0·15	0·33
2	3	Control	0·87		
		CCl_4	0·29		
		DPPD	0·94		
		DPPD + CCl_4	0·49		
		SKF	0·83		
		SKF + CCl_4	0·55		
3	1	Control	0·45		
		CCl_4	0·35		
4	2	Control	100%		
		CCl_4	50%		

References: 1, Sesame *et al.* (1968); 2, Castro *et al.* (1968); 3, Slater and Sawyer (1969); 4, Smuckler *et al.* (1967).

non-radical products which are responsible for the localised damage to the cytochrome P_{450} region.

The decrease in drug-metabolising activity reported above after the administration of carbon tetrachloride is much more rapid than the concomitant decrease in cytochrome P_{450} content. This suggests that metabolites of carbon tetrachloride are destroying the rate-limiting factor X_1 of the NADPH–cytochrome P_{450} chain (see Figure 10.3) preferentially to the cytochrome P_{450} itself. If factor X_1 is a distinct component of the NADPH–cytochrome P_{450} chain, and is not identical with the more distal binding-site proteins by which drugs are anchored close to the P_{450} activation site, then this may explain why substances like SKF 525A (which interact with the NADPH chain at the binding site) are not effective in preventing the inhibitory action of carbon tetrachloride on drug metabolism (Castro *et al.*, 1968; Dingell and Heimberg, 1968). The suggestion mentioned above is favoured to some extent by the finding that although there are probably a number of different binding sites around the P_{450} locus, carbon tetrachloride appears to affect all of them in that it decreases the metabolism of all drugs so far examined. (Chloroform, however, decreases the metabolising activity with respect to some drugs but not to others.) The simplest explanation is that it does so by damaging the X_1-site.

The data of Dallner *et al.* (1965) have already been mentioned briefly in connection with the rate limiting component X_1 (see Figure 10.3). Further evidence that points to the existence of such a factor proximal to cytochrome P_{450} is provided by the data of Gilbert (1969). He found that feeding rats with two substituted phenols led in each case to an increase in drug metabolism but no significant change in the content of cytochrome P_{450} (Table 10.19). The activity of the first component of the NADPH–P_{450} chain, the NADPH–cytochrome *c* reductase, is known to be greatly in excess of the normal rate of drug metabolism (Kato and

Table 10.19. Effects of administration of 2,6-di-t-butyl-4-methylphenol and 2,6-di-t-butyl-4-methoxymethylphenol to rats *in vivo* on the microsomal drug-metabolising system and cytochrome P_{450} content (data from Gilbert 1969). Both drugs were administered orally in a single dose of 1·5 mmol/kg body wt.; analyses were performed 24 h later.
Abbreviations: BHT, 2,6-di-t-butyl-4-methylphenol; BMP, 2,6,-di-t-butyl-4-methoxy-methylphenol.

Treatment	BHT oxidase μmol/h per g of liver	Cytochrome P_{450} nmol/g of liver
Controls	0·26 ± 0·04	9·2 ± 0·7
BHT	0·78 ± 0·07	10·5 ± 0·8
BMP	1·15 ± 0·02	12·4 ± 0·7

Gillette, 1965). These results suggest the presence of a rate-limiting factor between the flavoprotein and cytochrome P_{450}. This rate-limiting factor could be a separate entity different to the NADPH–flavoprotein and cytochrome P_{450} with its associated binding-site protein; or it could reflect a conformational property of the protein components of the P_{450} zone.

10.12 Hepatotoxicity of chloroform, fluorotrichloromethane and bromotrichloromethane

It has been known for a long time that chloroform produces a similar type of liver damage to that given by carbon tetrachloride. The mechanism by which chloroform produces this damage has not been studied, however, as extensively as for the latter agent. Chloroform has a much *slower* action on several liver parameters than has been found with carbon tetrachloride. This may reflect not only a more pronounced effect in lowering body temperature, which thereby slows up effects dependent on metabolism, but an increased bond-dissociation energy for homolytic reactions (see Section 9.1). The difference in bond-dissociation energy for the C—H bond in chloroform compared with the C—Cl bond in carbon tetrachloride may be expected to cause a fivefold difference in homolytic processes in favour of carbon tetrachloride (see Section 9.1).

There is no doubt that chloroform is less hepatotoxic than carbon tetrachloride on an equimolar basis; comparisons based on equal *doses* of the two compounds fail to take into account the much greater solubility of chloroform in tissue. This is clearly shown by the data of Reynolds and Lee (1967) summarised in Table 10.20. Although the ED_{50} for liver disturbance with carbon tetrachloride is much less than for chloroform (Plaa *et al.*, 1958), the LD_{50} for chloroform is lower than for carbon tetrachloride, owing to a more pronounced narcotic effect.

In view of the close chemical and physical similarities of chloroform and carbon tetrachloride it is not unreasonable to expect them to damage the liver by broadly similar mechanisms although specific features may be different in the two cases. From the difference in bond-dissociation energies we may expect, as already mentioned, that chloroform will act

Table 10.20. Concentrations of carbon tetrachloride and chloroform in rat liver at various times after dosing (data of Reynolds and Lee, 1967).

Agent	Dose (ml/100 g body wt.)	Liver concentration (μmol/g wet wt.)		
		1 h	2 h	3 h
CCl_4	0·25	0·74	2·15	2·21
$CHCl_3$	0·21	2·88	3·47	2·83
Ratio $\frac{CHCl_3}{CCl_4}$		3·9	1·6	1·3

more slowly than carbon tetrachloride in any disturbance that is dependent on homolytic bond scission. From the discussion presented for carbon tetrachloride we can expect therefore that chloroform will stimulate lipid peroxidation and produce an increased diene absorption peak, a decreased liver NADPH, a decreased content of microsomal 'antioxidant', and a decreased glucose 6-phosphatase activity. All of these changes that occur in intoxication with carbon tetrachloride have been related to homolysis of the toxic agent. In fact, chloroform does stimulate peroxidation and decrease glucose 6-phosphatase activity *in vitro*, but to a considerably less extent than does carbon tetrachloride, as shown in Table 10.7. These effects found with chloroform *in vitro* are inhibited by promethazine so that they are analogous to the changes seen with carbon tetrachloride *in vitro*. When the situation *in vivo* is considered, however, certain difficulties arise. Although chloroform decreases total liver NADPH (as expected this change is slower than with carbon tetrachloride, Table 9.3) there is no apparent increase in microsomal diene conjugation (Klaassen and Plaa, 1969) and no decrease in glucose 6-phosphatase activity (Reynolds and Lee, 1968; Klaassen and Plaa, 1969; Cawthorne *et al.*, 1971; T. F. Slater and B. C. Sawyer, unpublished data). How can these findings be reconciled with the concept that homolysis of chloroform is necessary for the development of necrosis and that this is accompanied, at least for carbon tetrachloride, by an increased diene absorption (Recknagel and Ghoshal, 1966a) and by a decreased glucose 6-phosphatase activity?

The diene absorption will be considered first. The data of Klaassen and Plaa (1969) clearly show that the increased diene-absorbing material produced by carbon tetrachloride is rapidly metabolised. Fitting a smooth curve to their data (Figure 10.9) allows approximate rates of accumulation and degradation of diene material to be calculated. Such calculations indicate that with carbon tetrachloride the production rate of diene material is approximately 70 arbitrary units h^{-1}, and the destruction rate is approximately 25 arbitrary units h^{-1}. If chloroform has approximately one-fifth the effectiveness of carbon tetrachloride in producing diene-rich material and if the rate of destruction is similar in both cases, then it can be expected that no net increase of diene material would occur with chloroform *in vivo*. This is indeed what Klaassen and Plaa (1969) have found.

The situation with glucose 6-phosphatase is less clear-cut. It is known that this enzyme is activated by hydroxyl ions and is strongly and irreversibly inhibited by exposure to pH values less than 6·5 (see Stetten and Burnett, 1966). It has been suggested previously that carbon tetrachloride reacts with the NADPH–flavoprotein by a process of electron capture from the fully reduced flavoprotein (FPH_3) as indicated in Reaction (10.14).

$$CCl_4 + FPH_3 \longrightarrow CCl_3^{\bullet} + Cl^- + FPH_2^{\bullet} + H^+ \qquad (10.15)$$

This process can be seen to involve the production of protons at the reactive site and can thus be imagined to be responsible for the decreased activity of glucose 6-phosphatase. The decreased activity has been closely related to the stimulation of lipid peroxidation produced by carbon tetrachloride (Ghoshal and Recknagel, 1965b) and is similarly affected by free-radical scavengers. Although similar mechanisms [Reactions (10.16) and (10.17)] may be written for

$$CHCl_3 + FPH_3 \longrightarrow CCl_3^{\bullet} + H^- + FPH_2^{\bullet} + H^+ \quad (10.16)$$

or

$$CHCl_3 + FPH_3 \longrightarrow CHCl_2^{\bullet} + Cl^- + FPH_2^{\bullet} + H^+ \quad (10.17)$$

chloroform these do not appear so probable on energetic considerations as Reaction (10.18).

$$CHCl_3 + FPH_2^{\bullet} \longrightarrow CCl_3^{\bullet} + FPH_3 \quad (10.18)$$

This latter reaction proceeds without the formation of protons and may account for the lack of effect of chloroform on glucose 6-phosphatase *in vivo*. Under conditions *in vitro* with an excess of reducing substrate present (NADPH) and no acceptor for electron flow other than chloroform in the medium it can be envisaged that the major flavin form present is FPH_3 and that under these conditions Reaction (10.16) is favoured, thereby producing the observed decrease in glucose 6-phosphatase activity. These mechanisms, which are of course highly speculative, at least suggest experimental approaches to an understanding of why chloroform shows certain differences from carbon tetrachloride in its action on the liver.

Fluorotrichloromethane is virtually non-toxic when administered to rats by mouth (Slater, 1965) or given by inhalation (Lester and Greenberg, 1950; Jenkins *et al.*, 1970). It has similar physicochemical properties to carbon tetrachloride and also reacts strongly with the cytochrome P_{450} in the endoplasmic reticulum. It does not give centrilobular necrosis nor does it produce any decrease in liver NADPH or in glucose 6-phosphatase activity. The conclusion that we may draw from this is that the increased bond-dissociation energy due to the presence of the electronegative fluorine atom retards homolytic fission relative to carbon tetrachloride and by about the same amount as for chloroform. The lack of effect of fluorotrichloromethane on early as well as late stages of the injury is probably due to the rapid excretion of this material through the lungs at 37°C.

Bromotrichloromethane is a very toxic material and when given orally usually produces death before liver damage can develop to a recognisable degree. The bond-dissociation energy of the C—Br bond is very low and there seems no doubt that trichloromethyl radicals would be produced in a wide variety of locations throughout the body and not merely in the endoplasmic reticulum of centrilobular cells of the liver. The action of bromotrichloromethane on rat liver *in vivo* has been reported recently by Calligaro *et al.* (1970), who found an early effect on the endoplasmic reticulum—changes were seen 15 minutes after a dose of 0·25 ml/100 g body wt. These authors also reported that the administration of bromotrichloromethane and, to a lesser extent, carbon tetrachloride produced an increased free radical signal in liver samples when studied by the electron spin resonance technique. Under conditions *in vitro* they found that bromotrichloromethane was more active in stimulating malonaldehyde production that carbon tetrachloride, a conclusion that confirms the results of Slater and Sawyer mentioned earlier in section 10.6. There seems no doubt that the more marked instability of bromotrichloromethane results in an accentuation of the homolytic processes observed with carbon tetrachloride under both *in vitro* and *in vivo* conditions.

10.13 Final remarks on the halogenomethanes

The last two Chapters have reviewed in some detail the actions of carbon tetrachloride and other halogenated methanes on the liver. The subject is one of considerable current interest and is developing rapidly. There is extensive controversy concerning the precise mechanisms by which carbon tetrachloride produces some of its effects on the liver, for example, the importance of lipid peroxidation, the interaction sites of carbon tetrachloride with the P_{450} chain, and the protective action of some antioxidants. Although I have tried to present a balanced account of the major areas of interest concerning the hepatotoxic action of carbon tetrachloride, the account of the more controversial areas is necessarily a somewhat biased one and represents in part an expression of personal opinions and speculations. This is, to some extent, inevitable when writing about a rapidly developing field in which one has been closely involved. One advantage of presenting such a monograph, in fact, is that it is possible to be more speculative and eclectic than normal.

The work on carbon tetrachloride over the last 10–15 years has been important for several major reasons. Firstly, the action of carbon tetrachloride on the rat liver provides a very good model for studying time sequences and mechanisms in a situation which results in extensive death of cells. Although the results are applicable in detail only to carbon tetrachloride and to its closely related halogenomethanes, they have a very broad importance to studies on necrosis in general. Secondly, such studies on carbon tetrachloride have led to the introduction of many new ideas into cell pathology and these have been associated with the stimulation of multidisciplinal attacks involving varied and novel techniques. Thirdly, the study of the hepatotoxicity of carbon tetrachloride has been one of the major areas of research that has stimulated appreciation of the role of free radical mechanisms in a number of diverse types of cell damage: the topic, that is of course, covered by this monograph. In the remaining Chapters we shall consider a variety of injurious situations ranging from acute alcoholic intoxication to chemical carcinogenesis.

References

Adam, S. E. I., Thorp, E., 1970, *Br. J. exp. Path.*, **51**, 394.
Agostini, C., Comi, P., 1968, *Boll. soc. ital. Biol. sper.*, **45**, 161.
Alexander, N. M., Scheig, R., Klatskin, G., 1967, *Biochem. Pharmac.*, **16**, 1091.
Alpers, D. H., Isselbacher, K. J., 1968, *Biochim. biophys. Acta*, **158**, 414.
Bangham, A. D., Rees, K. R., Shotlander, V., 1962, *Nature, Lond.*, **193**, 754.
Bhattacharyya, K., 1965, *J. Path. Bact.*, **90**, 151.
Bizzi, A., Tacconi, M. T., Veneroni, E., Garattini, S., 1966, *Nature, Lond.*, **209**, 1025.
Brauer, R. W., 1965, in *The Biliary System*, Ed. W. Taylor (Blackwell, Oxford), p.101.
Bray, H. G., Thorpe, W. V., Vallance, D. K., 1952, *Biochem. J.*, **51**, 193.
Butler, T. C., 1961, *J. Pharmac. exp. Ther.*, **134**, 311.
Calligaro, A., Congiu, L., Vannini, V., 1970, *Rass. med. sarda*, **73**, 365.
Cameron, G. R., Karunaratne, W. A. E., 1936, *J. Path. Bact.*, **42**, 1.
Castro, J. A., Sesame, H. A., Sussman, H., Gillette, J. R., 1968, *Life Sci.*, **7**, 129.
Cawthorne, M. A., Bunyan, J., Sennitt, M. V., Green, J., Grasso, P., 1970, *Br. J. Nutr.*, **24**, 357.
Cawthorne, M. A., Palmer, E. D., Bunyan, J., Green, J., 1971, *Biochem. Pharmac.*, **20**, 494.
Christie, G. S., Judah, J. D., 1954, *Proc. R. Soc. B.*, **142**, 241.
Claridge, R. F. C., Willard, J. E., 1965, *J. Am. Chem. Soc.*, **87**, 4992.
Clark, J. B., Pinder, S., 1969, *Biochem. J.*, **114**, 321.
Clifford, J. I., Rees, K. R., 1966, *J. Path. Bact.*, **91**, 215.
Clower, B. R., Douglas, B. H., Carrier, O., 1968, *Eur. J. Pharmac.*, **2**, 276.
Comporti, M., Saccocci, C., Dianzani, M. U., 1965, *Enzymologia*, **29**, 185.
Conney, A. H., Gillette, J. R., Inscoe, J. K., Trams, E. R., Posner, H. S., 1959, *Science*, **130**, 1478.
Csallany, A. S., Chiu, M., Draper, H. H., 1970, *Lipids*, **5**, 63.
Dallner, G., Siekevitz, P., Palade, G., 1965, *Biochem. biophys. Res. Commun.*, **20**, 135.
Dawkins, M. J. R., 1963, *J. Path. Bact.*, **85**, 189.
Di Luzio, N. R., 1966, *Life Sci.*, **5**, 1467.
Dianzani, M. U., 1954, *Biochim. biophys. Acta*, **14**, 514.
Dingell, J. V., Heimberg, M., 1968, *Biochem. Pharmac.*, **17**, 1269.
Fischler, F., 1913, *Mitt. Grenzgeb. Med. Chir.*, **26**, 553.
Fowler, J. S. L., 1969, *Brit. J. Pharmac. Chemother.*, **37**, 733.
Fox, C. F., Dinman, B. D., Frajola, W. J., 1962, *Proc. Soc. exp. Biol. Med.*, **111**, 731.
Frunder, H., 1968, *Vop. med. Khim.*, **14**, 130.
Frunder, H., Blume, E., Thielmann, K., Börnig, H., 1961, *Hoppe-Seyler's Z. physiol. Chem.*, **325**, 146.
Gallagher, C. H., 1960, *Aust. J. exp. Biol. med. Sci.*, **38**, 251.
Gallagher, C. H., 1962, *Aust. J. exp. Biol. med. Sci.*, **40**, 241.
Garner, R. C., McLean, A. E. M., 1969, *Biochem. Pharmac.*, **18**, 645.
Ghoshal, A. K., Recknagel, R. O., 1965a, *Life Sci.*, **4**, 1521.
Ghoshal, A. K., Recknagel, R. O., 1965b, *Life Sci.*, **4**, 2195.
Gibb, J. W., Brody, T. M., 1967, *Biochem. Pharmac.*, **16**, 2047.
Gilbert, D., 1969, *Biochem. J.*, **115**, 59P.
Glaumann, H., Kuylenstierna, B., Dallner, G., 1969, *Life Sci.*, **8**, 1309.
Glende, E. A., Recknagel, R. O., 1969, *Expl. molec. Path.*, **11**, 172.
Gordis, E., 1969, *J. clin Invest.*, **48**, 203.
Graham, E. A., 1915, *J. exp. Med.*, **22**, 48.
Green, J., Bunyan, J., Cawthorne, M. A., Diplock, A. T., 1969, *Br. J. Nutr.*, **23**, 297.
Greenberger, N. J., Cohen, R. B., Isselbacher, K. J., 1965, *Lab. Invest.*, **14**, 264.
Hashimoto, S., Glende, E. A., Recknagel, R. O., 1968, *N. Engl. J. Med.*, **14**, 1082.
Heppel, L. A., Porterfield, V. T., 1948, *J. biol. Chem.*, **176**, 763.

Higgins, G. M., Cragg, R. W., 1937, *Proc. Staff. Meet. Mayo Clin.*, **12**, 582.
Hill, H. A. O., Röder, A., Williams, R. J. P., 1969, *Biochem. J.*, **115**, 59P.
Himsworth, H. P., 1950, *The Liver and its Diseases* (Blackwell, Oxford), p.32.
Hine, J., 1956, in *Physical Organic Chemistry* (McGraw-Hill, New York), p.131.
Hochstein, P., Ernster, L., 1963, *Biochem. biophys. Res. Commun.*, **12**, 388.
Holtcamp, D. E., Hill, R. M., 1951, *Archs. Biochem. Biophys.*, **34**, 216.
Horecker, B. L., 1950, *J. biol. Chem.*, **183**, 593.
Horning, M. G., Earle, M. J., Maling, H. M., 1962, *Biochim. biophys. Acta*, **56**, 175.
Horton, A. A., Packer, L., 1970, *Biochem. J.*, **116**, 19P.
Hove, E. L., 1953, *J. Nutr.*, **51**, 609.
Icén, A. L., Huovinen, J. A., 1959, *Acta path. microbiol. scand.*, **47**, 297.
Jenkins, L. J., Jones, R. A., Coon, R. A., Siegel, J., 1970, *Toxic. appl. Pharmac.*, **16**, 133.
Kamin, H., Master, B. S., Gibson, Q. H., Williams, C. H., 1965, *Fedn Proc. Fedn Am. Socs exp. Biol.*, **24**, 1164.
Kaplan, N. O., 1968, *J. Vitam.*, **14**, 103.
Kato, R., Gillette, J. R., 1965, *J. Pharmac. exp. Ther.*, **150**, 279.
Kitabchi, A. E., Williams, R. H., 1968, *J. biol. Chem.*, **243**, 3248.
Klaassen, C. D., Plaa, G. L., 1969, *Biochem. Pharmac.*, **18**, 2019.
Klinger, W., Neugebauer, A., Splinter, F. K., 1968, *Arch. Tox.*, **23**, 178.
Knecht, M., 1966, *Z. Naturf.*, **21**, 799.
Larson, R. E., Plaa, G. L., 1965, *J. Pharmac. exp. Ther.*, **147**, 103.
Lee, I. P., Yamamura, H. I., Dixon, R. L., 1968, *Biochem. Pharmac.*, **17**, 1671.
Lester, D., Greenberg, L. A., 1950, *Archs. ind. Hyg.*, **2**, 335.
Liebecq-Hutter, S., Bacq, Z. M., 1958, *Arch. int. Physiol. Biochim.*, **66**, 469.
Lin, C. C., Chen, C., 1969, *Biochim. biophys. Acta*, **192**, 133.
Long, C., 1961, *Biochemists Handbook* (Spon, London).
Lu, A. Y. H., Strobel, H. W., Coon, M. J., 1969, *Biochem. biophys. Res. Commun.*, **36**, 545.
Mager, J., Halbreich, A., Bornstein, S., 1965, *Biochem. biophys. Res. Commun.*, **18**, 576.
Marchand, C., McLean, S., Plaa, G. L., Traiger, G., 1971, *Biochem. Pharmac.*, **20**, 869.
May, H. E., McCay, P. B., 1968a, *J. biol. Chem.*, **243**, 2288.
May, H. E., McCay, P. B., 1968b, *J. biol. Chem.*, **243**, 2296.
May, H. E., Poyer, J. L., McCay, P. B., 1966, *Fedn Proc. Fedn Am. Socs exp. Biol.*, **25**, 301.
McLean, A. E. M., 1967a, *Biochem. Pharmac.*, **16**, 2030.
McLean, A. E. M., 1967b, *Br. J. exp. Path.*, **48**, 632.
McLean, A. E. M., Marshall, W. J., 1971, *Biochem. J.*, **123**, 28P.
McLean, A. E. M., McLean, E. K., 1966, *Biochem. J.*, **100**, 564.
Mellors, A., Tappel, A. L., 1966, *J. biol. Chem.*, **241**, 4353.
Morrison, G. R., Brock, F. E., Karl, I. E., Schank, R. E., 1965, *Archs. Biochem. Biophys.*, **111**, 448.
Müller, R., 1911, *Z. exp. Path. Ther.*, **9**, 103.
Neubert, D., Maibauer, D., 1959, *Arch. exp. Path. Pharmak.*, **235**, 291.
Nordenbrand, K., Hochstein, P., Ernster, L., 1964, *6th Int. Congr. Biochem., New York*, **8**, 76.
Nordlie, R. C., Arion, W. J., 1965, *J. biol. Chem.*, **240**, 2155.
Nyquist, S. E., Barr, R., Morré, D. J., 1970, *Biochim. biophys. Acta*, **208**, 532.
Oettingen, W. F. van, 1955, *The Halogenated Hydrocarbons, Toxicity and Potential Dangers* (US Department of Health, Education and Welfare, Public Health Service Publication number 414, Washington DC).

Orrenius, S., Ernster, L., 1964, *Biochem. biophys. Res. Commun.*, **16**, 60.
Paul, B. B., Rubenstein, D., 1963, *J. Pharmac. exp. Ther.*, **141**, 141.
Petrelli, M., Stenger, R. J., 1969, *Expl. molec. Path.*, **10**, 115.
Pette, D., Brandau, H., 1966, *Enzym. biol. Clin.*, **6**, 79.
Plaa, G. L., Evans, E. A., Hine, C. H., 1958, *J. Pharmac. exp. Ther.*, **123**, 224.
Placer, Z., Veselkova, A., Rath, R., 1965, *Experientia*, **21**, 19.
Platt, D. S., Cockrill, B. L., 1967, *Biochem. Pharmac.*, **16**, 2257.
Platt, D. S., Cockrill, B. L., 1969, *Biochem. Pharmac.*, **18**, 445.
Priest, R. E., Smuckler, E. A., Iseri, O. A., Benditt, E. P., 1962, *Proc. Soc. exp. Biol. Med.*, **111**, 50.
Rao, K. S., Glende, E. A., Recknagel, R. O., 1970, *Expl. molec. Path.*, **12**, 324.
Rao, K. S., Recknagel, R. O., 1968, *Expl. molec. Path.*, **9**, 271.
Rao, K. S., Recknagel, R. O., 1969, *Expl. molec. Path.*, **10**, 219.
Rappaport, A. M., 1963, in *The Liver*, volume I, Ed. C. Rouiller (Academic Press, New York), p.265.
Recknagel, R. O., Anthony, D. D., 1959, *J. biol. Chem.*, **234**, 1052.
Recknagel, R. O., Ghoshal, A. K., 1965, *Fedn Proc. Fedn Am. Socs exp. Biol.*, **24**, 299.
Recknagel, R. O., Ghoshal, A. K., 1966a, *Nature, Lond.*, **210**, 1162.
Recknagel, R. O., Ghoshal, A. K., 1966b, *Lab. Invest.*, **15**, 132.
Recknagel, R. O., Ghoshal, A. K., 1966c, *Expl. molec. Path.*, **5**, 108.
Rees, K. R., Spector, W. G., 1961, *Nature, Lond.*, **190**, 821.
Reynolds, E. S., 1963, *Fedn Proc. Fedn Am. Socs exp. Biol.*, **22**, 370.
Reynolds, E. S., 1967, *J. Pharmac. exp. Ther.*, **155**, 117.
Reynolds, E. S., Lee, A. G., 1967, *Lab. Invest.*, **16**, 591.
Reynolds, E. S., Lee, A. G., 1968, *Lab. Invest.*, **19**, 273.
Schenkman, J. B., Remmer, H., Estabrook, R. W., 1967, *Mol. Pharmac.*, **3**, 113.
Seawright, A. A., McLean, A. E. M., 1967, *Biochem. J.*, **105**, 1055.
Serratoni, F. T., Schnitzer, B., Smith, E. B., 1969, *Archs Path.*, **87**, 46.
Sesame, H. A., Castro, J. A., Gillette, J. R., 1968, *Biochem. Pharmac.*, **17**, 1759.
Siekevitz, P., 1955, *J. biophys. biochem. Cytol.*, **1**, 477.
Sigel, B., Baldia, L. B., Dimbiloglu, M. E., 1967, *Nature*, **213**, 1258.
Slater, T. F., 1965, *Biochem. Pharmac.*, **14**, 178.
Slater, T. F., 1966a, *Nature, Lond.*, **209**, 36.
Slater, T. F., 1966b, *Biochem. J.*, **101**, 16P.
Slater, T. F., 1968, *Biochem. J.*, **106**, 155.
Slater, T. F., 1969, in *Lysosomes in Biology and Pathology*, volume I, Eds. J. T. Dingle, H. Fell (North-Holland, Amsterdam), p.469.
Slater, T. F., Delaney, V. B., 1970, *Biochem. J.*, **116**, 303.
Slater, T. F., Greenbaum, A. L., 1965, *Biochem. J.*, **96**, 484.
Slater, T. F., Sawyer, B. C., 1969, *Biochem. J.*, **111**, 317.
Slater, T. F., Sawyer, B. C., 1971, *Biochem. J.*, **123**, 805.
Slater, T. F., Sawyer, B. C., Strauli, U. D., 1966, *Biochem. Pharmac.*, **15**, 1273.
Slater, T. F., Strauli, U. D., Sawyer, B. C., 1964, *Biochem. J.*, **93**, 260.
Smuckler, E. A., Arrhenius, E., Hultin, T., 1967, *Biochem. J.*, **103**, 55.
Smuckler, E. A., Hultin, T., 1966, *Expl. molec. Path.*, **5**, 504.
Smuckler, E. A., Koplitz, M., Striker, G. E., 1968, *Lab. Invest.*, **19**, 218.
Stammberger, K., Richter, G., Frunder, H., 1968, *Acta biol. med. germ.*, **20**, 443.
Staudinger, H., Kerekjarto, B., Ullrich, V., Zubrzycki, Z., 1965, in *Oxidases and Related Redox Systems*, volume 2, Eds. T. E. King, H. S. Mason, M. Morrison (John Wiley, New York), p.815.
Steacie, E. W. R., 1954, in *Atomic and Free Radical Reactions* (Reinhold, New York), p.682.

Stetten, M. R., Burnett, F. F., 1966, *Biochim. biophys. Acta,* **128**, 344.
Stoner, H. B., 1956, *Br. J. exp. Path.,* **37**, 176.
Striker, G. E., Smuckler, E. A., Kohnen, P. W., Nagle, R. B., 1968, *Am. J. Path.,* **53**, 769.
Strobel, H. W., Lu, A. Y. H., Heidema, J., Coon, M. J., 1970, *J. Biol. Chem.,* **245**, 4851.
Thielmann, K., Schulz, M., Kramer, P., Frunder, H., 1963, *Hoppe-Seyler's Z. physiol. Chem.,* **332**, 204.
Torrielli, M. V., Slater, T. F., 1970, *Br. J. exp. Path.,* in the press.
Ugazio, G., Torrielli, M. V., 1969, *Biochem. Pharmac.,* **18**, 2274.
Wattenberg, L. W., Leong, J. L., 1962, *J. Histochem. Cytochem.,* **10**, 412.
Wells, H. G., 1908, *Archs intern. Med.,* **1**, 589,
Wigglesworth, J. S., 1964, *J. Path. Bact.,* **87**, 333.
Wills, E. D., 1969, *Biochem. J.,* **113**, 333.
Wirtschafter, Z. T., Cronyn, M. W., 1964, *Environ. Health,* **9**, 186.
Witting, L. A., 1969, *Archs. Biochem. Biophys.,* **129**, 142.
Wyard, S. J., 1969, *Solid State Biophysics* (McGraw-Hill, London), p.278.

11

Hepatotoxic effects of alcohol

11.1 Introductory remarks

In this Chapter the effects of alcohols on the liver will be examined in some detail. By 'alcohols' in this context are meant the short-chain monohydric aliphatic alcohols in general and **ethanol** in particular. By far the most attention will be paid to ethanol[1] because of the widespread habit of drinking it and the clinical significance of alcoholism. However, chronic alcoholism itself will not be discussed in detail, since most of the studies connecting the biological effects of alcohol with homolytic reactions have been concerned with the results of a single acute dose of alcohol. There are adequate background reviews concerning chronic alcoholism that may be consulted (e.g. Jellinek, 1960; Williams, 1959). Except where clearly stated to the contrary, the term 'alcohol' will be used synonymously with 'ethanol' from this point on.

It is neither possible nor necessary to give a comprehensive account of the chemical, pharmacological and biochemical properties of ethanol (see Goodman and Gilman, 1955) but it will be useful, however, to provide at least some background material before discussion of the mechanisms that have been suggested to be responsible for the alcohol-induced fatty liver. Such considerations will lead on to recent work in which homolytic reactions have been implicated, and where protection against the development of the fatty liver has been achieved *in vivo* by prior dosing of the experimental animals with free-radical scavengers.

11.2 Background information

Ethanol is a colourless liquid at room temperature and has a burning taste; its density is 0·79 g cm^3 at 15°C, it boils at 78°C, and is freely miscible with water. Ethanol and its related homologues have many industrial and commercial uses: for example they are important as solvents in the chemical industry, being used in the resin and varnish trades, in organic syntheses and in perfumery and in pharmaceutics (as lotions and colognes). Alcohol for industrial use is made unfit for consumption by the addition of various toxic agents that include pyridine bases, naphtha and benzene. Ethanol is made industrially by the sulphonation of ethylene followed by hydrolysis:

$$C_2H_4 + H_2SO_4 \longrightarrow C_2H_5{-}HSO_4(+H_2O) \longrightarrow C_2H_5OH + H_2SO_4 \qquad (11.1)$$

The major amount of ethanol that is used annually is consumed in intoxicating beverages; the ethanol is produced for this purpose by

[1] Some information relevant to the toxicities of methanol and of allyl alcohol is given at the end of the Chapter.

biological fermentation[2]. In this process glucose is fermented by yeast: the complex sequence of enzyme-catalysed reactions is identical as far as the pyruvate stage with the sequence that yields lactate in human muscle. Whereas in muscle pyruvate is reduced to lactate by lactate dehydrogenase, in yeast the pyruvate is decarboxylated to acetaldehyde, which is then reduced by NADH to ethanol. The last-named reaction utilises **alcohol dehydrogenase** and this will be dealt with in more detail later. Figure 11.1 summarises the reaction pathway and lists the names of the enzymes concerned.

Many alcoholic beverages contain a variety of materials that are major contributors to flavour and bouquet, and which are potentially of pharmacological interest. Such components formed during fermentation and subsequent distillation processes include higher alcohols such as isoamyl alcohol, propyl and isobutyl alcohol, aldehydes such as furfuraldehyde, and organic acids such as tartaric, succinic, lactic, acetic and malic acids.

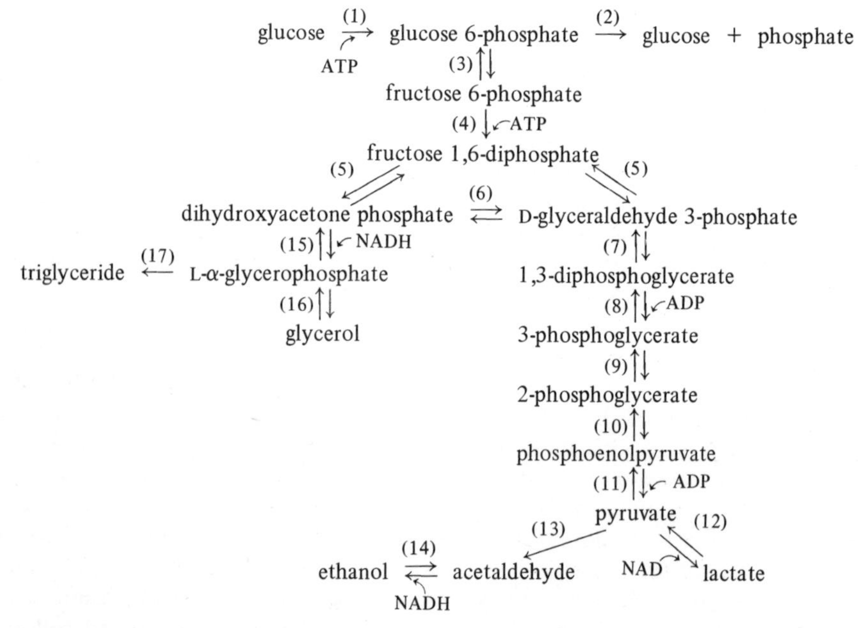

Figure 11.1. Enzymes involved in the conversion of glucose into ethanol or lactate. (1) Glucokinase and hexokinase; (2) glucose 6-phosphatase; (3) phosphoglucose isomerase; (4) phosphofructokinase; (5) aldolase; (6) triose phosphate isomerase; (7) glyceraldehyde phosphate dehydrogenase; (8) phosphoglycerate kinase; (9) phosphoglyceromutase; (10) enolase; (11) pyruvate kinase; (12) lactate dehydrogenase; (13) pyruvate decarboxylase; (14) alcohol dehydrogenase; (15) L-glycerol 3-phosphate oxidoreductase; (16) L-glycerol 3-phosphate phosphatase.

[2] Ethanol also occurs normally in mammalian tissues in trace amounts; it is formed apparently through the decarboxylation of pyruvate to acetaldehyde, which then undergoes reduction to ethanol (McManus *et al.*, 1966).

In general, however, because of the low concentrations of these components in the commercially prepared beverages, the principal pharmacological action of the alcoholic beverage on the body results from the ethanol content. This may not be true in some home-made beverages where purification and quality control are not rigorously supervised[3].

The ethanol content of wines and spirits is often expressed in terms of a percentage of *proof spirit*; this contains 57·1% (*v*/*v*) of ethanol. For example, whisky is often sold in England at 75% proof and contains 42% (*v*/*v*) ethanol. Fortified wines such as sherry and port contain approximately 20% (*v*/*v*) ethanol and English draught beer contains 3–6% (*v*/*v*) ethanol. The ethanol contents of standard English measures may easily be calculated. A single nip of whisky ($\frac{1}{6}$ gill or 23 ml) contains 8 g of ethanol; a pint of draught beer (3%, *v*/*v*, alcohol) contains 12 g of ethanol. The calories contained in such drinks are approximately: nip of whisky, 80 kcal; pint of English bitter beer, 250 kcal. Other components (e.g. carbohydrates) contribute to the calorie content of alcoholic drinks; ethanol itself has a calorific value of 7 kcal g^{-1}.

Alcohol is often mistakenly called a stimulant, but pharmacologically it is the opposite: it is a *depressant* of the central nervous system that, as blood concentration rises, produces an irregularly descending depression of the central nervous system, resembling the effects produced by general anaesthetics. The depressant action increases with the chain length of the aliphatic alcohol and increases with the degree of chain branching. This is illustrated by t-amyl alcohol (**amylene-hydrate**), which is used clinically as a hypnotic.

In excessive amounts ethanol produces the same sequence of events as seen in anaesthetic toxicity: analgesia, unconsciousness, anaesthesia and death from respiratory failure. Thus ethanol is a powerful depressant drug; Hutchinson (1969) in his well-known book on nutrition calls it a narcotic but this term has come to have a rather different everyday meaning from that used strictly in pharmacology. The social value of ethanol in moderation is usually said to derive from its ability to override many cerebral inhibitions and reticence. ("Speo donare novas largus amaraque curarum eluere efficax"—Horace: mighty to inspire new hopes and powerful to drown the bitterness of cares.)

Alcohol causes venous dilatation, leading to increased heat loss and decreased body temperature; it produces a marked diuresis. In high concentrations it is used as a local astringent; this property may produce damage to the oral and pharyngeal mucosa, and gastritis when neat spirits

[3] An example of the *toxic* effects of some home-made alcoholic beverages concerns cider production. During this process the organic acids in the apple extract may abstract heavy metals from the vats used in the preparation and storage. Heavy metal poisoning may then result when quantities of the cider are drunk. A classic of medical detective work concerned just such a situation. The Devon physician, George Baker, traced the cause of widespread colic to lead poisoning from contaminated cider in 1767.

are drunk in large quantities on an empty stomach. Ethanol has marked antiseptic qualities: the optimum strength for this purpose is said to be about 70% (*v*/*v*). The use of an ethanol aerosol has recently been proposed for sterilising artificial lung respirators but in the presence of oxygen there is an explosion risk.

11.3 Metabolism of ethanol

In normal adults ethanol can be metabolised at the rate of 10–15 ml/h.[(4)] The alcohol concentration in blood rises rapidly after the oral ingestion of alcoholic beverages (Haggard *et al.*, 1941); in a group of normal Americans the ingestion of 44 g of whisky (approximately five English nips) or about 4 pints of bitter beer on an empty stomach gave blood values of 67–92 mg/100 ml and 41–49 mg/100 ml respectively; after a meal the corresponding values were 30–53 mg/100 ml and 23–29 mg/100 ml respectively. In England at the present time it is illegal for a car driver having a blood ethanol concentration of 80 mg/100 ml (18 mM) or above to drive on the public highway. With an excessive intake of ethanol the concentration in blood reaches very high values and stays high for a very long period. In samples *post-mortem* from patients who died from alcoholic intoxication brain concentrations of ethanol of 270–510 mg/100 g and blood concentrations of 500–800 mg/100 ml have been reported.

Ethanol is metabolised in the liver by three main routes, of which the one utilising alcohol dehydrogenase has been known and well studied for many years; a second involves **catalase** and, although recognised for a considerable period, it has only recently been studied in detail. Recently, a third mechanism involving the endoplasmic reticulum has been discovered. These mechanisms will be discussed briefly in turn.

11.3.1 Alcohol dehydrogenase

Two main types of alcohol dehydrogenase have been studied: the enzymes from yeast and from liver. *Yeast* alcohol dehydrogenase has a molecular weight of 151 000. It has four catalytic sites per molecule and contains zinc; this is of interest in view of an old treatment for chronic alcoholism based on the administration of zinc oxide (Marcet, 1860). *Liver* alcohol dehydrogenase has a molecular weight of 73 000 and requires free thiol groups for activity.

[(4)] Above a blood alcohol concentration of about 10 mg/100 ml the rate of metabolism is independent of the blood value since the alcohol dehydrogenase system in liver is working at its full rate. Similar findings have been reported for the intact rat. Gordon (1968) has found with isolated perfused rat livers, however, that when the perfusing fluid is gassed with oxygen + carbon dioxide (95:5) the rate of utilisation of ethanol rises with its concentration. The mechanism of this effect is unknown. Alcohol is metabolised considerably more rapidly in the rat than in the human. For example, Lester *et al.* (1968) report a utilisation in the rat of about 300 mg/kg body wt./h; this is approximately 2–3 times the rate quoted above for the human.

The mechanism of alcohol dehydrogenase activity has been studied in considerable detail, particularly by Theorell and colleagues (Theorell, 1967). One interesting feature of the reaction is the stereospecificity of the hydrogen transfer from NADH. By using the 4-monodeuterated NADH, which contains a new centre of asymmetry, it has been found that NAD-linked dehydrogenases can distinguish between the conformation of the deuterium atom in the two diastereoisomers at the 4-position. One class of dehydrogenase reacts with the hydrogen on one side of the pyridine ring (side α); the other class reacts with the hydrogen on the other (β) side. Alcohol dehydrogenases are type α, as are lactate and malate dehydrogenase of heart muscle; β specificity is shown by glutamate dehydrogenase, cytochrome *c* reductase and α-glycerophosphate dehydrogenase of muscle (Figure 11.2).

In absolute quantities alcohol dehydrogenase is mainly located in liver although high specific activities may be found in other tissues such as retina, where the conversion of retinaldehyde into retinol is catalysed by an alcohol dehydrogenase.

Krebs *et al.* (1969) have reported on the alcohol dehydrogenase activity of a number of rat tissues (see Table 11.1). In liver the enzyme has a fairly wide specificity, reacting with other aliphatic alcohols (such as propyl alcohol and vitamin A), with intermediates in terpene and cholesterol synthesis (geraniol, mevalonic acid), but only very slowly with *methyl* alcohol. The liver enzyme is present mainly in the soluble fractions of tissue homogenates and histochemical examination shows that it is preferentially located in *portal* areas: this intralobular distribution is more marked in the human than in the rat (Greenberger *et al.*, 1965).

The product of alcohol dehydrogenase activity with ethanol as substrate is acetaldehyde. This is normally converted into acetate[5] by aldehyde dehydrogenases, which can be flavin-linked and coupled with the respiratory chain. The metabolic reactions involved in the conversion of ethanol into acetyl-coenzyme A are shown in Figure 11.3. Pig liver flavin-linked aldehyde dehydrogenase contains flavinadenine dinucleotide, molybdenum

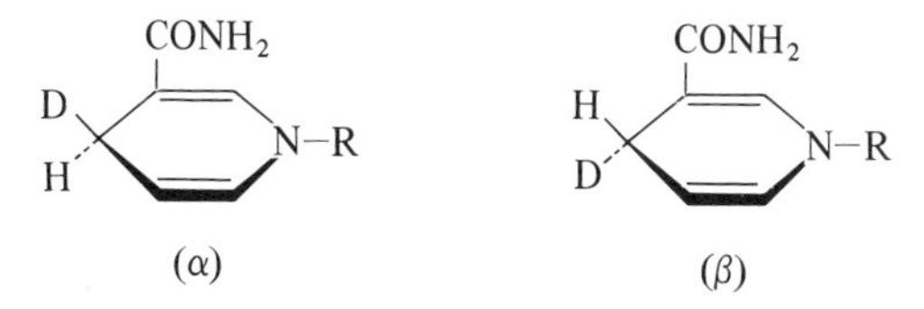

Figure 11.2. Diastereoisomers of NADH involving the hydrogen and deuterium atoms at C-4 on the pyridine ring. The thick edge of the ring is to be imagined facing the reader; in the left-hand illustration the bond to the deuterium atom is pointing upwards.

[5] Blood acetate concentration rises during ethanol intoxication, showing that the production of acetate exceeds its utilisation in liver; for example by reaction with thiokinase and coenzyme A to yield acetyl-coenzyme A.

Table 11.1. Alcohol dehydrogenase activity in normal rat tissues (data summarised from Krebs *et al.*, 1969).

Tissue	Activity (μmol of substrate utilised/min per g wet wt. of tissue)
Liver	20·3
Kidney cortex	2·1
Gastric mucosa, fundus	1·5
body	0·4
Duodenum	0·4
Testis	0·27
Uterus	0·19
Brain cortex, Pancreas, Lung, Spleen	0·0

and an iron porphyrin, and can be coupled with the reduction of cytochrome *c*; it is completely inhibited by 100 μM-quinacrine. In pig liver, aldehyde dehydrogenase has an activity of approximately 10 μmol of substrate utilised/g per h. In a survey of aldehyde dehydrogenases in rat, rabbit, dog and monkey tissues, Dietrich (1966) found most activity in the liver. For example, in the rat the V_{max} (nmol/mg of protein per min) for liver was 49 in female rats and 29 in male rats; the next most active tissue was kidney (9 and 4·1 for females and males respectively), followed by uterus (0·4). Rabbit liver aldehyde dehydrogenase was about three times as active as rat liver and most of the activity was in the soluble fraction of the liver, with the remainder in the mitochondria.

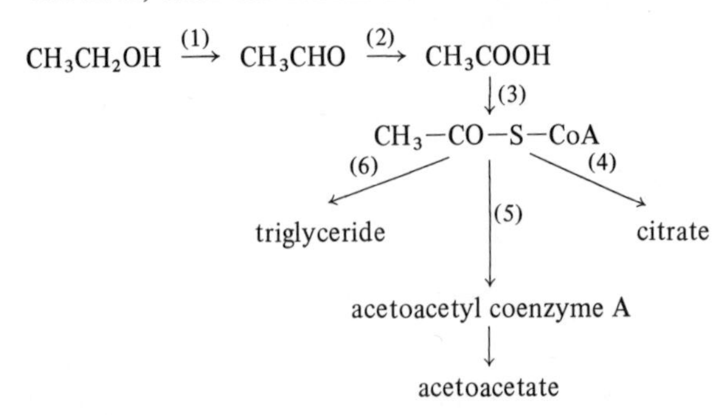

Figure 11.3. Metabolic routes for the breakdown of ethanol in the liver. The enzymes involved are:
(1) alcohol dehydrogenase; (2) aldehyde dehydrogenase; (3) thiokinase; (4) citrate synthetase; (5) acetyl-CoA-acetyl CoA, C-acetyl transferase (acetyl-CoA acetyl transferase); (6) fatty acid synthesis system.

11.3.2 Catalase

The second mechanism of oxidising ethanol is that dependant on catalase, a ferriprotoporphyrin-IX enzyme. Catalase normally catalyses the breakdown of hydrogen peroxide to water and has an extremely high turnover number. Short-chain alkyl hydroperoxides are also utilised by this enzyme. In addition catalase catalyses *peroxidative* type reactions (Figure 11.4).

If hydrogen peroxide is produced in the vicinity of catalase, for example by a flavoprotein-linked oxidation of substrate (SH_2), then ethanol can be oxidised as is shown in Figure 11.5.

In many species so far examined catalase, peroxidase, uricase and L-hydroxy acid dehydrogenase are localised in liver in cytoplasmic particles that have been called **peroxisomes** (see De Duve and Baudhuin, 1966). In the guinea pig, however, only some 18% of total catalase is particulate in nature, compared with 73% in the mouse; Feinstein *et al.* (1953) also found that only a small amount of catalase was present in the peroxisome fraction of monkey liver (see Makar and Mannering, 1968). Hydrogen peroxide is generated in the peroxisomes by the oxidation of the L-hydroxy acids as shown in Figure 11.5.

(a) $$HO{-}OH + \begin{matrix} AH_2 \\ 2X^{n+} \end{matrix} \longrightarrow 2H_2O + \begin{matrix} A \\ 2X^{(n+1)+} \end{matrix}$$

(b) $$HO{-}OH + H_2O_2 \longrightarrow 2H_2O + O_2$$

(c) $$BH_2 + O_2 \xrightarrow{FP} B + H_2O_2 \xrightarrow{catalase} O_2 + 2H_2O$$

(d) $$SH_2 + AH_2 + O_2 \longrightarrow S + A + 2H_2O$$

Figure 11.4. Breakdown of hydrogen peroxide by catalase (a) and (b); the peroxidation type reaction is (a) and the catalatic reaction is (b). In reaction (d) catalase and a flavoprotein dehydrogenase (FP) combine to oxidise the two substrates SH_2 and AH_2 utilising the peroxidative rate shown in (a). In reaction (c) the catalatic reaction is utilised.

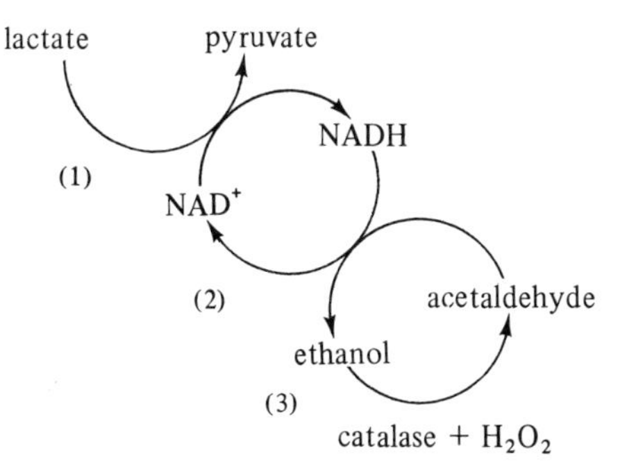

Figure 11.5. Oxidation of lactate by coupling lactate dehydrogenase (1) with alcohol dehydrogenase (2), catalase (3) and a source of hydrogen peroxide. The net result of all three reactions operating is

$$H_2O_2 + \text{lactate} \longrightarrow \text{pyruvate} + 2H_2O\,.$$

The role of catalase in the metabolism of ethanol is controversial. Several authors have claimed that it is not of major significance in ethanol metabolism, whereas recent work with methanol (see later) shows that the catalase pathway is a major one in the rat and is proportional to the amount of catalase in the peroxisomes. Further, Wartburg and von Röthlisberger (1961) have shown an increased catalase activity in rat liver in *chronic* alcohol intoxication, a finding that may be relevant to the tolerance known to develop to alcoholic intoxication in man. The increased tolerance, which may mean that three to four times as much alcohol can be taken to produce the same degree of intoxication as in a normal person, has been said to be unaccompanied by increased ethanol metabolism; tolerance is known to be long-lasting for several months after ending chronic ethanol ingestion and may reflect a genuine central nervous system adaptation to the effects of ethanol on the higher centres. Chronic alcoholics also show cross-tolerance to volatile anaesthetics such as ether[6].

In contrast to the above-mentioned view that alcoholic tolerance does not involve increased metabolism of the alcohol are the experiments reported by Isselbacher and Greenberger (1964). It was shown very clearly that chronic alcoholics could consume relatively large quantities of alcohol before the concentration of alcohol in their blood rose above 80 mg/100 ml. Whether this effect is due largely to an increased rate of metabolism or an increased rate of excretion, or of storage in peripheral depots, is not clear. Certainly, evidence is now available from recent studies that ethanol ingestion in rats (and man) causes proliferation of endoplasmic reticulum and an increased activity of drug-metabolising enzymes, as will now be summarised before the third mechanism of alcohol metabolism is discussed.

11.4 Ethanol and induction of liver enzymes

Many substances are now known that stimulate (**induce**) the activity of the enzyme system in liver that is concerned with the metabolism of foreign compounds (for review see Conney, 1967). This electron chain, which is shown diagrammatically in Figure 10.3, will be referred to as the drug-metabolising sequence. Ethanol has recently been added to the list of inducers.

Rubin *et al.* (1968) showed that ethanol administered to rats[7], particularly if on a diet deficient in choline, casein, methionine and cystine, markedly stimulated microsomal aniline hydroxylase and cytochrome P_{450} activity; nitroreductase was increased much less (Table 11.2).

(6) The effects of ethanol in increasing the activity of the NADPH-cytochrome P_{450} chain in liver endoplasmic reticulum are considered in Section 11.4.

(7) The liquid diet used contained 36% of total calories as ethanol and was given over a 15 day period; this is a very considerable intake of alcohol and is roughly equivalent to about a daily intake of one-half a bottle of whisky in an adult human.

Table 11.2 The effect of ethanol administration on rat liver cytochrome P_{450} and drug-metabolising enzyme activities. Ethanol was included in liquid diets and comprised 36% of total calorie intake. The diets were either (a) an adequate diet or (b) an inadequate diet with respect to choline and sulphur-containing amino acids. Rats were kept on the diets for 15 days. Data from Rubin *et al.* (1968).

Treatment	Aniline hydroxylase (nmol of *p*-aminophenol formed/g of fat-free liver per 15 min)	Nitroreductase (nmol of *p*-aminobenzoic acid formed/g of fat-free liver per 15 min)	Cytochrome P_{450} (%)	Liver triglyceride (mg/g wet wt.)
Adequate diet	97	1·2	100	11
+ ethanol	755	1·7	139	44
Inadequate diet	127	1·6	100	28
+ ethanol	1358	1·9	180	108

The report for the rat was followed by a further communication by Rubin and Lieber (1968b) on the response of both rat and human liver to ingestion of ethanol. The ethanol dose to the three human volunteers was 42% of total calories as ethanol for 12 days. Ethanol greatly stimulated aniline hydroxylase in the rat as previously reported and there were considerably smaller increases in pentobarbital hydroxylase and benzpyrene hydroxylase. In the human volunteers, although pentobarbital hydroxylase doubled in activity after ethanol ingestion, benzpyrene hydroxylase was unchanged.

The effects of high ethanol intake on liver morphology in human volunteers was studied by Rubin and Lieber (1968a). There was a rapid accumulation of fat and an increased amount of smooth endoplasmic reticulum when blood alcohol values were maintained at 20–80 mg/100 ml. The authors point out that the amount of alcohol consumed by many 'social' drinkers is sufficient to cause recognisable disturbances in the liver, and that one need not get drunk to sustain alcohol-induced hepatic injury.

The action of ethanol in increasing the activity of drug-metabolising enzymes in liver has been used by Waltman *et al.* (1969) in attempts to lower plasma bilirubin values in neonates. The rationale is that neonatal liver has a very low activity of the microsomal enzyme glucuronyl transferase that catalyses Reaction (11.2).

$$\text{Bilirubin} + \text{UDP-glucuronic acid} \longrightarrow \text{bilirubin glucuronide} + \text{UDP} \quad (11.2)$$

Bilirubin cannot be excreted from the body in the bile unless converted first into the glucuronide[8]. Thus bilirubin may reach very high and toxic

[8] But see the report by Kuenzle (1970) that bilirubin is mainly secreted in human bile as conjugates other than the simple glucuronide.

concentrations in the neonatal period owing to the slowness of biliary excretion. The enzyme glucuranyl transferase also shows an increase of activity after treatment of the liver with inducers such as phenobarbital, which increase the drug-metabolising sequence to a considerably greater degree. Waltmann *et al.* (1969) have found that ethanol given to mothers at delivery (118 g of ethanol by slow intravenous administration) had a significant effect in decreasing the plasma bilirubin concentrations in the newborn infants (Table 11.3). Similar findings have been reported for more powerful inducers (for example phenobarbital on neonatal bilirubin concentrations and dicophane on bilirubin content in a patient with the Crigler–Najjar syndrome).

Table 11.3. Effect of administration of ethanol to pregnant women just before delivery on the plasma bilirubin concentrations of neonates. Data from Waltman *et al.* (1969).

Days post-partum	Bilirubin (mg 100 ml^{-1})	
	control group	ethanol group
1	4·6 ± 0·4	2·8 ± 0·4
2	6·4 ± 0·5	3·8 ± 0·4
3	6·4 ± 0·8	3·7 ± 0·4
4	5·7 ± 0·8	3·5 ± 0·6

Ethanol not only produces an increased amount of the drug-metabolising enzyme sequence but also increases the activity of an enzyme system in microsomes that oxidises ethanol to acetaldehyde (Lieber and De Carli, 1968). This system was first described by Orme-Johnson and Ziegler (1965) and has recently been studied by Roach and colleagues (1969). The system is dependent on NADPH and oxygen; it is inhibited by carbon monoxide, cyanide and azide, but not by SKF 525A. Partial inhibition (40–50%) of the system by 3-amino-1,2,4-triazole (a powerful inhibitor of catalase) suggests that the oxidation of ethanol is at least partially dependent on hydrogen peroxide. This is to be expected, however, from the discussion given in Section 11.3.2, since the microsomal fraction used in all the above-quoted reports is likely to be contaminated with peroxisomes.

Isselbacher and Carter (1970) also report that the microsomal fraction may contain some alcohol dehydrogenase but controversy exists concerning the actual contributions of this enzyme and of catalase to the overall metabolism of ethanol by the microsomes. The effects of various inhibitors both *in vitro* and *in vivo* on the microsomal ethanol oxidising system (MEOS) have been studied by Lieber and De Carli (1970a, b); Lieber *et al.* (1970); Isselbacher and Carter (1970); and Khanna *et al.* (1970). Generally, the MEOS system appears insensitive to SKF 525A (as mentioned previously) but is partially inhibited by carbon monoxide, azide, and cyanide. The *in vivo* administration of inhibitors of alcohol dehydrogenase and catalase (pyrazole; 3-amino-1,2,4-triazole) produced varying degrees of inhibition, or no effect at all, on MEOS. Surface active agents like cholate or deoxycholate (but not digitonin) decreased the MEOS activity, but Lieber *et al.* (1970) have shown such action to be unspecific and to affect a variety of other microsomal enzymes. Isselbacher and Carter (1970) suggest that the MEOS activity can be explained by a combination of NADPH oxidase activity in the microsomes which produces hydrogen peroxide; by a catalytic oxidation of ethanol in the presence of hydrogen peroxide; and by the presence of alcohol dehydrogenase in the microsomes that has some small activity with respect to NADPH. This latter point seems rather suspect since the reactivity of liver ADH with NADPH is very low and the amount of ADH contamination of microsomal pellets is generally small. The involvement of NADPH oxidase and catalase seems quite likely but whether this is sufficient to account for the total activity of MEOS is not clear. This point is discussed in the reports of Lieber and De Carli (1970a, b) and of Lieber *et al.* (1970) who found that the *in vivo* administration of pyrazole or 3-amino-triazole had very much more marked effects on catalase and ADH than on MEOS. It is possible that there is a route for ethanol oxidation in liver microsomes which is independent of catalase and ADH but the significance of its contribution to the overall metabolism of ethanol *in vivo* is at present uncertain.

Lieber and De Carli (1970c) have found that MEOS activity *in vitro* is stimulated by a prior treatment of the rats with ethanol, phenobarbital, butylated hydroxytoluene, or 3-methyl cholanthrene, all inducers of liver endoplasmic reticulum. Pre-treatment with high calorie intakes of ethanol also led to increased clearance rates of ethanol from the blood (Lieber and De Carli, 1970a). However, Tephly *et al.* (1969) and Khanna and Kalant (1970) have shown that ethanol metabolism *in vivo* is largely independent of prior feeding with phenobarbital or with SKF 525A in contrast to the stimulatory action of phenobarbital on MEOS *in vitro* (Lieber and De Carli, 1970c). Moreover, Tephly *et al.* (1969) have calculated that MEOS contributes less than 7% towards the total metabolism of ethanol *in vivo*. Perhaps the difference between the *in vitro* and *in vivo* findings lies in the relative excess of NADPH and oxygen under *in vitro* conditions, and in the altered composition of the fluid in which the microsomes are suspended *in vitro* compared to the cell supernatant *in vivo*. The latter change entails that the microsomes *in vitro* are exposed to a different antioxidant status than within the intact cell. Certainly, antioxidants affect the consequences of acute ethanol intoxication as will be discussed later (Section 11.8). Moreover, the increased liver triglyceride that results from acute ethanol intoxication *in vivo* is partially prevented by prior treatment with phenobarbital and SKF 525A (Vincenzi *et al.*, 1967;

Wooles, 1968) as well as by free radical scavengers like diphenyl-*p*-phenylene diamine (Di Luzio, 1963). Lieber and De Carli (1970b) have made the interesting suggestion that one possible effect of chronic ethanol dosing would be to induce NADPH oxidase that would lead to an increased production of hydrogen peroxide. In the presence of transitional metals this could initiate lipid peroxidation and damage to the endoplasmic reticulum; antioxidants would be expected to have a protective action against such damage.

It is apparent that more data are required before the significance of MEOS activity to the overall metabolism of ethanol and to its hepatotoxic action can be properly assessed. It seems likely, however, that research in this area is likely to be fruitful in terms of evaluating the damaging action of ethanol on the liver.

11.5 Nucleotide changes

A major event in the oxidation of ethanol in the liver is the disturbance in the $NAD^+/NADH$ ratio. In the rat, the ratio in *whole* liver samples falls from 3·5 to 0·9 within 30 min after dosing (Figure 11.6) with a single intraperitoneal injection of 3 ml/kg body wt. as a 20% solution in saline. The dose is a large one even for the rat, which has a very rapid metabolism; on a body-weight basis, ingestion of this dose by the rat is equivalent to a man consuming a bottle of whisky at a sitting (the LD_{50} for the rat has been reported to be 14 ml/kg body wt.). It is interesting that the ratio $NADPH/NADP^+$ shows much less change than the ratio $NAD^+/NADH$, indicating that *transhydrogenases* catalysing Reaction (11.3) are not particularly effective in ethanol intoxication, where the NADH is produced mainly in the cytoplasm.

$$NADH + NADP^+ \longrightarrow NAD^+ + NADPH \qquad (11.3)$$

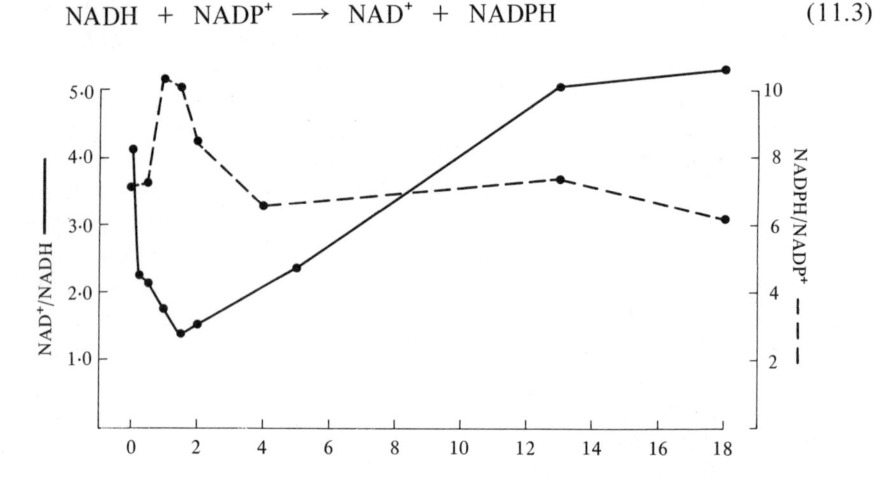

Figure 11.6. Changes in the ratios $NAD^+/NADH$ and $NADPH/NADP^+$ in rat liver after the administration of ethanol (2·4 g/kg body wt.) at time 0. Data of Slater *et al.* (1964).

Similar changes to those outlined above in the ratio $NAD^+/NADH$ after ethanol administration have been found by other authors (Forsander *et al.*, 1958; Smith and Newman, 1959; Lieber and Schmid, 1961; Reboucas and Isselbacher, 1961; Jouany, 1965; Fazekas and Rengei, 1969). These reports have been concerned with the overall changes in nucleotide content in whole liver samples. It has become increasingly evident, however, that relatively low-molecular-weight coenzymes and substrates are compartmented in the liver in much the same way as has been known for a long time for enzymes. Thus relatively minor changes in whole tissue samples may conceivably hide much larger changes in an intracellular compartment. It is difficult to obtain precise measurements of the intracellular distributions of small molecules since the normal methods of cell fractionation involve centrifuging of homogenates, and even the simplest scheme for separating mitochondria and nuclei from microsomes and supernatant takes approximately 15 min from killing of the rat; in this time appreciable redistribution between different fractions may occur. An experimental technique that solves this problem at least partially has been developed by Delaney and Slater (1970), in which the tissue homogenate is rapidly passed through Millipore filters that retain the mitochondria and nuclei and yet allow most of the smaller fragments of the endoplasmic reticulum and all of the soluble fraction to pass through. In this way the soluble fraction can be separated from the large-particle fraction within 2 min of killing the animal. The differences found in the intracellular distribution of several coenzymes with this technique compared with the results found by classical centrifuging analyses are shown in Table 11.4. Here it can be seen that most of the liver ATP and NADPH is in the mitochondrial fraction and most of the NAD^+ is in the cell sap. Rapid changes occur in this distribution after liver injury as discussed in Chapter 10 for carbon tetrachloride. After administration of ethanol the main changes in the liver nucleotides concern the ratio $NAD^+/NADH$. Since NAD^+ and NADH occur largely in a single compartment (the cell sap) the whole liver values obtained correspond fairly closely to the changes in the total nucleotide

Table 11.4. Percentage distributions of ATP and of $NADP^+$ + NADPH in the cell sap fraction of rat liver homogenates as measured by a rapid centrifuging technique (Slater, 1967) or by the microfiltration procedure (Delaney and Slater, 1970). Values are given as the percentage of nucleotide in cell sap compared with the content in whole homogenate.

	ATP	$NADP^+$ + NADPH	NAD^+
Centrifuging	72 ± 3	49 ± 5	84[a]
Filtration	49 ± 3	40 ± 2	92

[a] Data of Glock and McLean (1956).

content of the cell sap. The word 'total' is used in the last sentence to imply the entire nucleotide content of the cell sap fraction in contrast to the nucleotide fraction bound as coenzyme to various dehydrogenases present in the cell sap. A method that allows the calculation of the *free* NAD^+/NADH in the cell sap (and in the mitochondria) in contrast to the total nucleotide content has been elaborated by Bücher and Klingenberg (1958). They have suggested that if a nucleotide-linked enzyme that is characteristic of the cell sap (i.e. lactate dehydrogenase) or the mitochondria (i.e. glutamate dehydrogenase) is considered then the equations shown in Figure 11.7 may be written. Determination of the concentrations of the respective pairs of substrates, e.g. lactate and pyruvate for Equation (a), together with a knowledge of the value of the equilibrium constant (K) allows calculation of the ratio NAD^+/NADH associated with the dehydrogenase (see Veech *et al.*, 1969). This approach gives values very

Cytoplasmic compartment

$$\text{Lactate} + NAD^+ \rightleftharpoons \text{pyruvate} + NADH$$

$$K = \frac{[\text{pyruvate}]}{[\text{lactate}]} \times \frac{[NADH]}{[NAD^+]} = 1{\cdot}11 \times 10^{-4}$$

Mitochondrial compartment

$$\text{Glutamate} + H_2O + NAD^+ \rightleftharpoons \text{2-oxoglutarate} + NH_3 + NADH$$

$$K = \frac{[\text{2-oxoglutarate}][NH_3]}{[\text{glutamate}]} \times \frac{[NADH]}{[NAD^+]} = 3{\cdot}87 \times 10^{-3}\ \text{mM}$$

Figure 11.7. Redox-couples (a) in liver cytoplasm and (b) in liver mitochondria. Values are calculated for pH 7·0, activity of water = unity and ionic strength 0·25. Values from Williamson *et al.* (1967).

Table 11.5. Effect of ethanol on the NAD^+/NADH ratios of liver cell fractions as estimated by the redox procedure (data of Baxter and Hensley, 1969). Ethanol was given to rats by stomach tube as an aqueous 25% (*w*/*v*) solution at a dose of 2·5 g/kg body wt. Rats were killed 45 min later and acid extracts of the frozen livers were analysed for substrates (lactate; pyruvate; glutamate; 2-oxoglutarate). NAD^+/NADH ratios in cytoplasm and in the mitochondria were determined from the equation

$$K = \frac{\text{substrate}_{\text{ox.}}}{\text{substrate}_{\text{red.}}} \times \frac{NADH}{NAD^+},$$

where $K = 1{\cdot}11 \times 10^{-4}$ for the lactate redox-couple and $K = 3{\cdot}87 \times 10^{-3}$ mM for the glutamate redox-couple.

	$\frac{\text{Lactate}}{\text{Pyruvate}}$	$\frac{NAD^+}{NADH}$ (cytoplasm)	$\frac{NAD^+}{NADH}$ (mitochondria)
Control	32	280	5·6
Ethanol	49	184	3·4

different from those found by direct analysis of total nucleotides in liver cell fractions (see Figure 11.6). By using the redox-couple method Baxter and Hensley (1969) have followed changes in the ratio $NAD^+/NADH$ after giving ethanol (2·5 g/kg^{-1}) to female rats. The ratios had changed in the cell sap from 280 to 184, and in the mitochondrial fraction from 5·6 to 3·4, by 45 min after dosing (Table 11.5).

During the oxidation of ethanol, NAD^+ becomes rate-limiting in many NAD-linked reactions. This follows from work by Theorell showing that the dissociation of NADH from alcohol dehydrogenase is relatively slow compared with the reduction of NAD^+ (for recent studies on the binding of NAD^+ and alcohol dehydrogenase see Theorell *et al.*, 1967); further, NADH produced in the cytoplasm cannot be oxidised directly by mitochondrial NADH-dehydrogenases but must be transported across the mitochondrial membrane by a comparatively slow shuttle process. Tygstrup *et al.* (1965) have suggested that to overcome this limiting availability of NAD, alternative substrates may be supplied to reoxidise the NADH produced by alcohol dehydrogenase in the cytoplasm. They suggest that fructose increases the oxidation rate of ethanol *in vivo* by such an effect in regenerating NAD^+ from NADH. Their scheme to explain the clinically useful process of intravenous fructose administration to patients in acute alcoholic intoxication is shown in Figure 11.8. In connection with the clinical use of fructose to restore the liver $NAD^+/NADH$ ratio to normal during acute ethanol intoxication some other recently described effects of fructose are noteworthy. A large parenteral dose of fructose given to rats has been found rapidly to decrease both liver **ATP** concentrations and liver **protein synthesis** (Mäenpää *et al.*, 1968; Raivio *et al.*, 1969; Burch *et al.*, 1969). A similar dose of fructose also produces a transient large decrease in **bile flow** in the rat (T. F. Slater and M. N. Eakins, unpublished work).

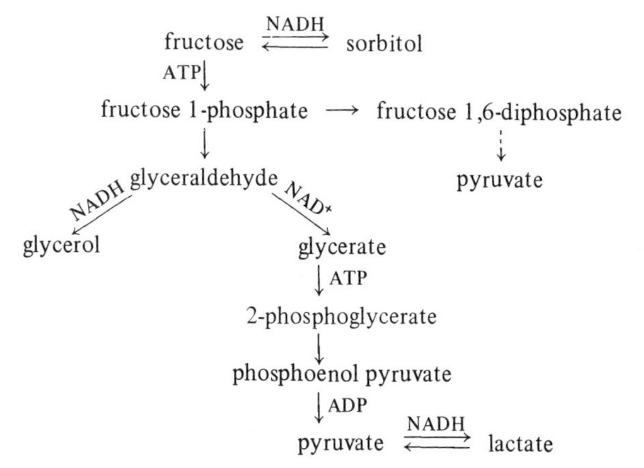

Figure 11.8. Metabolism of fructose and its ability to re-oxidise the NADH produced by ethanol oxidation. From Tygstrup *et al.* (1965).

11.6 Ethanol intoxication and liver injury

Two main aspects of the liver injury resulting from ethanol intoxication in the rat have been particularly studied. The first aspect concerns the acute injury that follows the administration of a single large dose of ethanol and where nutritional factors in general are not relevant to the ensuing injury. The second aspect concerns chronic intoxication by ethanol where the nutritional status of the animal during the course of a long study may be of major importance, and where controversy exists over the most suitable model diets to be used in experimental studies. This latter aspect of ethanol intoxication is of course closely connected with our understanding of the tissue disturbances that are associated with chronic alcoholism in man. In this Chapter we shall concentrate mainly on the liver disturbances that occur in acute ethanol intoxication, and only brief reference will be made to the chronic state.

11.7 Acute liver injury after administration of alcohol

The administration of a large single dose of ethanol to rats was shown by Mallov and Bloch (1956) to result in a fatty liver with little or no associated necrosis. The increase in lipid was demonstrated by Di Luzio (1958) to be mainly **triglyceride** and this may increase by more than 10-fold during the acute stage of the intoxication. Horning *et al.* (1960) showed that the fatty acids associated with the increased liver triglyceride originated mainly from the adipose tissue; this has been confirmed by Isselbacher's (1964) group by using gas chromatography to identify the component fatty acids that occurred in the accumulated triglyceride in the liver. These studies show that the increased accumulation of triglyceride mainly utilises preformed fatty acids from the depots and does not result from an increased synthesis *de novo* of fatty acids in the liver. The rate of accumulation of liver triglyceride has been studied by several authors and Wooles (1966) gives clear data (Figure 11.9) that show the rapidity of

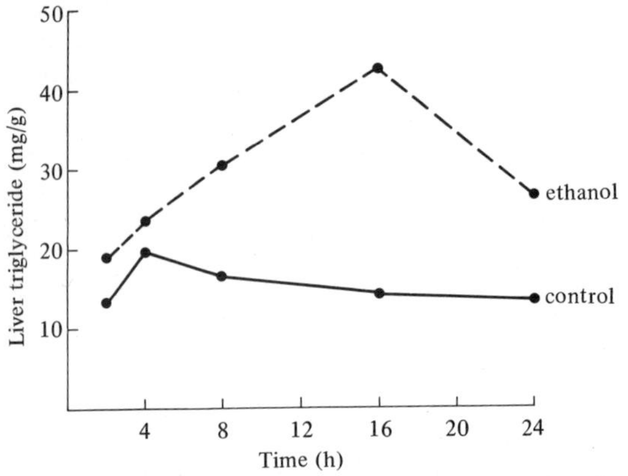

Figure 11.9. Increase in liver triglyceride after ethanol administration (6 g/kg body wt.). Data from Wooles (1966).

the process. Increases in liver triglyceride can be observed within 30 min of oral administration of ethanol. The possible mechanisms that could result in such an increased liver triglyceride have been summarised in Figure 9.2, which is modified from the diagram of Isselbacher and Greenberger (1964).

It can be seen that possible mechanisms leading to an increased liver triglyceride include: (1) an increased synthesis of triglyceride in the liver involving (a) an increased synthesis of fatty acids and (b) an increased coupling of fatty acid with phosphatidic acid to yield triglyceride rather than phospholipid; (2) a decreased oxidation of fatty acids in the liver; (3) an increased release of free fatty acids from the fat depots to the liver that would favour the increased synthesis of triglyceride; (4) a decreased output of lipoprotein by the liver leading to an accumulation of fat, as occurs in intoxication with carbon tetrachloride. It is difficult to summarise briefly the numerous and conflicting results that have appeared in this connection, but it is probably true to say that convincing evidence for a major role for any of the above mechanisms has not yet been obtained. Detailed references to this work are given in a review by Isselbacher and Greenberger (1964).

An interesting finding that warrants further study is the effect of ethanol in increasing the microsome-catalysed incorporation of fatty acid into diphosphatidic acid; this may be of relevance to the stimulatory action of ethanol on other microsomal enzyme systems (see Section 11.4). As mentioned above, the reported effects of ethanol on other possible routes leading to increased triglyceride deposition in the liver are confusing in their complexity. Much of this confusion arises from the use of widely varying experimental conditions by the investigators (for example: dose of ethanol administered; sex and age of rat; dietary regime; length of treatment etc.). The position is therefore unsatisfactory for determining the metabolic events involved although fortunately advances in two new directions have offered the hope of a better understanding of the basic lesions. These advances have concerned gluconeogenesis and the protective role of antioxidants.

Studies on gluconeogenesis *in vivo* are complicated by hormonal interactions and consistent results have proved difficult to obtain. There were several early reports, however, that administration of ethanol decreased gluconeogenesis *in vivo*, although the mechanism of this effect was not understood (see Isselbacher and Greenberger, 1964). Krebs (1968) has reported that reproducible conditions are obtainable with the isolated perfused liver and, as a consequence, has been able to develop a hypothesis to explain the effects of ethanol on gluconeogenesis. Krebs and coworkers found that ethanol at quite low concentrations (5 mM) in the perfusing medium caused a pronounced decrease in gluconeogenesis with lactate as substrate but no comparable effect when pyruvate was the substrate.

As Krebs points out, 5 mM is a concentration quite likely to be obtained in the human. Krebs' finding suggests that the critical step inhibited by ethanol is the conversion of lactate into pyruvate, which requires NAD^+; this reaction is thus in competition with alcohol dehydrogenase for NAD^+ [Reactions (11.5) and (11.6)].

$$\text{Lactate} + NAD^+ \longrightarrow \text{pyruvate} + NADH \qquad (11.4)$$

$$\text{Ethanol} + NAD^+ \longrightarrow \text{acetaldehyde} + NADH \qquad (11.5)$$

We have already seen that ethanol intoxication rapidly produces a change in the ratio $NAD^+/NADH$ in favour of the reduced form. Thus the conversion of lactate into pyruvate becomes limited by the very low concentration of NAD^+ available. Krebs pointed out that this would lead to a decrease in the pyruvate concentration in the soluble compartment of the liver and this would affect the rate of the enzyme pyruvate carboxylase reaction, since the K_m for the reaction is probably high (0·44 mM in chicken liver). Since the normal concentration of pyruvate in liver is at the most 0·1 mM, a substantial decrease in pyruvate concentrations would cause a pronounced decrease in pyruvate carboxylase activity, the first enzyme in the conversion of pyruvate into glucose (see Figure 11.10). This inhibition of gluconeogenesis from lactate may largely explain the long-known association of *hypo*glycaemia with chronic alcoholism, particularly in those who are in a poor state of nutrition, as the main conversion of lactate (e.g. from muscle) into blood sugar occurs in the liver.

Krebs (1968) also discusses the effects of the disturbed $NAD^+/NADH$ ratio on other NAD-linked enzyme reactions. In particular, α-glycerophosphate dehydrogenase is inhibited and α-glycerophosphate accumulates in the liver. Since α-glycerophosphate is a key intermediate in the

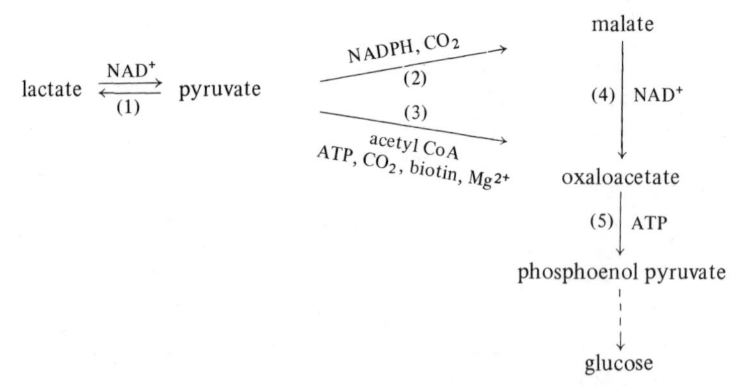

Figure 11.10. Metabolic conversions of pyruvate into phosphoenolpyruvate utilising: (1) lactate dehydrogenase; (2) malic enzyme; (3) pyruvate carboxylase; (4) malate dehydrogenase; (5) phosphoenol pyruvate carboxykinase.

synthesis of triglyceride Krebs has suggested that the fatty liver resulting from ethanol intoxication arises basically from the $NAD^+/NADH$ control of the α-glycerophosphate dehydrogenase step (see Figure 11.1). However, as pointed out by Isselbacher, an increased *synthesis* of triglyceride is by no means a universal finding in ethanol-induced fatty liver, and an interference with secretory processes may also be of considerable importance.

The hypothesis suggested by Krebs is attractive in many respects and explains several of the major features of acute ethanol intoxication. However, there are some aspects of the overall liver disturbance that require further discussion, particularly the ideas that have arisen from recent work involving the protective action of antioxidants.

11.8 Antioxidants and accumulation of liver triglyceride

Observations concerning the effects of treatment with antioxidants on the ethanol fatty liver have come principally from Di Luzio's laboratory (for summary, see Di Luzio and Hartman, 1967). Di Luzio (1963) reported that the increased liver triglyceride after ethanol dosing was alleviated by pretreating rats with α-tocopherol or *NN′*-diphenyl-*p*-phenylenediamine by *intraperitoneal* injection at −48, −24 and −2 h before the *oral* dosing with ethanol (Table 11.6). In 1964 he showed that the concomitant *oral* administration of an antioxidant mixture (containing butylated hydroxytoluene, butylated hydroxyanisole and propyl gallate) with the ethanol also protected against the increased liver fat that was measured 16 h later. A similar result with *NN′*-diphenyl-*p*-phenylenediamine given *orally* with the ethanol was reported by Di Luzio and Costales (1965). This effect obtained with a variety of known antioxidants led Di Luzio to the conclusion that ethanol was damaging the liver by a process of lipid peroxidation. Kalish and Di Luzio (1966) reported that ethanol intoxication resulted in an

Table 11.6. Effects of treatment with various antioxidants on the triglyceride content of liver after ethanol administration (data from Di Luzio, 1964, and Di Luzio and Costales, 1965). Ethanol was given as a single oral dose of 6 g/kg body wt. for (a) and (c) and 3 g/kg body wt. for (b). In (a) *NN′*-diphenyl-*p*-phenylenediamine was given by intraperitoneal injections at −48, −24 and −2 h before ethanol (600 mg/kg body wt.); (b) G-50R mixture was given with the ethanol (7·5 ml/kg body wt.); (c) α-tocopherol acetate was given by intraperitoneal injections at −48, −24 and −2 h before ethanol (100 mg/kg body wt.).

Treatment	Liver triglyceride (mg/g)		
	(a)	(b)	(c)
Control	16	18	10
+ drug	11	11	10
Ethanol	36	33	47
+ drug	20	22	30

increased **peroxide** content of liver lipid extracts, and that this was prevented (as was the increased liver triglyceride) by dosing with ubiquinone-4 at −24 h and also at 0 h together with the ethanol (Table 11.7). Rather similar findings had also been obtained in Hartcroft's laboratory (Porta and Hartcroft, 1965), where it was found that α-tocopherol was effective in preventing the increased liver triglyceride developing after an acute dose of ethanol and that chronic ethanol intoxication resulted in an increased liver **thiobarbituric acid** reaction. Comporti *et al.* (1967) found that ethanol added to liver homogenates *in vitro* stimulated malonaldehyde production to a small extent, but the concentration of ethanol required for this effect was critical and higher concentrations decreased malonaldehyde production. The antioxidant effect was further studied by Di Luzio (1968), who found that lipid-soluble antioxidants were more effective than the water-soluble antioxidants. Further, although antioxidants prevented the increased liver triglyceride they had no effect on *blood ethanol concentrations* and therefore presumably did not increase the rate of ethanol metabolism. This was in agreement with an earlier finding of Di Luzio (1964) that an antioxidant mixture given orally had no effect on the *clinical signs* of ethanol intoxication although protecting against the increased liver triglyceride. The conclusion may thus be reached that the antioxidants were reacting with some inter-hepatic material produced as a result of ethanol metabolism, or by correcting some metabolic disturbance (e.g. the altered NAD^+/NADH ratio).

The role of *lipid peroxidation* in the ethanol-induced fatty liver has been disputed by Bunyan *et al.* (1969) and by Hashimoto and Recknagel (1968), who found no evidence for an increased **diene** conjugation in microsomal, mitochondrial or nuclear fractions after oral administration of ethanol. Further, these workers found no increase in *lipid peroxides* in lipid extracts of microsomes, mitochondria or whole liver suspensions obtained 1 or 18 h after ethanol administration; their interpretation of the diene-conjugation data has been disputed in turn by Di Luzio (1968).

Table 11.7. Effect of administration of ethanol (6 g/kg body wt.) on liver triglyceride and peroxide contents. The Table also includes the protective action of two intravenous doses of ubiquinone-4 (500 mg/kg body wt.) given to the female rats 24 h previously and also at the time of administering the ethanol. Data from Kalish and Di Luzio (1966). The peroxide content of lipid extracts was measured by an iodometric method.

Treatment	Liver triglyceride (mg/g)	Liver peroxide (μequiv./g of liver lipid)
Control	16 ± 2	34 ± 2
Ubiquinone-4 + glucose	12 ± 2	28 ± 3
Ethanol	36 ± 4	58 ± 2
Ethanol + ubiquinone-4	21 ± 3	24 ± 3

The relationships of free-radical scavengers, ethanol metabolism and triglyceride accumulation has been made considerably more complex by the observations of Vincenzi *et al.* (1967) and Wooles (1968). The former group of investigators found that *phenobarbital* given at −50, −26 and −2 h before oral ethanol and SKF 525A given at −2 h were both partially effective in decreasing the expected rise in liver triglyceride measured 16 h after giving ethanol. Pretreatment with phenobarbital is known to increase liver drug-metabolising enzyme activity and SKF 525A is known to be a powerful blocking agent of drug metabolism during the first few hours after administration, although later a stimulation of enzyme activity has been observed. Wooles (1968) found that *promethazine* given at −6 and −2 h before ethanol dosing was partially effective in preventing the increased liver triglyceride: a 40% protection was observed under these conditions. How is it possible to interpret these data that appear so convincing and yet are apparently so contradictory?

First, the results with promethazine are to be expected in view of its powerful free-radical scavenging action, discussed extensively in Chapter 10. Promethazine, however, has a relatively short half-life in the rat and, in general, a second dose of promethazine is required after administration of a toxic agent for suppression of an effect *in vivo* for longer than about 8 h. Repeated prior injections of promethazine are known to increase the rate of metabolism *in vivo* of a dose given subsequently; under these conditions the half-life of promethazine is considerably decreased. Such behaviour probably accounts for the finding of Wooles that pretreatment with promethazine had no effect on accumulation of liver triglyceride measured 18 h after the last dose of promethazine. It might be expected that pretreatment with phenobarbital would give a similar result in terms of accumulation of triglyceride. However, as mentioned above, phenobarbital pretreatment is quite effective in attenuating the ethanol-induced fatty liver. A possible explanation of this concerns the known stimulatory action of phenobarbital on the microsome-catalysed oxidation of ethanol (Roach *et al.*, 1969). Pretreatment with phenobarbital may thus alleviate the accumulation of triglyceride by increasing the metabolic removal of ethanol by some microsomal-linked route. Whether promethazine also stimulates the microsomal-catalysed oxidation of ethanol is not known. The action of SKF 525A on the ethanol-induced fatty liver is likewise clouded with uncertainty. It is known that the microsome-catalysed oxidation of ethanol requires NADPH but not cytochrome P_{450}. SKF 525A may act therefore by blocking the normal NADPH–cytochrome P_{450} chain and making available more NADPH for ethanol oxidation.

The evidence presented above shows that free-radical scavengers may be considerably effective in attenuating the increased triglyceride content in liver resulting from ethanol administration. The question naturally arises: what mechanism is responsible for the protective action of free-radical scavengers on the increased liver triglyceride?

Ethanol itself may enter free-radical reactions relatively easily (see Schönberg, 1968). The formation of the ethoxy radical ($C_2H_5O^\bullet$) by interaction with some endogenous radical $R^\bullet$ introduces a strongly reducing component into the system, which, in the presence of a suitable oxidant, is converted into acetaldehyde [Reaction (11.7)].

$$R^\bullet + C_2H_5OH \longrightarrow RH + C_2H_5O^\bullet \tag{11.6}$$

$$C_2H_5O^\bullet + X \longrightarrow C_2H_4O + XH^\bullet \tag{11.7}$$

An illustration of the free-radical behaviour of ethanol is the photochemical reaction between ethanol and bromotrichloromethane which results in the appearance of chloroform, bromine and other products.

The NADPH–cytochrome P_{450} chain is known to involve several endogenous radical species (see Chapter 10), and an interaction between one of these and ethanol could result in the homolytic cleavage of the alcohol with production of the reducing ethoxyl radical. The production of radicals from ethanol in the endoplasmic reticulum might result in several deleterious reactions that affect triglyceride metabolism. For example, there might be an inhibition of lipoprotein excretion, as found in intoxication with carbon tetrachloride, or damage to the fatty acid ω-oxidation system that involves cytochrome P_{450}.

In the author's opinion the results with antioxidants and the attractive hypothesis of Krebs appear at present to be incompatible. What is needed to clarify the situation is the answer to such questions as: (a) does antioxidant treatment alter the ratio $NAD^+/NADH$ in a way more favourable to α-glycerophosphate dehydrogenase? (b) if the antioxidant alters the ratio $NAD^+/NADH$ what is the mechanism? Is (b) a direct reaction between antioxidant and an NADH-dependent system or is there some more complex reaction involved? If the antioxidant treatment does not restore the $NAD^+/NADH$ ratio to a more normal value in the presence of ethanol then obviously Krebs' hypothesis concerning the mechanism responsible for the increased triglyceride will need reassessing. The answers to such questions are of great interest and importance to our understanding, not only of the pathogenesis of liver damage in acute alcoholic intoxication in the rat but also to human liver disease in chronic alcoholism.

11.9 Chronic ethanol intoxication

"Alcohol consumption may seem to provide a pleasant warmth but if alcoholism develops this is like a slow fire which tortures and eventually destroys its victim" (Williams, 1966).

The problems associated with chronic alcoholism will not be dealt with here in other than a cursory manner. The experimental model is more complex than with acute ethanol intoxication, and the data from human clinical studies are voluminous and relatively unhelpful so far as the mechanisms of injury are concerned.

It has long been believed that there is a connection between a chronically heavy intake of ethanol, fatty degeneration and liver cirrhosis. Such a suggestion is given qualitative significance by the studies of Lelbach (1968), which show that a high incidence of cirrhosis can be related both to a high intake of alcohol and to the length of time over which a patient has continually ingested large quantities of alcohol.

For experimental models of chronic alcoholism it is difficult to choose the most suitable diet and many variants have been tried. For instance, if a high intake of ethanol is to be given over a prolonged period, should the controls receive an isocaloric amount of (for example) glucose? Should the calories from the administered ethanol be additional to the normal calories consumed or should the normal diet be adjusted so that together with ethanol the total calories given are unchanged? These are questions that have provoked much argument and have no clear and unequivocal answers.

However, to summarise very briefly some of the experimental data that have appeared, it has been found that rats accumulate fat in the liver only when the calorie intake from alcohol is more than 35% of the total calorie intake. The effects of antioxidants on the increased liver triglyceride in chronic alcoholism have been studied and the results appear to depend on the proportion of dietary calories supplied by fat. For example with high fat content little protection was obtained with antioxidant treatment but considerable protection has been found when the fat intake was decreased (Di Luzio and Hartman, 1967).

The relationship between ethanol intake and **cirrhosis** is at present too obscure to warrant much discussion. Not only do we know very little about the aetiology of cirrhosis in general but better experimental models for studying this aetiology are available than chronic alcohol intoxication in experimental animals[(9)]. Liver disease in chronic alcoholism is of course further complicated by the frequent ingestion of minor components of the alcoholic beverages used, for example higher alcohols, aldehydes and metals, which in small amounts may be quite innocuous but when taken in large amounts for prolonged periods may well be hepatotoxins of considerable importance. McDonald's work on the intake of iron in alcoholic beverages and the incidence of **haemochromatosis** is an example (see Chapter 15). Perhaps relevant to the long-term action of ethanol on the liver and the association of chronic alcoholism with cirrhosis is the finding that ethanol *decreases* bile flow in isolated perfused rat liver preparations (T. F. Slater and M. N. Eakins, unpublished work); similar

[(9)] Porta *et al.* (1969) have recently succeeded in producing cirrhosis in rats by using a liquid intake containing 32% (*w*/*v*) alcohol and 25% (*w*/*v*) sucrose; the solid diet administered was relatively deficient in lipotropic factors (less than 17 mg of choline/100 kcal). The alcohol intake accounted for approximately 37% of the total calorie intake. A review of the recent work concerning the effects of alcohol on the liver is provided by Porta *et al.* (1970).

results have been reported by Kotelanski *et al.* (1969). An interesting study of chronic alcoholism in terms of an impaired and inadequate glucose metabolism of the brain has been developed by Williams and co-workers (Roach and Williams, 1966). The studies have indicated that the addiction to alcohol can be overcome by substituting an alternative metabolite that can cross the blood–brain barrier and can then supply the brain with metabolic energy; glutamine has been tried with some success in this respect[10].

11.10 Alcohols and necrosis

So far in this Chapter the acute effects of ethanol in liver have been considered solely in terms of its action in increasing liver triglyceride and indeed this is the almost universally observed effect with little or no associated necrosis. A most interesting extension of alcohol-induced liver injury to include the production of necrosis has been made by Lelbach (1969). Since rat liver can metabolise ethanol very rapidly, Lelbach argued that ethanol, even when ingested in massive amounts per day, probably never reached a concentration sufficient to cause a more severe cell damage than triglyceride accumulation. He therefore pretreated rats with an inhibitor of alcohol dehydrogenase (1,2-pyrazole) and found that blood concentrations of ethanol were very much higher than in the rats not treated with 1,2-pyrazole. The combination of daily ethanol (3–4 g/day) with 1,2-pyrazole (31 mg/kg body wt.) proved fatal to male rats in 5–19 days and the liver was found at autopsy to have extensive *centrilobular necrosis*[11]. These studies have been extended by Lester and Benson (1970) who have shown that pyrazole also blocks the oxidation of methanol, propanol and other higher alcohols. Additionally a number of oximes and amides were also found to be effective inhibitors of ethanol oxidation in rats. Blomstrand and Theorell (1970) have shown that 4-methylpyrazole is an effective inhibitor of ethanol oxidation in man.

Liver necrosis is well known to follow the ingestion of allyl alcohol ($CH_2{=}CH{-}CH_2OH$) and this has been demonstrated by Rees and Tarlow (1967) to result from oxidation of the alcohol by alcohol dehydrogenase to **acrolein** ($CH_2{=}CH{-}CHO$). The necrotic lesions are mainly located around the portal tracts where alcohol dehydrogenase is concentrated (see Section 11.3.1). Schwarzmann *et al.* (1967) have demonstrated that administration of ethanol decreases the toxicity of allyl alcohol in the liver, presumably by competing for the dehydrogenase site.

[10] Davis and Walsh (1970) have suggested a biochemical basis for ethanol addiction. They suggest that acetaldehyde formed by the oxidation of ethanol augments the formation of the *alkaloid* tetrahydropapaveroline, itself derived from dopamine.

[11] For dietary conditions that predispose to liver necrosis after a single dose of ethanol see Takeuchi *et al.* (1969).

It has already been mentioned that liver alcohol dehydrogenase reacts only slowly with methanol[12], which is, however, metabolised by the catalase-dependent pathway described in Section 11.3.2. The product of both routes is formaldehyde. It is known that the retina contains a very active alcohol dehydrogenase but studies on the catalase pathway in retina have not been reported. In primates, methanol is extremely toxic to the retina *in vivo* and this is generally agreed to be due to the production of formaldehyde. Which pathway of methanol oxidation is used is not clear, however. Cooper and Marchesi (1959) have shown with retina that the glycolytic enzyme hexokinase is especially sensitive to formaldehyde *in vitro*; an 84% inhibition is obtained with a concentration of 0·01 M-formaldehyde. Since the retina has the highest aerobic glycolytic rate among mammalian tissues it is possible that inhibition of the initial enzyme (hexokinase) in the sequence by the product of methanol oxidation (formaldehyde) is responsible for the severe ocular toxicity that occurs.

References

Baker, G., 1767, *An Essay Concerning the Cause of the Endemial Colic of Devonshire* (Private publication, London).
Baxter, R. C., Hensley, W. J., 1969, *Biochem. Pharmac.*, **18**, 233.
Blomstrand, R., Theorell, H., 1970, *Life Sci.*, **9** (II), 631.
Bücher, T., Klingenberg, M., 1958, *Angew. Chem.*, **70**, 552.
Bunyan, J., Cawthorne, M. A., Diplock, A. T., Green, J., 1969, *Br. J. Nutr.*, **23**, 309.
Burch, H. B., Max, P., Chyu, K., Lowry, O. H., 1969, *Biochem. biophys. Res. Commun.*, **34**, 619.
Comporti, M., Hartman, A., Di Luzio, N. R., 1967, *Lab. Invest.*, **16**, 616.
Conney, A. H., 1967, *Pharmac. Rev.*, **19**, 317.
Cooper, J. R., Marchesi, V. T., 1959, *Biochem. Pharmac.*, **2**, 313.
Davis, V. E., Walsh, M. J., 1970, *Science*, **167**, 1005.
De Duve, C., Baudhuin, P., 1966, *Physiol. Rev.*, **46**, 323.
Delaney, V. B., Slater, T. F., 1970, *Biochem. J.*, **116**, 299.
Dietrich, R. A., 1966, *Biochem. Pharmac.*, **15**, 1911.
Di Luzio, N. R., 1958, *Am. J. Physiol.*, **194**, 453.
Di Luzio, N. R., 1963, *Physiologist*, **6**, 169.
Di Luzio, N. R., 1964, *Life Sci.*, **3**, 113.
Di Luzio, N. R., 1968, *Expl. molec. Path.*, **8**, 394.
Di Luzio, N. R., Costales, F., 1965, *Expl. molec. Path.*, **4**, 141.
Di Luzio, N. R., Hartman, A. D., 1967, *Fedn Proc. Fedn Am. Socs exp. Biol.*, **26**, 1436.
Fazekas, I. G., Rengei, B., 1969, *Enzymologia*, **36**, 59.
Feinstein, R. N., Hampton, M., Cotter, G. J., 1953, *Enzymologia*, **16**, 219.
Forsander, O., Räihä, N., Suomalainen, H., 1958, *Hoppe-Seyler's Z. physiol. Chem.*, **312**, 243.
Glock, G. E., McLean, P., 1956, *Expl. Cell Res.*, **11**, 234.
Goodman, L. S., Gilman, A., 1955, in *The Pharmacological Basis of Therapeutics*, second edition (Macmillan, New York), p.103.
Gordon, E. R., 1968, *Canad. J. Physiol. Pharmac.*, **46**, 609.

[12] Methanol oxidation in monkey liver utilises mainly the alcohol dehydrogenase route, in contrast to the peroxidative system that is of major importance in the rat (see Makar *et al.*, 1968).

Greenberger, N. J., Cohen, R. B., Isselbacher, K. J., 1965, *Lab. Invest.*, **14**, 264.
Haggard, H. W., Greenberg, L. A., Lolli, G., 1941, *Q. Jl. Stud. Alcohol*, **1**, 684.
Hashimoto, S., Recknagel, R. O., 1968, *Expl. molec. Path.*, **8**, 225.
Horning, M. G., Williams, E. A., Maling, H. M., Brodie, B. B., 1960, *Biochem. biophys. Res. Commun.*, **3**, 635.
Hutchison, R., 1969, *Food and the Principles of Nutrition*, twelfth edition, revised by H. M. Sinclair and D. F. Hollingsworth (Arnold, London).
Isselbacher, K. J., Carter, E. A., 1970, *Biochem. biophys. Res. Commun.*, **39**, 530.
Isselbacher, K. J., Greenberger, N. J., 1964, *New Engl. J. Med.*, **270**, 351, 402.
Jellinek, E. M., 1960 *The Disease Concept of Alcoholism* (College and University Press, New Haven, Conn.).
Jouany, J. M., 1965, *Annls. Biol. clin.*, **23**, 1115.
Kalish, G. H., Di Luzio, N. R., 1966, *Science*, **152**, 1390.
Khanna, J. M., Kalant, H., 1970, *Biochem. Pharmac.*, **19**, 2033.
Khanna, J. M., Kalant, H., Lin, G., 1970, *Biochem. Pharmac.*, **19**, 2493.
Kotelanski, B., Groszmann, R. J., Kendlar, J., Zimmerman, H. J., 1969, *Proc. Soc. exp. Biol. Med.*, **132**, 715.
Krebs, H. A., 1968, in *Stoffwechsel der Isoliert perfundierten Leber*, Eds. W. Staib, R. Scholz (Springer-Verlag, Berlin), p.216.
Krebs, H. A., Freedland, R. A., Hems, R., Stubbs, M., 1969, *Biochem. J.*, **112**, 117.
Kuenzle, C. C., 1970, *Biochem. J.*, **119**, 411.
Lelbach, W. K., 1968, *Germ. med. Mon.*, **13**, 31.
Lelbach, W. K., 1969, *Experientia*, **25**, 816.
Lester, D., Benson, G. D., 1970, *Science*, **169**, 282.
Lester, D., Keokosky, W. A., Felzenberg, F., 1968, *Q. Jl. Stud. Alcohol*, **29**, 449.
Lieber, C. S., De Carli, L. M., 1968, *Science*, **162**, 917.
Lieber, C. S., De Carli, L. M., 1970a, *J. Biol. Chem.*, **245**, 2505.
Lieber, C. S., De Carli, L. M., 1970b, *Science*, **170**, 78.
Lieber, C. S., De Carli, L. M., 1970c, *Life Sci.*, **9** (II), 267.
Lieber, C. S., Rubin, E., De Carli, L. M., 1970, *Biochem. biophys. Res. Commun.*, **40**, 858.
Lieber, C. S., Schmid, R., 1961, *J. clin. Invest.*, **40**, 394.
Mäenpää, P. H., Raivio, K. O., Kekomäki, M. P., 1968, *Science*, **161**, 1253.
Makar, A. B., Mannering, G. J., 1968, *Mol. Pharmac.*, **4**, 484.
Makar, A. B., Tephly, T. R., Mannering, G. J., 1968, *Mol. Pharmac.*, **4**, 471.
Mallov, S., Bloch, J. L., 1956, *Am. J. Physiol.*, **184**, 29.
Marcet, W., 1860, *On Chronic Alcoholic Intoxication* (Churchill, London).
McManus, I. R., Brotsky, E., Olson, R. E., 1966, *Biochem. biophys. Acta*, **121**, 167.
Orme-Johnson, W. H., Ziegler, D. M., 1965, *Biochem. biophys. Res. Commun.*, **21**, 78.
Porta, E. A., Hartcroft, W. S., 1965, in *Therapeutic Agents and the Liver*, Eds. N. McIntyre, S. Sherlock (Blackwell, Oxford), p.145.
Porta, E. A., Koch, O. R., Hartcroft, W. S., 1969, *Lab. Invest.*, **20**, 562.
Porta, E. A., Koch, O. R., Hartcroft, W. S., 1970, *Expl. molec. Path.*, **12**, 104.
Raivio, K. O., Kekomäki, M. P., Mäenpää, P. H., 1969, *Biochem. Pharmac.*, **18**, 2615.
Reboucas, G., Isselbacher, K. J., 1961, *J. clin. Invest.*, **40**, 1355.
Rees, K. R., Tarlow, M. J., 1967, *Biochem. J.*, **104**, 757.
Roach, M. K., Reese, W. N., Creaven, P. J., 1969, *Biochem. biophys. Res. Commun.*, **36**, 596.
Roach, M. K., Williams, R. J., 1966, *Proc. natn. Acad. Sci. U.S.A.*, **56**, 566.
Rubin, E., Hutterer, F., Lieber, C. S., 1968, *Science*, **159**, 1469.
Rubin, E., Lieber, C. S., 1968a, *New Engl. J. Med.*, **278**, 869.
Rubin, E., Lieber, C. S., 1968b, *Science*, **162**, 690.

Schönberg, A., 1968, *Preparative Organic Photochemistry* (Springer-Verlag, Berlin).
Schwarzmann, V., Infante, R., Raisonnier, A., Caroli, J., 1967, *C. r. Séanc. Soc. Biol.*, **161**, 2425.
Slater, T. F., 1967, *Biochem. J.*, **104**, 833.
Slater, T. F., Sawyer, B. C., Strauli, U. D., 1964, *Biochem. J.*, **93**, 267.
Smith, M. E., Newman, H. W., 1959, *J. biol. Chem.*, **234**, 1544.
Takeuchi, J., Takada, A., Kanayama, R., Ohata, N., Okumura, Y., 1969, *Lab. Invest.*, **21**, 398.
Tephly, T. R., Tinelli, F., Watkins, W. D., 1969, *Science*, **166**, 627.
Theorell, H., 1967, *Harvey Lect.*, **61**, 17.
Theorell, H., Ehrenberg, A., de Zolenski, C., 1967, *Biochem. biophys. Res. Commun.*, **27**, 309.
Tygstrup, N., Winkler, K., Lundquist, F., 1965, *J. clin. Invest.*, **44**, 817.
Veech, R., Eggleston, L. V., Krebs, H. A., 1969, *Biochem. J.*, **115**, 609.
Vincenzi, L., Meldolesi, J., Morini, M. T., Bassan, P., 1967, *Biochem. Pharmac.*, **16**, 2431.
Waltman, R., Bonura, F., Nigrin, G., Pipat, C., 1969, *Lancet*, ii, 108.
Wartburg, J. P., Röthlisberger, M. von, 1961, *Helv. physiol. pharmac. Acta*, **19**, 30.
Williams, R. J., 1959, *Alcoholism: The Nutritional Approach* (University of Texas Press, Texas).
Williams, R. J., 1966, Speech at *12th Int. Institute on Prevention and Treatment of Alcoholism, Prague* (see also *Proc. natn. Acad. Sci. Wash.*, **56**, 566).
Williamson, D. H., Lund, P., Krebs, H. A., 1967, *Biochem. J.*, **103**, 514.
Wooles, W. R., 1966, *Life Sci.*, **5**, 267.
Wooles, W. R., 1968, *Toxic. appl. Pharmac.*, **12**, 186.

12

Free-radical damage to biological membrane systems

12.1 Introduction

The cell membrane delineates the living cell: it separates the highly ordered functional activities of metabolism from the relative disorder of the external environment. In so doing it has need of specialised properties; among these is a necessary resistance to injurious attacks by environmental agents with which it may come into contact. Another important requirement in differentiated multicellular organisms is the mechanism by which cells recognise one another. This property, which is basic to immune reactions, to organogenesis and to cancer probably involves specific receptor sites on the exterior cell membrane. Neville (1968), for example, has isolated a tissue-specific protein from liver cell membranes that was not present in a poorly differentiated liver tumour.

Within the cell are a number of distinctive structures, each bounded by membranes of broadly similar composition to the exterior membrane. The occurrence of such intracellular membranes enables particular metabolic activities to be compartmented away from others that might compete for a common substrate, or which might even degrade the enzymes themselves. An obvious example is the localisation of acid hydrolases within the confines of the lysosomal fraction where digestion of endogenous cell material or ingested exogenous food or other material can occur without derangement of other aspects of the cell's behaviour. Numerous similar examples could be given and the interested reader is referred to reviews on intracellular localisation by De Duve (1964), De Duve and Baudhuin (1966), De Duve and Wattiaux (1966), and Lehninger (1964). An indication of the wide variety of membrane-limited structures within a mammalian cell has already been given in Figure 1.1. Such pictures should not mislead one into thinking of the membrane systems as essentially static systems. The time-lapse cinematographic work of Frederic and Chèvremont (1952) has shown clearly that the mitochondria, for example, undergo rapid changes in shape and size in living cells. More recent work on phagocytosis illustrates how dynamic is the interaction between particular membrane-limited particles, such as the fusion of primary lysosomes with a phagosome.

All of the membranes mentioned above in relation to intracellular structures are lipoprotein in character and often have haem components or flavin groups built into the basic structure. The lipid component invariably contains unsaturated fatty acid residues. Such membranes are bathed in fluid that contains oxygen and trace metals. In these circumstances it might seem unlikely that the membranes would survive at all under such favourable conditions for peroxidation. Of course, protective mechanisms are operative; for example, there is a relative abundance of radical scavengers such as glutathione, vitamin E etc. (see

Chapter 5). Even so it may be that membrane material turns over rather quickly through oxidative degradation, with a subsequent rapid enzymic removal of the toxic lipid peroxides so produced, perhaps by the mechanisms discussed in Section 4.2.3.

Before examples of peroxidative attack on biological membranes are given it will be necessary to discuss the lipid composition of such membranes in some detail. This Chapter will consider membrane damage that arises exclusively from the initiation of lipid peroxidation. As a consequence the major component of the membrane that is of direct relevance is the unsaturated fatty acid fraction.

12.2 Membrane composition

The membranes that are dealt with in later Sections of this chapter include the erythrocyte membrane, the liver cell membrane and membranes of liver mitochondrial, lysosomal and microsomal fractions. The discussion on membrane composition will therefore be limited to these types of membranes.

12.2.1 Erythrocyte membrane composition

The membrane of the lysed erythrocyte ('erythrocyte ghost') has been extensively studied by a number of investigators; the results are summarised by Maddy (1966) and Korn (1969). The 'ghost membrane' has a rather high protein/lipid ratio and a very high (cholesterol/polar lipid) ratio in relation to the other membranes considered here. The latter ratio, for example, is 0·9–1·0 in erythrocyte ghosts compared with 0·3–0·5 in liver cell membranes and 0·03–0·08 in microsomal membranes. The major polar lipids reported (Maddy, 1966) for the ghost membrane are: sphingomyelin, phosphatidylethanolamine, phosphatidylcholine and phosphatidylserine. The major *fatty acids* of the erythrocyte membrane lipid are summarised by Maddy (1966). The percentage composition of the fatty acid content of human erythrocyte membranes was found to be: palmitic, 31; stearic, 17–24; oleic, 16; linoleic, 6–10; linolenic, trace; arachidonic, 8–10. It can be seen that the membrane contains appreciable quantities of poly-unsaturated fatty acids, in particular arachidonic acid, potentially capable of peroxidative degradation.

12.2.2 Liver cell membrane composition

The detailed lipid composition of the rat liver cell membrane has been given by Skipski *et al.* (1965). In percentages of total extractable lipid material the composition of the membrane was: cholesterol, 16; total phospholipids, 39; total neutral lipid, 35; free fatty acids, 7; phosphatidylcholine, 14; sphingomyelin, 7; phosphatidylethanolamine, 6; total lipid not accounted for by known components, 26. Although the membrane lipids were relatively enriched in saturated fatty acids in comparison with the whole liver lipid fraction, there were substantial amounts of 18:1 (oleic), 18:2 (linoleic) and 20:4 (arachidonic) acids present in the cell membrane lipids.

12.2.3 Mitochondrial and microsomal membrane composition

The lipid composition of these membranes has been described by Getz *et al.* (1962). The major polar lipids in the mitochondrial are diphosphatidylglycerol, phosphatidylcholine, phosphatidylethanolamine, plasmalogen; in microsomal membranes they are phosphatidylcholine, phosphatidylethanolamine, sphingomyelin. Most of the **cardiolipin** in liver is found in the mitochondrial membranes.

The major fatty acids found in mitochondrial membrane lipid in percentages of total extracted lipid were: palmitic, 16; stearic, 15; oleic, 12; linoleic, 23; arachidonic, 17. The corresponding values for microsomes were 21, 18, 13, 17 and 17.

May *et al.* (1965) have incubated rat liver microsome fractions with and without ADP and a source of NADPH, and have analysed the resultant changes in lipid content. In the presence of ADP and NADPH there was an accelerated production of malonaldehyde (see Section 4.2.2) and a marked loss of phosphatidylethanolamine and double bonds in fatty acids. The loss of unsaturated linkages was found to be due mainly to oxidation of arachidonic acid and docosahexaenoic acid.

12.2.4 Lysosomal membrane composition

The composition of lysosome membranes has been reviewed by Tappel (1969) and Lucy (1969). The study is complicated by the heterogeneity of this fraction, which includes cytolysomes containing inclusions of

$CH_3(CH_2)_4.CH{=}CH.CH_2CH{=}CH.CH_2.CH{=}CH.CH_2CH{=}CH(CH_2)_3.COOH$

14 11 8 5

Arachidonic acid ($C_{20}H_{32}O_2$)

$CH_3(CH_2)_4.CH{=}CH.CH_2.CH{=}CH(CH_2)_7.COOH$

12 9

Linoleic acid ($C_{18}H_{32}O_2$)

$CH_3{-}CH_2{-}CH{=}CH{-}CH_2{-}CH{=}CH{-}CH_2{-}CH{=}CH{-}CH_2{-}CH{=}CH{-}CH_2{-}CH{=}CH{-}CH_2{-}CH{=}CH{-}CH_2{-}CH_2{-}COOH$

(double bonds numbered 19, 16, 13, 10, 7, 4; chain printed vertically from CH = 7 CH – CH_2 – CH = 4 CH – CH_2 – CH_2 – COOH)

4,7,10,13,16,19-Docosahexaenoic acid ($C_{22}H_{32}O_2$)

Figure 12.1. Structures of arachidonic acid, linoleic acid and docosahexaenoic acid.

mitochondria and endoplasmic reticulum. The phospholipid fraction extracted from lysosomes showed no qualitative differences compared with mitochondrial and microsomal phospholipids. The percentage fatty acid composition was: palmitic, 18; stearic, 22; oleic, 8; linoleic, 22; arachidonic, 23.

The discussion outlined above has shown that in each class of membranes considered there is a considerable content of unsaturated fatty acids that may undergo peroxidative change under suitable conditions. The results of May *et al.* (1965) have indicated that the largest changes can be expected to occur in arachidonic acid (20:4) and docosahexaenoic (22:6) acid (for structures see Figure 12.1). In the following Sections a number of examples will be considered where peroxidative attack on membrane-limited systems is associated with severe malfunction of the cells or cell organelles concerned.

For general surveys of membrane function and structure the reader is referred to Dowben (1969), Ling (1969), Lucy (1969) and Weiss (1969).

12.3 Changes in the erythrocyte membrane

Procedures that stimulate lipid peroxidation can result in an increased haemolysis of erythrocytes *in vitro* and, under certain conditions, *in vivo*. It can be easily demonstrated that direct exposure to a source of free radicals (e.g. ferrous ions + vitamin C, or ionising radiation) leads to the rapid breakdown of the erythrocyte membrane *in vitro*. Another example that has been well studied is the haemolysis *in vitro* by dialuric acid (hydroxybarbituric acid) in vitamin E deficiency (Tsen and Collier, 1960; Bunyan *et al.*, 1961). Bunyan and colleagues showed that the haemolysis produced by dialuric acid was proportional to the extent of lipid peroxidation and that agents known to act as antioxidants (for example α-tocopherol, butylated hydroxytoluene and thyroxine) prevented haemolysis.

The question whether lipid peroxidation produces haemolysis under conditions *in vivo* has been considered particularly by Mengel and colleagues. The question has become specially relevant in view of the increasing uses of *hyperoxia* in various clinical treatments. For example, oxygen partial pressures greater than atmospheric are used in certain radiotherapeutic procedures, in flushing out wounds suspected of harbouring *Clostridium welchii*, in treating arterial occlusion and pure oxygen at low pressure is used in space capsules. Mengel and colleagues have considered the possible complications of lipid peroxidation in such clinical applications of hyperoxia since lipid peroxidation is stimulated by increasing the oxygen partial pressure[1]. Mengel and Kann (1966) subjected mice, that were

[1] In view of such results it may be clinically valuable to administer an antioxidant drug (for example intravenous vitamin E) to patients receiving hyperoxygenation as therapy for hypoxic shock. The variable interval of hypoxia preceding treatment may render the erythrocytes more susceptible to subsequent hyperoxia given as treatment.

either on a normal diet or on a diet deficient in vitamin E, to oxygen at high pressure. They found that extensive lipid peroxidation (as measured by the thiobarbituric acid reaction of isolated washed erythrocytes) and haemolysis *in vivo* occurred only in association with vitamin E deficiency. Oxygen at high pressure under conditions *in vitro* was shown to give similar results to that produced by the addition of hydrogen peroxide or by prolonged irradiation of erythrocytes with ultraviolet light. No significant changes were observed in erythrocyte glutathione concentrations or in the activity of glucose 6-phosphate dehydrogenase under these conditions.

Merriwether and Mengel (1967) have examined the problem of **paroxysmal nocturnal haemoglobinuria**. In this syndrome the patients' erythrocytes are characteristically more fragile during periods of sleep so that the morning urine samples are distinctly red in colour. As the name implies, attacks are periodic rather than continuous in nature. Erythrocytes from patients with this condition are particularly liable to haemolysis by agents that induce lipid peroxidation and, under standard conditions of peroxidation, lipid obtained from their erythrocytes produced more peroxide than lipid obtained from normal erythrocytes. Mengel and colleagues noted that after incubating *normal* erythrocytes with glutathione there occurred changes in the erythrocyte membrane so that they then resembled erythrocytes from the patients with paroxysmal nocturnal haemoglobinuria. For example, after ultraviolet irradiation lipid extracted from normal erythrocytes produced 87 nmol of malonaldehyde; the corresponding value obtained from erythrocytes taken from the patients was 126; normal erythrocytes incubated with glutathione gave a value of 145. The patients had normal serum vitamin E concentrations and normal erythrocyte glucose 6-phosphate dehydrogenase, catalase and glutathione peroxidase activities. Mengel and co-workers suggest that the tendency to an increased lipid peroxidation is a basic *intracorpuscular* defect in this disease.

In an interesting report Goldstein and Balchum (1967) have shown that quite low concentrations of **ozone** in air (40 p.p.m.) produced increased lipid peroxidation in a human erythrocyte suspension, and this was associated with an increased osmotic fragility (Table 12.1). The increased peroxidation was detectable within 10 min of exposure to ozone. The authors suggested that as chronic exposure to ozone produces a generalised ageing syndrome (Stokinger, 1965), this may result from a radical-motivated mechanism similar to that discussed in Section 12.6. It may be speculated that some degenerative lung changes (for example emphysema), which are most common in urban surroundings, may be a result of chronic contact with atmospheric free radicals (see Section 13.5) or with agents that initiate free-radical processes in the lungs.

The classical studies on photosensitisation of erythrocytes that led to an appreciation of the role of oxygen in such systems should be mentioned.

Table 12.1. Effect of exposure to ozone (40 p.p.m.; 2 h) on the thiobarbituric acid-reactive material in erythrocyte suspensions *in vitro*. Data from Goldstein and Balchum (1967).

Treatment	Malonaldehyde (nmol/g of haemoglobin)
Non-exposed control	12·2
Exposed to room air	11·8
Exposed to ozone	51·1

If freshly obtained human erythrocytes are suspended in an iso-osmotic buffered solution and irradiated with light of wavelength greater than approximately 350 nm then no haemolysis occurs during the course of several hours. If, however, a small concentration (0·01–0·1 mM) of Eosin, Rose Bengal, Neutral Red or some similar dye is included in the buffer then haemolysis occurs within a few minutes of irradiating the suspension[(2)].

This photosensitised reaction has been found to have all of the properties required for a mechanism based on a free-radical peroxidative attack on membrane lipid. The action spectrum for haemolysis is coincident with the absorption spectrum of the dyestuff present in solution. There is a requirement for oxygen and free-radical scavengers inhibit the reaction[(3)]. There is an appreciable lag phase after commencement of irradiation and the first signs of haemolysis, which then rapidly increases in extent; such a lag phase is typical of chain reactions in general. The light-radiation is only necessary for the initiation process and does not have to be continued throughout the remainder of the experiment, although cutting off the light after the original exposure increases the lag phase. Finally the process is largely independent of temperature; the $Q_{10°}$ is close to unity. Such properties clearly show the homolytic nature of the underlying mechanism.

In addition to the photosensitised haemolysis outlined above, many of the dyes also produce haemolysis in the absence of light when present in much higher concentrations (1–10 mM). This so-called dark reaction, however, involves completely different mechanisms. For example, it is independent of oxygen, dependent on temperature and may well be a property of the surface activity (i.e. detergent-like action) of the dye.

(2) A wide variety of other membrane-bounded structures behave similarly to erythrocytes in this respect. For references see Blum (1964).

(3) An example of a similar type of reaction is the protective action of β-carotene against lethal photosensitisation produced by haematoporphyrin in mice (Mathews, 1964).

12.4 Changes in the mitochondrial membrane

Waldschmidt *et al.* (1968) have shown that freshly prepared rat liver mitochondria contain appreciable numbers of free radicals, as evidenced by electron-spin-resonance signals. The number of radical-centres increased with the age of the parent tissue up to a plateau value of approximately 10^{13} radicals mg^{-1} of mitochondrial protein. Changes in the number of radical-centres were observed when mitochondrial metabolism was disturbed. For example starving the rats for 24 h before isolation of mitochondria *increased* the number of free radicals by 150%, whereas the addition of respiratory inhibitors such as malonate, cyanide or 2,4-dinitrophenol decreased the number of radical-centres.

Many reports have discussed the mechanisms that underly mitochondrial **swelling**. Lehninger (1962) has reviewed the subject in a comprehensive manner. Isolated mitochondria react as osmometers to changes in the ionic strength of the suspending medium. This seems largely due to the freely permeable nature of the outer membrane of the mitochondria to small ions. Some materials, however, cause mitochondrial swelling quite out of proportion to their ionic strength. For example, 1 mM-inorganic phosphate, 10 μM-thyroxine and 10 μM-reduced glutathione all cause rapid and extensive swelling that is dependent on a continued mitochondrial respiration; this swelling is called *high-amplitude swelling*. Inhibition of respiration by cyanide or uncoupling of associated phosphorylation by 2,4-dinitrophenol abolishes the swelling action normally seen with phosphate, thyroxine and glutathione. In addition to high-amplitude swelling, which is dependent on respiration, there is a *low-amplitude swelling* behaviour shown by tightly coupled mitochondria and which appears limited to ionic shifts during oxidative phosphorylation.

It has been shown particularly by the work of Hunter and colleagues (Hunter *et al.*, 1963; Schneider *et al.*, 1964; McKnight and Hunter, 1966) that reduced glutathione or ferrous ions produce a mitochondrial swelling that is associated with an increased lipid peroxidation of the membrane (see Figure 12.2). Such swelling causes irreversible damage to the mitochondria. This increase in lipid peroxidation is strongly inhibited by inorganic phosphate or arsenate and by very low concentrations of thyroxine. Recent work indicates that these inhibitions operate at least partially through a binding of mitochondrial iron. However, the stoicheiometry of the reactions indicates that other (scavenger?) mechanisms operate as well.

Cash *et al.* (1966) have shown that the swelling action of thyroxine is markedly potentiated by low concentrations of calcium; this illustrates the difficulties that abound in this field, for many reagents contain sufficient calcium as impurity to affect the thyroxine stimulation.

Lehninger and Beck (1967) have studied the mechanism of mitochondrial swelling by phosphate, thyroxine and reduced glutathione in considerable detail. They found that swelling induced by glutathione was not reversed by additional ATP, unlike the swelling caused by phosphate and thyroxine. Mitochondrial swelling by glutathione is accompanied by the loss of a number of soluble materials from the mitochondria and Lehninger showed that one of these materials was a protein that could restore the ability of the mitochondria to contract when mixed with ATP. This protein was later shown to be **glutathione peroxidase** (for the role of glutathione peroxidase in the metabolism of fatty acid peroxides see Section 4.2.3).

Coupland *et al.* (1969) have studied the effects of hyperbaric oxygen on mitochondrial function in organ cultures and in suspensions. At 3 atmospheres pressure of oxygen there were marked decreases in the activities of various mitochondrial enzymes and a large increase in malonaldehyde production. Although the production of malonaldehyde (from lipid peroxides) was largely prevented by the addition of **EDTA** to the medium this did *not* affect the decreased enzyme activities.

Mellors and Tappel (1966) have studied the peroxidation of mitochondrial lipid induced by irradiation; both ubiquinol-6 and α-tocopherol were effective as antioxidants. In other peroxidising systems tested at oxygen concentrations equivalent to those present *in vivo* Mellors and Tappel found that ubiquinol-6 was as active as α-tocopherol and was present in

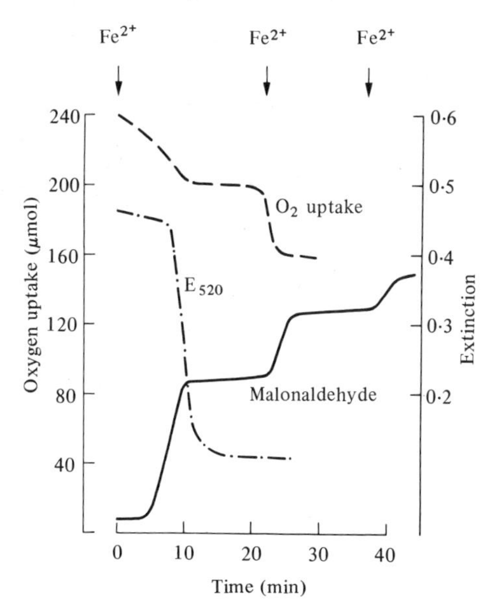

Figure 12.2. Effects of added ferrous ions to a rat liver mitochondrial suspension on production of malonaldehyde, mitochondrial swelling as measured by the extinction of the suspension at 520 nm, and on oxygen uptake. Data of Schneider *et al.* (1964). The mitochondria were suspended in potassium chloride-tris buffer, pH 7·4, at a concentration of approximately 150 μg of protein ml^{-1}; ferrous ammonium sulphate (20 μM) was added as indicated by the arrows.

the mitochondria in about 10 times the concentration. In the report of Cash *et al.* (1966) it was shown that thyroxine was also about as active as α-tocopherol on a molar basis as an antioxidant in the iron-induced lipid-peroxidation reaction in liver mitochondria. The antioxidant activity of thyroxine *in vitro* could be observed at 0·5 μM, which is about 10 times the estimate of its normal concentration in body water. Increases of concentrations to 100–1000-fold are required to produce swelling of mitochondria.

Placer *et al.* (1966) have distinguished between non-enzymic and enzymic lipid peroxidation in mitochondrial suspensions. Non-enzymic lipid peroxidation is induced by ferrous iron, glutathione or vitamin C and involves the formation of $OH^\bullet$, $FeO_2^\bullet$ or $HO_2^\bullet$ radicals. *Enzymic* lipid peroxidation requires a reducing source such as NADPH and is blocked in mitochondria by cyanide or azide. *Partial* enzymic lipid peroxidation proceeds via a glutathione peroxidase route, which yields reduced glutathione, which then results in a non-enzymic type of lipid peroxidation.

12.5 Changes in the endoplasmic reticulum membrane

We have already seen that the membranes of the endoplasmic reticulum are rich in $C_{20:4}$ and $C_{22:6}$ unsaturated fatty acids and that there are several endogenous radical sources in rat liver endoplasmic reticulum. Since these membranes are bathed in an oxygen-rich cell sap (partial pressure of oxygen approximately 16 Torr) it may be expected that peroxidation processes can readily occur. In fact it is well known that incubation of liver, brain or kidney microsomes in buffer at approximately neutral pH produces lipid peroxides with an associated formation of malonaldehyde and the uptake of oxygen. An indication of the rate of this reaction at 0°C and at 37°C is given in Figure 12.3.

Lipid peroxidation in washed microsomes can be greatly stimulated by the addition of a number of substances. Gillette *et al.* (1957) found that the NADPH mixed-function oxidase in *rabbit* liver microsomes forms lipid peroxide when incubated in the absence of drug substrate. In 1963 Beloff-Chain *et al.* found that NADPH oxidation by *rat* liver microsomes was itself greatly stimulated by the addition of ADP. This was closely followed by a report of Hochstein and Ernster (1963) that this oxidation of NADPH stimulated by ADP was linked to the production of lipid peroxide as evidenced by the production of malonaldehyde. These authors showed that the lipid peroxidation was blocked by a metal chelator (EDTA), antioxidants (α-tocopherol), *p*-chloromercuribenzoate and SKF 525A (see Table 12.2). It was soon shown that ferrous iron was necessary for the reaction; this was present in the original solutions as a contaminant of the ADP samples. A survey of other diphosphates confirmed the results of Beloff-Chain *et al.* that other nucleoside diphosphates could replace ADP, and that inorganic pyrophosphate was even more effective over a low concentration range (up to 0·1 μM) but less effective at higher concentrations than observed with ADP. Similar systems were found in rat brain and rat kidney microsomes although drug-metabolising activity and cytochrome P_{450} concentrations in these tissues were much lower than in liver, if present at all. The mechanisms responsible for the increased lipid peroxidation in the presence of ADP and ferrous ions appeared to be associated with the NADPH–cytochrome P_{450} chain, however, since Orrenius *et al.* (1964) reported that the lipid peroxidation was inhibited by two drugs, codeine and aminopyrine, that undergo lipid oxidative demethylation and thus compete with the lipid peroxidation pathway for electron flow from NADPH.

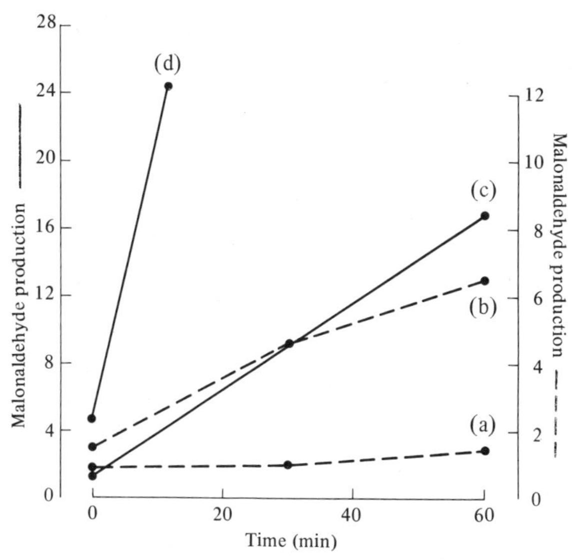

Figure 12.3. Malonaldehyde production by suspensions of rat liver microsomes in 0·15 M potassium chloride incubated for different times at (a) 0°C and (b) 37°C. The production of malonaldehyde by a suspension of microsomes plus cell sap in 0·25 M sucrose at 37°C is shown in (c). The effect of adding a mixture of ferrous ions and ADP (2·6 and 1·6 μmoles respectively) to (c) on the production of malonaldehyde is shown in (d).

Table 12.2. The effects of various drugs *in vitro* on the NADPH–ADP–Fe^{2+} lipid-peroxidation system. Data from Slater (1968) and Hochstein and Ernster (1963).

Drug	Concentration (μM)	Inhibition (%)
Promethazine	25	50
Propyl gallate	5	50
EDTA	10	97
NN'-Diphenyl-*p*-phenylenediamine	0·25	98
SKF 525A	100	44

May *et al.* (1965, 1966) have shown that lipid peroxidation coupled to NADPH oxidation in the presence of ADP and ferrous ions is increased by performing the incubation in atmospheres enriched with oxygen. At 5 atmospheres pressure there were extensive changes in the lipids of the endoplasmic reticulum, as already described. May and McCay (1968) extended these studies and showed that the main fatty acid oxidised was arachidonic acid; the peroxidation was accompanied by an increased polarity of the membrane lipids, possibly due to the formation of hydroxy fatty acids as shown in model systems by Christopherson (1968). The lipid peroxidation was dependent on a continued source of NADPH; the formation of malonaldehyde was correlated with the loss of unsaturated fatty acids and with oxygen uptake as shown in Figure 12.4.

The role of iron has been studied by Marks and Hecker (1967), who showed that with low concentrations of iron the system was enzymic and had a requirement for NADPH, whereas high iron concentrations initiated a non-enzymic reaction.

Slater (1968) has shown that the NADPH–ADP–ferrous iron system has a very low activity in liver microsomal suspensions prepared from newborn rats or from adult chickens, compared with preparations from adult rat liver (see Table 10.4). Dosing adult rats *in vivo* with the free-radical scavenger promethazine caused a marked decrease in adult rat liver NADPH–ADP–ferrous iron lipid-peroxidation activity. Table 12.2 illustrates these results and also shows the effects of various other inhibitors on the NADPH–ADP–ferrous iron system. These studies

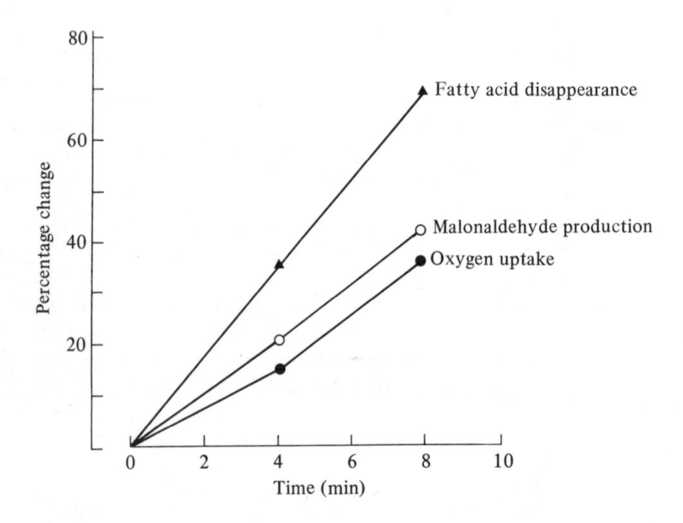

Figure 12.4. Correlation between oxygen uptake, destruction of polyunsaturated fatty acid and malonaldehyde production in microsomal suspensions containing a source of NADPH, ADP and ferric ions. Data of May and McCay (1968).

strengthen the conclusion that the pathway of lipid peroxide formation in the presence of ADP and iron salts is as shown in Figure 10.3. The precise position of coupling at the terminal end of the electron-transport chain is uncertain. Cytochrome P_{450} appears unnecessary since, firstly, NADPH–ADP–ferrous iron lipid peroxidation is demonstrable in brain microsomes where cytochrome P_{450} is not detectable. Secondly, Placer *et al.* (1966) have shown that carbon monoxide, which binds to reduced cytochrome P_{450}, stimulates the lipid peroxidation coupled to NADPH oxidation; in addition some drugs that react with the cytochrome P_{450} end of the electron-transport chain to undergo metabolism (aminopyrine and codeine) inhibited lipid peroxidation. These data suggest that the side-reaction leading to lipid peroxidation in the presence of ADP and ferrous iron diverges at the site where drugs are bound in a suitable conformation relative to cytochrome P_{450}.

It is interesting, in view of the location for the lipid-peroxidation diversion outlined above, that Utley *et al.* (1967) suggested a similar location for the non-enzymic lipid peroxidation that is produced by the addition of ferrous or **mercurous** ions. In their system microsomes in phosphate buffer[4] peroxidised relatively rapidly when ferrous sulphate, mercurous chloride or *p*-chloromercuribenzoate were added; the system was insensitive to heating but was sensitive to carbon monoxide or EDTA. The authors suggest that only lipid around the terminal microsomal Fe_x is peroxidised by mercurous ions whereas ferrous ions peroxidise a wider area of lipid; the situation with mercurous ions is analogous to that described for carbon tetrachloride, where it is suggested that only lipid around the proximal NADPH–flavoprotein is attacked (Section 10.7).

Microsomes also peroxidise when exposed to vitamin C, especially in the presence of added ADP–ferrous ions. This system is essentially insensitive to heat and to *p*-chloromercuribenzoate, but is partially sensitive to SKF 525A and very sensitive to α-tocopherol and EDTA (Hochstein and Ernster, 1963). This effect of vitamin C on lipid peroxidation complicates studies on the biosynthesis of this vitamin for, as Chatterjee and McKee (1965) have shown, lipid peroxidation causes a decrease in the activity of L-gulonolactone oxidase, which is on the biosynthetic route to vitamin C. Low concentrations of EDTA, however, prevent this peroxidation and allow accurate measurements of vitamin C synthesis to be made (see Figure 12.5). The system is also responsive to hyperbaric oxygen; McCay *et al.* (1960) showed that L-gulonolactone oxidase was increased in activity under such conditions and this, by increasing the synthesis of vitamin C, *increased* lipid peroxidation, causing a *decrease* in the unsaturated fatty acid content of rat liver microsomes.

[4] Phosphate buffer may contain quite significant quantities of pyrophosphate as impurity.

L-gulonic acid $\xrightarrow[\text{gulonic oxidase}]{NAD^+}$ L-gulonolactone $\longrightarrow$ 2-oxo-L-gulonolactone $\longrightarrow$ L-ascorbic acid

```
  COOH            O=C ──┐          O=C ──┐          O=C ──┐
   |               |    |           |    |           |    |
HO-CH          HO-CH    |          O=C   |        HO-C    |
   |               |    O           |    O           ||   O
HO-CH          HO-CH    |       HO-CH    |        HO-C    |
   |               |    |           |    |           |    |
  HC-OH           HC ───┘          HC ───┘          HC ───┘
   |               |                |                |
HO-CH          HO-CH            HO-CH            HO-CH
   |               |                |                |
  CH2OH          CH2OH            CH2OH            CH2OH
```

Figure 12.5. Conversion of L-gulonic acid into L-ascorbate by liver enzymes.

12.6 Changes in the lysosomal membrane

The effects of lipid peroxidation on the lysosomal membrane are described in Chapters 14 and 15 in connection with effects of trace metals and radiation. In this Section we shall consider briefly two aspects: 'age-pigment' and ageing.

Many cells accumulate a lipid residue inside membrane-bounded particles; since this appears to increase in some cases with the age of the organism it has been called 'age-pigment' or 'wear and tear' pigment. It is also called **lipofuschin**. The particles appear to be derived from lysosomes, as shown by histochemical staining reactions for acid phosphatase; there is also some evidence that as the amount of pigment increases the amount of acid phosphatase decreases. These particles have been studied particularly by Björkerud (1963), who isolated purified fractions from horse heart. The general development is possibly as follows: lysosomes ingest lipid-rich material in relative excess. This slowly peroxidises and is partially degraded; the composite particle gradually loses acid-phosphatase activity and perhaps other lysosomal enzyme activities. The peroxidised lipid accumulates, imparting the characteristic colour and substructure to the particle. Goldfischer *et al.* (1966) have suggested that peroxidation is helped by the ingestion of mitochondria (and endoplasmic reticulum?) containing haem, that is known to be a potent pro-oxidant. An excessive number of lipofuschin-rich particles is an indication that a considerable amount of membrane engulfment has occurred (e.g. by a stimulated autophagy).

With regard to the ageing process itself, Harman has stressed the degradative role of free-radical-motivated reactions in this process. In a proposed mechanism analogous to that outlined above, Harman (1956, 1962) suggested that continuous production of reactive free radicals over a long time causes a chronic and gradual increasing damage to the structure-dependent function of intracellular membranes. In an extension of this idea Harman and Piette (1966) have considered that blood (which contains high concentrations of ferrous ions, copper ions and haem groups, high serum lipoprotein and a high oxygen partial pressure) could produce a chronic intravascular radiation syndrome through free-radical-motivated reactions. This could lead to fibrosis and a decreased nutritional supply to the underlying tissue. In an attempt to find evidence for this hypothesis, they carried out electron-spin-resonance studies on serum in the presence or absence of a number of naturally occurring materials. Quite large signals were obtained when serum was incubated with **adrenalin** or **vitamin C** in the presence of oxygen. It will be interesting to know if blood-vessel endothelium has a protective mucoprotein action that can protect against this potential free-radical hazard; if so does the nature of this coating change with age? There is a similarity between the effects suggested by Harmanand Piette and the arteriole-capillary fibrosis that is produced by radiation or during normal ageing (Casorett, 1960).

In a different approach from that of Harman regarding the role of free radicals in ageing, other workers have fed animals with antioxidants from an early age to see if this increases life-span. The results suggest that antioxidant feeding may prolong life-span in such situations, and further work on this question should be encouraged. However, the precise interpretation of such chronic experiments is very difficult. The feeding with such agents, which are powerful enzyme inducers (e.g. butylated hydroxytoluene), can produce large changes in hormone patterns, liver size, reaction to other agents and so on, thereby complicating any conclusion that the actions of such substances are simply those of antioxidants. The question of antioxidants and ageing is considered in some detail by Tappel (1968). In a recent study showing an effect of the antioxidant ethoxyquin on the life span in mice (Comfort *et al.*, 1971) the authors have also given a discussion of the complications involved in interpreting data obtained from such prolonged experiments.

12.7 Abnormal amounts of antioxidants

In this Section examples of tissue disturbances that result from abnormalities in dietary intake of antioxidants will be considered. These examples cover (a) excessive intakes (for example, acute experiments involving antioxidants used in low concentrations in the food industry) and (b) dietary deficiencies (for example, of vitamin E).

12.7.1 Excessive intakes

Antioxidants are a common additive to many food preparations and their use is strictly controlled by Government statute. Numerous toxicological studies have been made in which widely varying concentrations of antioxidants have been administered to experimental animals. It is not proposed to go into these experiments in any detail; a few brief examples will suffice to illustrate special points.

Antioxidants that undergo metabolic conversion *in vivo* by reaction with the NADPH-cytochrome P_{450} electron-transport chain in liver endoplasmic reticulum may act as very potent inducing agents. As a consequence the ability of the liver to metabolise other drugs[5] may be increased and liver size may also be increased. An example of this is the chronic administration of the antioxidant butylated methoxymethyl phenol. Gilbert *et al.* (1969) found that its administration to rats (380 mg/kg body wt.) for 10 days increased liver weight by approximately 60% (see Figure 12.6). With acute doses of antioxidants several reports have stressed the occurrence of disturbances of the central nervous system[6]. For example, Di Luzio (1964) mentions the effect of a large dose of an antioxidant mixture containing butylated hydroxytoluene and propyl gallate to rats; similar disturbances have been reported after the administration of large doses of propyl gallate alone. Orten *et al.* (1949) have studied the acute and chronic toxicity of propyl gallate to rats, guinea pigs and dogs. The LD_{50} in the rat was 3·8 g/kg body wt. after oral dosing and 0·38 g/kg body wt. after intraperitoneal dosing. Death was preceded by obvious signs of cardiorespiratory failure and convulsions. In contrast to the experiments just discussed, Hillman (1957) reported that a normal adult human male ingested 296 g of α-tocopherol over a period of 93 days with a subsequent high plasma concentration of over 2 mg 100 ml^{-1}. No clinical signs of toxicity were observed. The normal daily intake of this vitamin is approximately 30 mg and normal plasma concentration is about 1 mg 100 ml^{-1}.

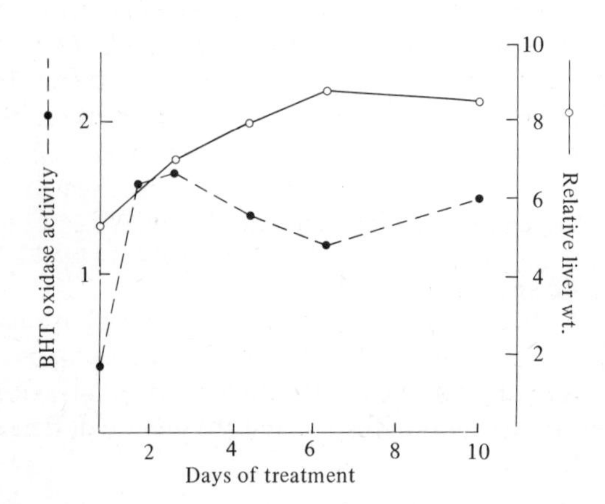

Figure 12.6. Effect of dosing rats with 2,6-di-t-butyl-4-methoxymethylphenol (380 mg/kg body wt.) on liver weight and the activity of BHT oxidase. Data of Gilbert *et al.* (1969).

[5] This may lead to increased toxicity in the presence of certain additional agents, as discussed in relation to carbon tetrachloride intoxication in Chapter 10.

[6] In this respect it is interesting that Polis *et al.* (1969) have found brain excitatory effects on injecting rabbits intravenously with a 'protein-free radical' combination. They suggest that the free radical transfers energy to receptor sites in the nervous tissue.

It has been found that local application of the carcinogen, dimethyl benzanthracene, in the hamster cheek pouch causes pseudopod formation in epidermal cells that is associated with a micro-invasion of the basement membrane; non-carcinogenic reagents tested were not effective in this respect (Woods and Smith, 1969). Riley and Seal (1968) have tested a number of substituted anisoles and have found some of them active in giving micro-invasion when applied locally in very strong solution (20%) to guinea-pig ear skin. Active anisoles in this respect included butylated hydroxyanisole, which is used as an antioxidant in the food industry. In contrast to the disturbances seen with dimethylbenzanthracene the micro-invasion observed with butylated hydroxyanisole (and with butylated hydroxytoluene) was readily reversible.

12.7.2 Vitamin E deficiency

The term 'vitamin E' covers a group of chemically related derivatives of a hypothetical compound 2-methyl-2-(4,8,18-trimethyltridecyl)-6-chromanol or tocol (Figure 12.7).

α-Tocopherol is the most common derivative of tocol present in foods; it is 5,7,8-trimethyltocol. β-Tocopherol and γ-tocopherol are the 5,8-dimethyl- and the 7,8-dimethyl-tocols respectively. The α-, β- and γ-tocopherols are viscous oils, insoluble in water, relatively stable to light in the absence of oxygen; they are, however, decomposed by exposure to ultraviolet radiation, oxidising agents or peroxidising fats. In Nature they are particularly concentrated in cereals: α- and β-tocopherol in wheat germ, barley and rye; γ-tocopherol in soyabean, groundnuts and corn. The biological activities of these contents are in the proportions (α:β:γ) 100:30:20.

The tocopherols undergo a number of metabolic transformations *in vivo*. For example, interaction of α-tocopherol with autoxidising lipid yields the α-tocopherylquinone (Figure 12.8).

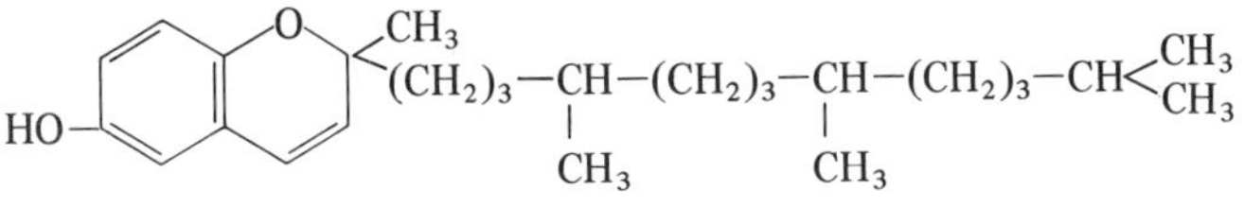

Figure 12.7. Structure of tocol.

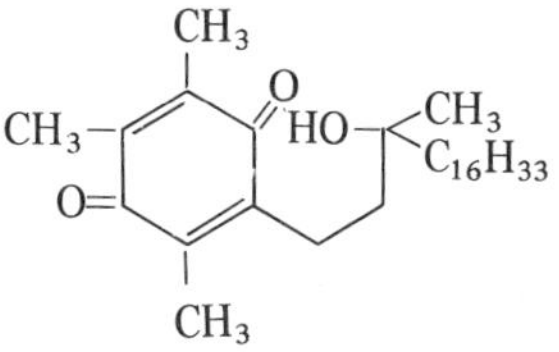

Figure 12.8. Structure of α-tocopheryl quinone.

Simon *et al.* (1956) have isolated products of α-tocopherol metabolism in urine: for example, the Simon metabolite (a) in Figure 12.9. Martius and Fürer (1963) believe that the active vitamin *in vivo* is the 2,5,6-trimethyl-(3-farnesyl-farnesyl-geranyl-geranyl)-1,4-benzoquinone (Figure 12.9b); more recent work has not supported this suggestion (see Draper and Csallany, 1969). Since α-tocopherol can function *in vitro* as a powerful antioxidant (see Tappel, 1962) and since many of the biological symptoms of α-tocopherol deficiency can be alleviated by the administration of a variety of other antioxidants, it has been generally assumed that the major function of α-tocopherol *in vivo* is as a biological antioxidant that is particularly concerned with maintaining the structure and integrity of lipid membranes. This view has been challenged (Green *et al.*, 1967; Bunyan *et al.*, 1967). An accepted settlement of this controversy has not yet been achieved but, in the author's opinion, it seems fair to conclude that both viewpoints contain elements of truth: some of the biological properties of α-tocopherol *in vivo* surely reflect its ability to function as a powerful lipid antioxidant. In other respects, however, additional properties of the molecule seem to be of considerable importance. The reaction of α-tocopherol or model compounds with free-radical-generating systems has been studied by various investigators in view of the probable action of vitamin E as an antioxidant *in vivo*. Skinner (1964) has shown, for example, that α-tocopherol reacts with azobis-isobutyronitrile (Figure 2.5) to yield two main products (Figure 12.10). The chromanoxy radical (Figure 12.10c) appears likely as an intermediate.

It will be useful to describe briefly the clinical manifestations of α-tocopherol deficiency before considering a few examples of disturbances that possibly involve free-radical mechanisms. In man, the deficiency is reported to be associated with xanthomatosis, cirrhosis of the gall bladder and creatinuria; in children, steatorrhoea and a tendency to erythrocyte haemolysis have been reported. In other species a wide variety of clinical symptoms involving a number of tissues have been described. In rats the failure to reproduce normally owing to male gonad atrophy and to placental

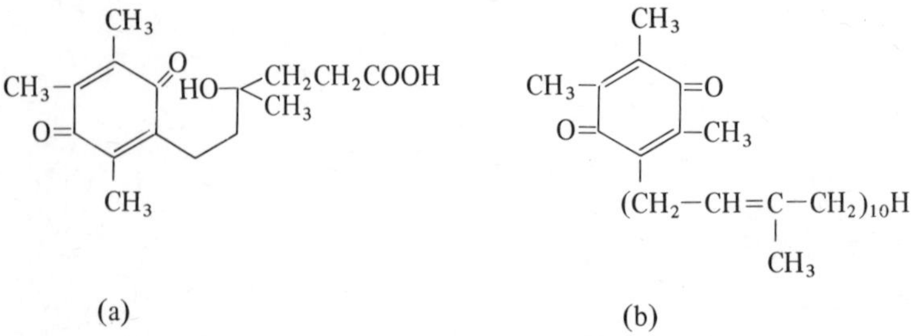

(a) (b)

Figure 12.9. Structures of (a) the Simon metabolite (Simon *et al.*, 1956) and (b) 2,5,6-trimethyl-3-(farnesyl-farnesyl-geranyl-geranyl)-1,4-benzoquinone (see Martius and Fürer, 1963).

changes led to the recognition of vitamin E in the early 1920s. In the dog, guinea pig, rabbit, chick and Rhesus monkey vitamin E deficiency leads to muscular dystrophy. In piglets there is a marked degeneration of skeletal and cardiac muscle that is associated with liver injury. Liver necrosis also develops in young rats on a diet deficient in vitamin E, provided that the intake of methionine and selenium is also restricted. In chicks that are vitamin E deficient there are neurological symptoms; exudative diathesis and encephalomalacia are common. Tubular degeneration has been seen in the rat and guinea pig kidneys during vitamin E deficiency; anaemia in the Rhesus monkey and an increased erythrocyte fragility in several species are other signs of the deficiency disease. Thus the signs of vitamin E deficiency are protean in their nature. The following discussion will be restricted to a few examples that illustrate points relevant to the general theme of this Chapter.

Rats fed with a diet deficient in vitamin E and low in sulphur-containing amino acids and factor 3 (a selenium compound) develop a fatal liver necrosis (Chernick *et al.*, 1955). Several weeks before the onset of this acute necrosis a disturbance of liver metabolism can be demonstrated *in vitro*; this is called **respiratory decline**. In this pre-necrotic period, liver *slices* from the affected lobes will show a progressive decline in oxygen uptake *in vitro* in comparison with the uptake by slices from control rats. The necrosis and respiratory decline can be prevented by the addition of vitamin E to the diet or by parenteral administration (Mertz and Schwarz, 1959) of the vitamin. The relative order of effectiveness in relation to the activity of DL-α-tocopherol when administered intraportally was D-α-tocopherol polyethylene glycol 1000 succinate, 277; DL-α-tocopherol,

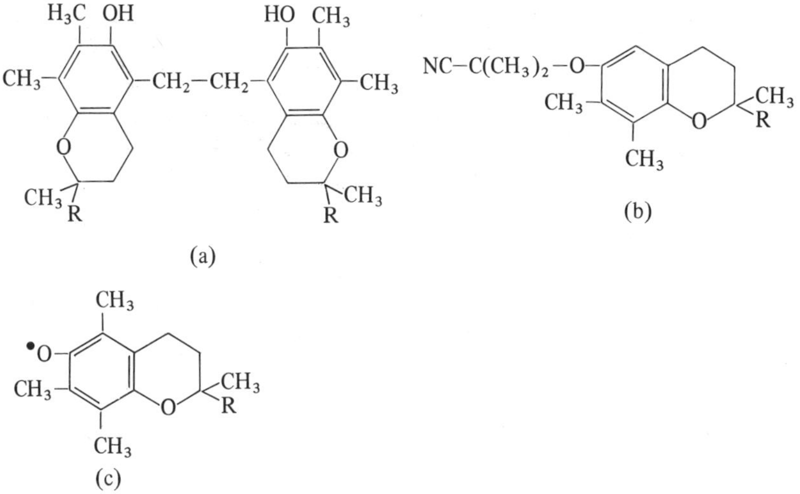

Figure 12.10. Reaction products (a, b) of α-tocopherol with azobisisobutyronitrile (Skinner, 1964). The chromanoxy radical (c) has been implicated in the reaction.

100; D-α-tocopherol, 83; D-β-tocopherol, 36. α-Tocopherol was not effective when added to liver slices *in vitro* although a tocopherol metabolite was effective. A series of antioxidants was tested for effectiveness in retarding the onset of necrosis *in vivo* and in preventing the phenomenon of respiratory decline *in vitro*; most were inactive but santoquin and *NN*-diphenyl-*p*-phenylenediamine prevented liver necrosis. The latter compound was also very active *in vitro* in overcoming respiratory decline. Methylene blue was inactive *in vivo* against necrosis but was very active *in vitro* in overcoming respiratory decline. Respiratory decline is not observed in liver homogenates fortified with NAD but with no added substrate, and this suggests that the basis of the lesion is the *leakage* of NAD from the mitochondria during incubation *in vitro*. This view is strengthened by the findings that many agents that decrease NAD leakage from the mitochondria also decrease respiratory decline; for example, ATP, EDTA, magnesium ions etc. Grove and Johnson (1967) have shown that respiratory decline is associated with a protein component of the microsomes, probably NAD-glycohydrolase. Destruction of NAD in the cytoplasm acts as a suction device, since the destruction of cytoplasmic NAD^+ affects the normal intercompartmental equilibrium thereby resulting in NAD^+ being drawn out of the mitochondria. In this situation there is an analogy with the necrosis produced by administering carbon tetrachloride, where destruction of NADPH in the endoplasmic reticulum causes a drainage of NADPH from the mitochondria by a similar mass-action effect. Both situations are responsive to certain antioxidants such as *NN*-diphenyl-*p*-phenylenediamine. Although Bunyan *et al.* (1963) have argued strongly against respiratory decline and liver necrosis in the vitamin E-deficient rat arising from a lipid peroxidation that is sensitive to vitamin E, the author thinks that a radical-motivated reaction is likely to be involved in the initiation of respiratory decline through changes in the organised structure of the endoplasmic reticulum.

As another example of the effects of vitamin E deficiency *in vivo* at the cellular level the report by Hess and Menzel (1968) can be cited. These authors showed that in rat kidney proximal tubular cells high oxygen pressures increased the number of centrioles when administered in association with a vitamin E deficiency. They suggest that an oxidative function is involved in centriolar formation and that the lesion of the reproductive tissues in vitamin E deficiency in the rat may be related to a protective role of vitamin E on the mitotic apparatus.

It is well known that liver homogenates from rats deficient in vitamin E will *peroxidise* faster *in vitro* than the control homogenates (Bieri and Anderson, 1960). Kitabchi and Williams (1968) have examined this effect in liver and have also extended the study to a variety of other tissues. They found that the adrenal was particularly sensitive to vitamin E deficiency in terms of peroxidation *in vitro*. Whereas liver homogenates from vitamin E-deficient rats peroxidise approximately eight times as fast as control liver suspensions, with adrenals the factor was approximately 20-fold. A significant increase in zero-time thiobarbituric acid value[7] in **adrenals** from vitamin E-deficient rats was also observed (Table 12.3) similar to that observed from livers of rats dosed previously with carbon tetrachloride (Section 10.8.1). A similar rise in zero-time thiobarbituric acid values was noted by Carpenter *et al.* (1959) with **muscle** tissue from rabbits that were deficient in vitamin E. The findings with the adrenals are particularly relevant to long-term experiments with vitamin E-deficient diets, for they suggest the possibility that some of the effects seen are the result of adrenal dysfunction consequent upon damage produced by the vitamin E deficiency.

Finally, and to illustrate yet another possible hazard in the investigation of the role of vitamin E in biological systems, the work of Detwiler and Nason (1966) can be mentioned. These authors found that α-tocopherol could reduce cytochrome *c* non-enzymically in the presence of unsaturated fatty acids as catalysts. In the course of this reaction the turbidity of the lipid suspension decreased. In this respect the tocopherol is acting analogously to a reduced coenzyme (e.g. NADH) in reducing a cytochrome. It would be of interest to know if the turbidity also decreases in the absence of oxygen. If not then it would suggest that the haem component of the cytochrome, oxygen and α-tocopherol are combining to produce hydroxy fatty acids that are known to be more hydrophilic than the parent unsaturated fatty acid (Christopherson, 1968; see Section 4.2.3).

Table 12.3. Effects of vitamin E-deficient diet on the malonaldehyde content of rat adrenal gland. Rats were kept on the diet for 10 weeks; the adrenals were then removed and immediately homogenised in trichloroacetic acid, followed by thiobarbituric acid determination. Data from Kitabchi and Williams (1968).

Diet	Malonaldehyde content (arbitrary units)
Control	0·122 ± 0·013
Vitamin E-deficient	0·260 ± 0·033

12.7.3 Ingestion of fatty acid peroxides and related materials

There are many reports in the literature that demonstrate the toxicity of oxidised fats (e.g. heated fatty foods) to animals. It is important to note, however, that several of the studies have used unusually severe heating conditions to produce the fatty acid oxidation (temperatures higher than 300°C) whereas fat is generally subjected to temperatures of less than

[7] By zero-time thiobarbituric acid value is meant the concentration of malonaldehyde found in tissue samples taken immediately at death (or with the animal under anaesthetic) and then being rapidly frozen. Malonaldehyde is then estimated in acid extracts of the frozen tissue by the thiobarbituric acid reaction.

350°F in normal cooking procedures. Harris (1963) comments on this in drawing attention to the carcinogenicity of fats heated to 300°C but the *non*-carcinogenicity of fats heated at 350°F. The author makes the interesting observation that the admixture of heated fats with ordinary diet causes a profound loss of vitamins E, A and K on storage, so that chronic experiments with diets enriched with such heated fats yield results not only dependent on the fat composition of the diet but also on avitaminosis. In his experiments rats died after 3 weeks on such dietary regimens, but where the diet was stored separately from the heated fat, only a mild diarrhoea was observed in the treated animals.

Oxidation of fats in foodstuffs can be expected to produce the hydroperoxides of unsaturated fatty acids, for example, linoleic acid hydroperoxide, linolenic acid hydroperoxide and other esters. Many reports have been concerned with the toxicity of these components of heated fats. Horgan *et al.* (1957) investigated the toxicity of a variety of organic peroxides and autoxidised linoleate and methyl linoleate after *intraperitoneal* injection into female mice. The LD_{50} values found (μmol/mouse) were: autoxidised linoleic acid, 4–8; autoxidised methyl linoleate, 45–60; disuccinoyl peroxide, 4–12; dibenzoyl peroxide, 20; di-t-butyl peroxide, 1080. Olcott and Dolev (1963) gave autoxidised crude methyl linoleate by *intraperitoneal* injection to rats and measured the mortality with and without the addition of α-tocopherol, given intraperitoneally 24 h before the injection of fatty acid ester. The injected fatty acid suspension was lethal at 150 μmol/100 g body wt.; α-tocopherol did not affect the LD_{50}. The major symptom in all of the rats that died was massive **ascites**. Kokatnur *et al.* (1966) have studied the toxicity of purified methyl linoleate hydroperoxide in rabbits. When given daily by *intravenous* injection for 10–14 days (dose 50 mg/day) there was a sharp increase in the creatine/creatinine ratio in the urine; the liver showed centrilobular fatty degeneration, scattered necrosis, some giant cells and bile-duct proliferation. When the hydroperoxide was given to rats that were deficient in vitamin E there was a high incidence of liver damage, a high creatine/creatinine ratio that was prevented by oral vitamin E, which did not, however, protect against the liver lesions. Continuous intravenous infusion of the hydroperoxide gave increased erythrocyte haemolysis. The dose used was 12 μmol/100 g body wt., which is much below the LD_{50} quoted above for rats and mice by intraperitoneal injection (approximately 150 μmol/100 g body wt.). Thus the lesions produced by intravenous methyl linoleate hydroperoxide were aggravated by vitamin E deficiency. Oral supplementation with vitamin E was protective against some but not all of the disturbances.

Nishida *et al.* (1960) found that the *intravenous* injection of 10 mg of methyl linoleate hydroperoxide into chicks gave cerebellar disorders similar to those seen in vitamin E deficiency. Thus it is possible that the parenteral administration of these hydroperoxides gives rise to destruction of vitamin E *in vivo* with subsequent appearance of disorders characteristic of vitamin E deficiency. The chemistry of the reaction between benzoyl peroxide and α-tocopherol has been studied (Goodhue and Risley, 1965). The reaction products were found to be strikingly dependent on the nature of the solvent used; in anhydrous alcohol the products were 8-α-alkoxy-α-tocopherones, homologues of tocopheroxide (Figure 12.11).

The reactivity of hydroperoxides with normal cell *components* has been discussed in Chapter 6 and will not be discussed further here.

A number of workers have investigated in detail the products formed by the autoxidation of unsaturated fatty acids. Schauenstein (1967), for example, studied the autoxidisation of polyunsaturated fatty acid esters in water suspension and separated the water-soluble products for biological testing. Various hydroperoxides tested were found to inhibit glycolysis and respiration of Ehrlich ascites tumour cells; **8-hydroperoxy-caprylic acid methyl ester**, the main active peroxide component, inhibited glycolysis at approximately 0·3 mM and respiration at approximately 2 mM *in vitro*. This effect seemed due to inhibition of the enzymes glyceraldehyde phosphate dehydrogenase and lactate dehydrogenase. Various non-peroxy components were also separated and several were hydroxyaldehydes. One active derivative tested was **4-hydroxyoct-2-en-1-al**, which also inhibited glycolysis and respiration; it caused a total inhibition of glyceraldehyde phosphate dehydrogenase at 0·1 mM and a 50% inhibition of lactate dehydrogenase at 8 mM. There were striking morphological changes in the ascites cells exposed to mM concentrations of the aldehyde; normal cells (normal monkey liver and kidney cells) showed no adverse reactions to the aldehyde. These observations seem of particular interest in view of Szent-Györgyi's views on the importance of aldehyde compounds to malignant transformation[8] (see Fodor *et al.*, 1967). Additional data on the water-soluble products of fatty acid oxidation are given by Baker and Wilson (1966).

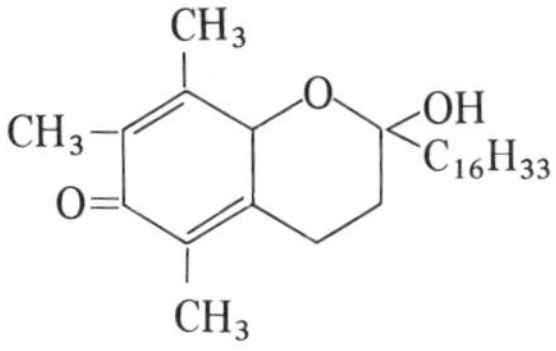

CH_3
CH_3 O OH
O $C_{16}H_{33}$
CH_3

Figure 12.11. Structure of α-tocopheroxide.

[8] The striking biological properties of hydroxy ocetenal (and hydroxy pentenal) reported by Schauenstein (1967, 1968) may be related to the activity of 4-hydroxy-2-oxo-butyraldehyde (Sparkes and Kenny, 1969). The latter substance has been found to be a bacterial growth inhibitor.

12.8 Bipyridylium compounds

To close this Chapter on membrane disturbances resulting from free-radical reactions, the bipyridylium compounds, widely used as herbicides (see Conning *et al.*, 1969), can be considered. The two substances that have been most studied for their biochemical mechanism of action are Diquat and Paraquat (Figure 12.12). Such substances are capable of reduction in solution to give a relatively stable free radical and this process is a prerequisite for herbicidal action. A scheme for action *in vivo* has been based on the reduction of the bipyridylium cation by chlorophyll within the chloroplast; the bipyridylium radical cation is then oxidised by molecular oxygen to yield hydrogen peroxide. After repeated cycles the concentration of hydrogen peroxide becomes toxic and damages the organised function of the chloroplast. Lipid peroxidation may also be involved in the overall injury produced by the bipyridylium radical ions.

The acute effects of Paraquat on mammals (for example, with accidental poisoning) are severe and often fatal. If death does not occur within a few days of ingestion of a large dose then extraordinary lung changes develop. There is often pulmonary oedema, with hyaline membrane formation and an inflammatory response. Later stages may be associated with a progressive fibrosis and death from respiratory failure. Diquat produces a different pathological picture, in which the major disturbances concern the intestinal tract.

Gage (1968) has found that rat liver microsomes with NADPH and in the absence of oxygen will catalyse the formation of Paraquat and Diquat radicals. Cyclic oxidation and reduction of the bipyridylium radical ions was found to lead to peroxidation of microsomal lipid. Further studies of the free-radical reactions resulting from the ingestion of Paraquat and Diquat by mammals are obviously of considerable interest and, in view of the widespread use of these compounds, of great importance.

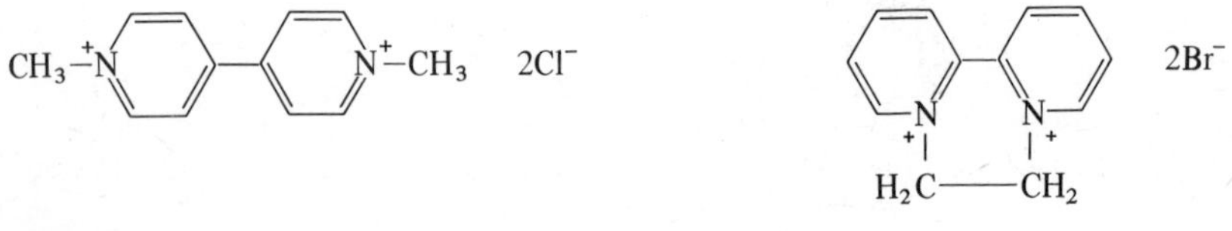

Figure 12.12. Structures of Paraquat dichloride and Diquat dibromide.

References

Baker, N., Wilson, L., 1966, *J. Lipid Res.*, **7**, 341, 349.
Beloff-Chain, A., Catanzaro, R., Serlupi-Crescenzi, G., 1963, *Nature, Lond.*, **198**, 351.
Bieri, J. G., Anderson, A. A., 1960, *Archs. Biochem. Biophys.*, **90**, 105.
Björkerud, S., 1963, *J. Ultrastruct. Res.*, supplement number 5.
Blum, H. F., 1964, *Photodynamic Action and Diseases Caused by Light* (Hafner, New York).
Bunyan, J., Green, J., Diplock, A. T., 1963, *Br. J. Nutr.*, **17**, 117.

Bunyan, J., Green, J., Edwin, E. E., Diplock, A. T., 1961, *Biochim. biophys. Acta,* **47**, 401.
Bunyan, J., Murrell, E. A., Green, J., Diplock, A. T., 1967, *Br. J. Nutr.,* **21**, 475.
Carpenter, M. P., Kitabchi, A. E., McCay, P. B., Caputto, R., 1959, *J. biol. Chem.,* **234**, 2814.
Cash, W. D., Gardy, M., Carlson, H. E., Ekong, E. A., 1966, *J. biol. Chem.,* **241**, 1745.
Casorett, G. W., 1960, in *The Biology of Ageing,* Ed. B. L. Strehler (American Institute of Biological Sciences, Washington), p.147.
Chatterjee, I. B., McKee, R. W., 1965, *Archs. Biochem. Biophys.,* **110**, 254.
Chernick, S. S., Moe, J. G., Rodnan, G. P., Schwarz, K., 1955, *J. biol. Chem.,* **217**, 829.
Christopherson, B. O., 1968, *Biochim. biophys. Acta,* **164**, 35.
Comfort, A., Youhotsky-Gore, I., Pathmanathan, K., 1971, *Nature,* **229**, 254.
Conning, D. M., Fletcher, K., Swan, A. A. B., 1969, *Br. med. Bull.,* **25**, 245.
Coupland, R. E., MacDougall, J. D. B., Myles, W. S., McCabe, M., 1969, *J. Path. Bact.,* **97**, 63.
De Duve, C., 1964, *J. theor. Biol.,* **6**, 33.
De Duve, C., Baudhuin, P., 1966, *Physiol. Rev.,* **46**, 323.
De Duve, C., Wattiaux, R., 1966, *A. Rev. Physiol.,* **28**, 435.
Detwiler, T. C., Nason, A., 1966, *Biochemistry, Easton,* **5**, 3936.
Di Luzio, N. R., 1964, *Life Sci.,* **3**, 113.
Draper, H. H., Csallany, A. S., 1969, *Fedn Proc. Fedn Am. Socs exp. Biol.,* **28**, 1690.
Dowben, R. M., 1969, *Biological Membranes* (Churchill, London).
Fodor, G., Sachetto, J., Szent-Györgyi, A., Együd, L. G., 1967, *Proc. natn. Acad. Sci. U.S.A.,* **57**, 1644.
Frederic, J., Chèvremont, M., 1952, *Archs Biol., Paris,* **63**, 109.
Gage, J. C., 1968, *Biochem. J.,* **109**, 757.
Getz, G. S., Bartley, W., Stirpe, F., Notton, B. M., Renshaw, A., 1962, *Biochem. J.,* **83**, 181.
Gilbert, D., Martin, A. D., Gangolli, S. D., Abraham, R., Golberg, L., 1969, *Fd. cosmetic Toxic.,* **7**, 603.
Gillette, J. R., Brodie, B. B., La Du, B. N., 1957, *J. Pharmac. exp. Ther.,* **119**, 532.
Goldfischer, S., Villaverde, H., Forschirm, R., 1966, *J. Histochem. Cytochem.,* **14**, 641.
Goldstein, B. D., Balchum, O. J., 1967, *Proc. Soc. exp. Biol. Med.,* **126**, 356.
Goodhue, C. T., Risley, H. A., 1965, *Biochemistry, Easton,* **4**, 854.
Green, J., Diplock, A. T., Bunyan, J., McHale, D., Muthy, I., 1967, *Br. J. Nutr.,* **21**, 69.
Grove, J. A., Johnson, R. M., 1967, *J. biol. Chem.,* **242**, 1623.
Harman, D., 1956, *J. Geront.,* **11**, 298.
Harman, D., 1962, *Radiat. Res.,* **16**, 753.
Harman, D., Piette, L. H., 1966, *J. Geront.,* **21**, 560.
Harris, R. J. C., 1963, *Rad. Res. Suppl.,* **3**, 209.
Hess, R. T., Menzel, D. B., 1968, *Science,* **159**, 985.
Hillman, R. W., 1957, *Am. J. clin. Nutr.,* **5**, 597.
Hochstein, P., Ernster, L., 1963, *Biochem. biophys. Res. Commun.,* **12**, 388.
Horgan, V. J., Philpot, J. S. L., Porta, B. W., Roodyn, D. B., 1957, *Biochem. J.,* **67**, 551.
Hunter, F. E., Gebicki, J. M., Hoffsten, P. E., Weinstein, J., Scott, A., 1963, *J. biol. Chem.,* **238**, 828.
Kitabchi, A. E., Williams, R. H., 1968, *J. biol. Chem.,* **243**, 3248.
Kokatnur, M. G., Bergan, J. G., Draper, H. H., 1966, *Proc. Soc. exp. Biol. Med.,* **123**, 254.
Korn, E. D., 1969, *Fedn Proc. Fedn Am. Socs exp. Biol.,* **28**, 6.

Lehninger, A. L., 1962, *Physiol. Rev.,* **42**, 467.
Lehninger, A. L., 1964, *The Mitochondrion* (Benjamin, New York).
Lehninger, A. L., Beck, D. P., 1967, *J. biol. Chem.,* **242**, 2098.
Ling, G. N., 1969, *Int. Rev. Cytol.,* **26**, 1.
Lucy, J. A., 1969, in *Lysosomes in Biology and Pathology,* volume 2, Eds. J. T. Dingle and H. B. Fell (North-Holland, Amsterdam), p.313.
Maddy, A. H., 1966, *Int. Rev. Cytol.,* **20**, 1.
Marks, V. F., Hecker, E., 1967, *Hoppe-Seyler's Z. physiol. Chem.,* **348**, 727.
Martius, C., Fürer, E., 1963, *Biochem. Z.,* **336**, 474.
Mathews, M. M., 1964, *Nature, Lond.,* **203**, 1092.
May, H. E., McCay, P. B., 1968, *J. biol. Chem.,* **243**, 2288.
May, H. E., Poyer, J. L., McCay, P. B., 1965, *Biochem. biophys. Res. Commun.,* **19**, 166.
May, H. E., Poyer, J. L., McCay, P. B., 1966, *Fedn Proc. Fedn Am. Socs exp. Biol.,* **25**, 301.
McCay, P. B., May, H. E., Kitabchi, A. E., Feinberg, R. H., Carpenter, M. P., Trucco, R. E., Caputto, R., 1960, *Biochem. biophys. Res. Commun.,* **3**, 441.
McKnight, R. C., Hunter, F. E., 1966, *J. biol. Chem.,* **241**, 2757.
Mellors, A., Tappel, A. L., 1966, *J. biol. Chem.,* **241**, 4353.
Mengel, C. E., Kann, H. E., 1966, *J. clin. Invest.,* **45**, 1150.
Merriwether, W. D., Mengel, C. E., 1967, *Nature, Lond.,* **216**, 85.
Mertz, W., Schwarz, K., 1959, *Proc. Soc. exp. Biol. Med.,* **102**, 561.
Neville, D. M., 1968, *Biochim. biophys. Acta,* **154**, 540.
Nishida, T., Tsuchiyama, H., Inoue, M., Kummerow, F. A., 1960, *Proc. Soc. exp. Biol. Med.,* **105**, 308.
Olcott, H. S., Dolev, A., 1963, *Proc. Soc. exp. Biol. Med.,* **114**, 820.
Orrenius, S., Dallner, G., Ernster, L., 1964, *Biochem. biophys. Res. Commun.,* **14**, 329.
Orten, J. M., Kuyper, A. C., Smith, A. H., 1949, *Fd. Technol., Champaign,* **2**, 308.
Placer, Z. A., Noel, D. A., Cushman, L. L., Johnson, B. C., 1966, *Fedn Proc. Fedn Am. Socs exp. Biol.,* **25**, 302.
Polis, B. D., Wyeth, J., Goldstein, L., Graedon, J., 1969, *Proc. natn. Acad. Sci. U.S.A.,* **64**, 755.
Riley, P. A., Seal, P., 1968, *Nature, Lond.,* **220**, 922.
Schauenstein, E., 1967, *J. Lipid Res.,* **8**, 417.
Schauenstein, E., Wöhl, W., Kramer, I., 1968, *Z. Naturforsch.,* **B23**, 530.
Schneider, A. K., Smith, E. E., Hunter, F. E., 1964, *Biochemistry, Easton,* **3**, 1470.
Simon, E. J., Eisengert, A., Sundheim, L., Milhorat, A. T., 1956, *J. biol. Chem.,* **221**, 807.
Skinner, W. A., 1964, *Biochem. biophys. Res. Commun.,* **15**, 469.
Skipski, V. P., Barclay, M., Archibald, F. M., Terebus-Kekish, O., Reichman, E. S., Good, J. J., 1965, *Life Sci.,* **4**, 1673.
Slater, T. F., 1968, *Biochem. J.,* **106**, 155.
Sparkes, B. G., Kenny, C. P., 1969, *Proc. Natl. Acad. Sci.,* **64**, 920.
Stokinger, H. E., 1965, *Archs envir. Hlth.,* **10**, 719.
Tappel, A. L., 1962, *Vitams Horm.,* **20**, 493.
Tappel, A. L., 1968, *Geriatrics,* **23**, 97.
Tappel, A. L., 1969, in *Lysosomes in Biology and Pathology,* volume 2, Eds. J. T. Dingle, H. B. Fell (North-Holland, Amsterdam), p.207.
Tsen, C. G., Collier, H. B., 1960, *Can. J. Biochem. Physiol.,* **38**, 957.
Utley, H. G., Bernheim, F., Hochstein, P., 1967, *Arch. Biochem. Biophys.,* **118**, 29.
Waldschmidt, M., Mönig, H., Schole, J., 1968, *Z. Naturf.,* **23b**, 798.
Weiss, L., 1969, *Int. Rev. Cytol.,* **26**, 63.
Woods, D. A., Smith, C. J., 1969, *J. invest. Derm.,* **52**, 259.

13

Free radicals and chemical carcinogenesis

13.1 Introduction

It has been recognised for almost 200 years that cancer may be induced by substances in the environment [for example, the recognition that scrotal cancer was particularly prevalent in chimney-sweeps (Percival Pott, 1775)], yet it is only some 40 years since the first pure carcinogens were isolated and tested by Kennaway and colleagues; these substances were polycyclic hydrocarbons such as 1,2,5,6-**dibenzanthracene** (Figure 13.1).

Soon after these early studies by Kennaway's group on dibenzanthracene, Cook, Hewett and Heiger isolated 3,4-**benzpyrene** from coal tar; this substance (Figure 13.2) produces malignant tumours when administered to animals.

In the last 30 years a wide variety of agents has been found that produce cancer when applied to or administered to animals. Such agents include polycyclic hydrocarbons, heterocyclic compounds, azo-dyestuffs, aromatic amines, nitrosamines, fungal toxins and inorganic materials[(1)].

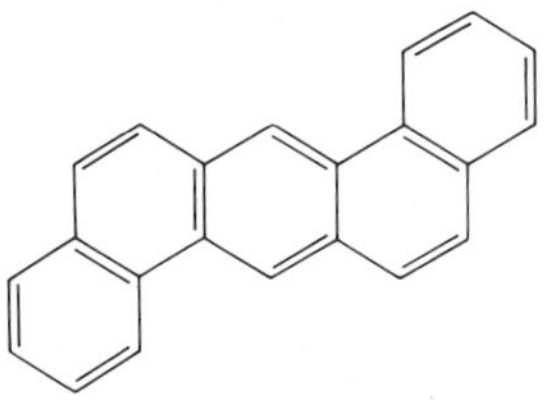

Figure 13.1. Structure of 1,2,5,6-dibenzanthracene.

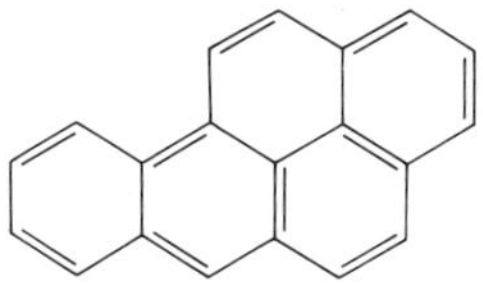

Figure 13.2. Structure of 3,4-benzpyrene.

13.2 Free radicals and carcinogenesis

Numerous hypotheses have been proposed in attempts to explain chemical carcinogenesis (see Pitot, 1966, for review); several of these hypotheses are directly relevant to the approach adopted in this Chapter. Kensler *et al.* (1942) studied the effects of **dimethylaminoazobenzene** (and related compounds) on glucose oxidation and found a correlation between the effect on glucose oxidation and the stability of the free-radical semiquinones formed by the metabolism of the azo-dyestuffs to substituted phenylene-diamines. Moreover, substances that produced free-radical semiquinones of relatively long half-life were also found to have a high carcinogenic action *in vivo*. Another study that indicated a relationship between free-radical

[(1)] It will be impossible to review even very briefly this complex field; instead, the emphasis will be on one specialised aspect: the possible role of free-radical reactions in chemically induced cancer. Reviews on chemical carcinogenesis that provide background material are by Carruthers (1961), Buu-Hoi (1964), Hueper and Conway (1964), Reid (1965), Miller (1970), Heidelberger (1970).

production and carcinogenicity was that by Lipkin *et al.* (1953); they suggested that polycyclic hydrocarbons having a strong tendency to abstract an electron from an alkali metal, thereby forming relatively stable free radicals, are powerful carcinogens. The relationship between electronic distribution and carcinogenicity in the polycyclic hydrocarbons has been studied particularly by the Pullmans in Paris (see Pullman and Pullman, 1955). They have observed that a high degree of carcinogenicity in the polycyclic hydrocarbons is associated with the occurrence in the molecule of high electron-density in the 9,10-position of the phenanthrene nucleus; this region is known as the **K-region** (from the German word, *Krebs*, for cancer). In later work the importance of the neighbouring L-region (the *meso*-positions in anthracene) was also recognised: as a consequence, it has been suggested that high carcinogenic activity is associated with a *high* π-electron density in the K-region and a *low* π-electron density in the L-region (Figure 13.3).

Direct suggestions that free-radical metabolites play a role in carcinogenesis were made by Gordy (1958). He pointed out that many environmental agents (for example, exhaust fumes, cigarette smoke, tobacco tars, charred foods) were known to contain material giving intense signals when studied by the electron spin resonance technique. These reactive entities may be trapped in the solid particles of material (as in the 'cage-effect') and may therefore be brought into intimate contact with epithelial cells during inhalation and digestion. Confirmatory evidence for the presence of free-radical signals in exhaust fumes, cigarette smoke and domestic chimney smoke was obtained by Lyons and Spence (1960). The hypothesis that environmental free radicals may contribute to carcinogenesis (and also to generalised ageing) was developed by Harman in an interesting article in 1962. Harman's speculations concerning the role of free radicals in ageing, and the effects of administered antioxidants on life-span in mice, are discussed in Section 12.6.

After such suggestions it was natural for studies to be made on the differences between tumours and the corresponding normal tissues by using the electron spin resonance technique. In 1963 Nebert and Mason described the electron spin resonance spectra of 29 types of mouse tumour tissue together with two normal mouse tissues; the tissue samples were

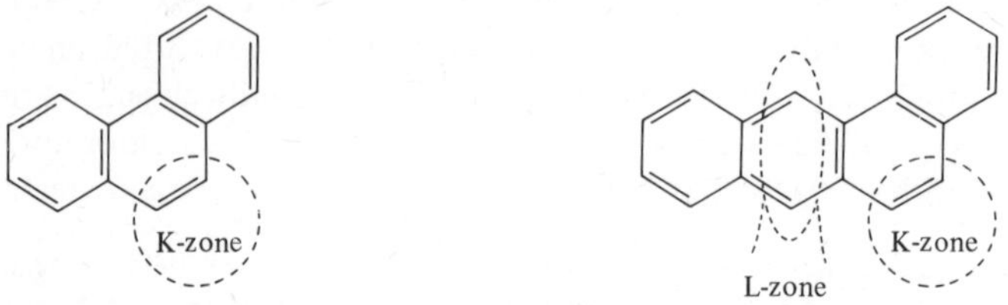

K-zone

L-zone

K-zone

Figure 13.3. Illustrating the positions of the K-region for phenanthrene, and the K- and L-regions for benz-(*a*)-anthracene.

examined in the frozen powdered state at −160°C. Strong signals were obtained from many tumour samples; a hepatoma studied (type BW 7756) showed *qualitative* differences from normal mouse liver. Large differences in the region of $g = 2 \cdot 25$ were observed to exist between the spectra obtained with microsomal fractions isolated from normal liver and from a hepatoma. As discussed in Section 6.3 the signal occurring in the $g = 2 \cdot 25$ region of liver microsomes is associated with the cytochrome P_{450} component; the decreased signal observed in liver tumours in the $g = 2 \cdot 25$ region is consistent with the generally observed decrease in drug metabolism in poorly differentiated liver tumours compared with normal liver tissue.

The work of Nebert and Mason was soon followed by studies on the changes in electron-spin-resonance spectra during the **development** of liver tumours after the administration to rats of a number of liver carcinogens. Vithayathil *et al.* (1965) used three such carcinogens: *p*-dimethylamino-azobenzene, **thioacetamide** and **2-acetylaminofluorene**, and observed the appearance of a *transitory* signal at $g = 2 \cdot 035 \pm 0 \cdot 002$ well before the development of recognisable tumours (Figure 13.4). In their experiments the tissue was *sliced* and suspended in 5% glucose medium; electron spin resonance was performed at 15°C with time-averaging of signals over 50

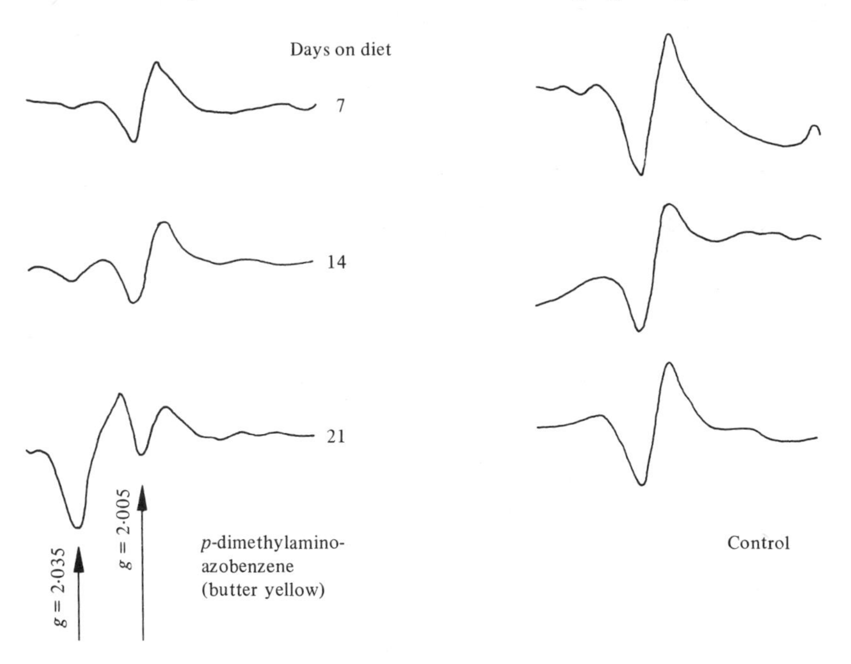

Figure 13.4. Changes in the electron spin resonance spectra of liver samples after the administration of butter yellow to riboflavin-deficient rats (data of Vithayathil *et al.*, 1965). The signal traces are averages of 50 successive sweeps; 50–100 mg wet wt. of liver at 15°C in a 5% glucose solution was used.

successive sweeps. Histological changes in the liver were minimal and non-specific during the periods when the signal at $g = 2 \cdot 035$ was observed (14, 40 and 6 days respectively for the three agents mentioned above). The material responsible for the signal was unstable to homogenising procedures that disrupted the integrity of the liver slices. Vithayathil *et al.* (1965) also examined biopsies of three **human** liver tumours and observed a decreased signal at $g = 2 \cdot 005$ compared with a human non-malignant liver sample. A similar decrease in signal at $g = 2 \cdot 005$ was also observed in liver tumours produced experimentally in rats with the carcinogen butter yellow. The component responsible for the transitory signal has been investigated more recently by Woolum and Commoner (1970), who have related the signal to the formation of an iron–nitric oxide free radical complex. In fact, normal liver was found to develop the signal at $g = 2 \cdot 035$ when incubated in phosphate buffer containing potassium nitrate. Since nitrate is reduced stepwise to nitric oxide, and iron plus nitric oxide is known to give a signal at $g = 2 \cdot 033$, the authors suggest that this compound in association with amino acid ligands is the source of the radical signal previously observed. Evidently the carcinogen stimulates the production of the radical *in vivo.* The iron–nitric oxide complex seems unrelated to the primary carcinogenic mechanisms, however, since rats fed carcinogens with distilled water free of nitrate and nitrite did not develop the characteristic signal at $g = 2 \cdot 0035$. In fact the presence of nitrite in the diet not only increased the strength of the radical signal but *decreased* the incidence of tumours. The authors suggest that the iron–nitric oxide complex may be functional in a process that inactivates the carcinogen.

An electron spin resonance spectrum strikingly different from that of the normal tissue was found with a tumour of mouse **spleen** by Brennan *et al.* (1966). Whereas normal tissue gave a single well-defined signal at $g = 2 \cdot 003$ the tumour material gave a **triplet** signal[(2)] and this result led the authors to suggest the occurrence of a new free-radical species in the tumour probably with the unpaired electron associated with a nitrogen atom.

The report by Brennan *et al.* suggested the possibility that if such *qualitative* differences exist between human tumours and their homologous normal counterparts it may be possible to use electron spin resonance techniques as a *diagnostic aid* in those cases where tumour material is readily accessible. Such relatively accessible tumours include those of the cervix and uterus, which exfoliate into the vaginal fluid and where cervical smears may contain large numbers of malignant cells. Preliminary studies to test the possible use of electron spin resonance in the diagnosis of such tumours have been made (Slater and Cook, 1969). It was found that

[(2)] The tumour type studied was a virus-induced reticular cell sarcoma of spleen; a similar triplet signal was observed with the JAX.C.1300 neuroblastoma but not with several other tumours studied.

samples of normal *human* cervix and uterus examined in the frozen powdered state at −180°C produced very large signals in the $g = 2{\cdot}003$ region. The material responsible for the signal was stable on storage at −196°C. A few cases of invasive carcinoma of the cervix and uterus were also examined and only very small signals were obtained in comparison with the samples of normal tissue (Figure 13.5). In this sense the results were analogous to the *quantitative* differences observed by Vithayathil *et al.* (1965) between liver tumours and normal liver tissue rather than to the *qualitative* differences observed by Brennan *et al.* (1966) with spleen tissue.

Certain tumour tissues or components (for example, ascites cell mitochondria; Utsumi *et al.*, 1965) do not peroxidise as readily as the corresponding normal material and this difference has been ascribed to the occurrence in the tumour of abnormally high concentrations of antioxidants (Schuster, 1955; Lash, 1966), and to the decreased content in tumour mitochondria of polyunsaturated fatty acids. The very much decreased signal observed in human tumours of the cervix and uterus may be due to one or other of the above causes, or to the absence of the material responsible for the signal in normal tissue. If the material that gives the signal in normal tissue can be identified then it should be possible to devise a histochemical test for its presence in cell-smears obtained from the cervix; such a test might then prove useful diagnostically.

Emanuel *et al.* (1969) have studied the electron spin resonance signals from certain transplantable tumour tissues and have compared them with the corresponding normal *rat* and *mouse* tissue samples. Hepatoma 22a, sarcoma 45 and Walker sarcoma all gave a **triplet** signal centred at $g = 2{\cdot}03$ and which was not observed in normal liver and muscle. In this respect their results are similar to those of Brennan *et al.* mentioned above.

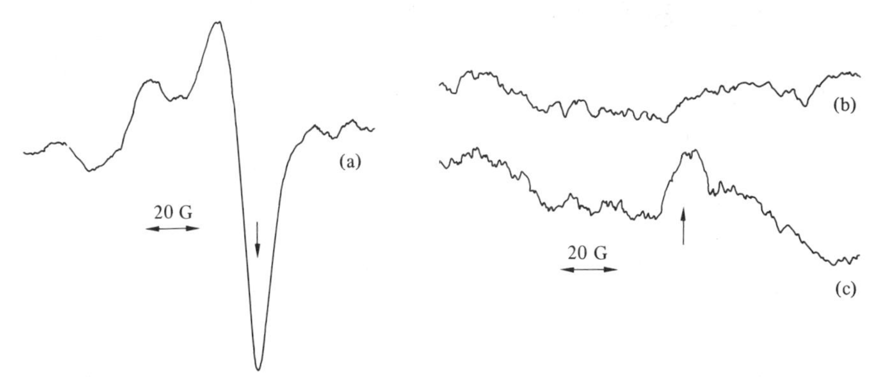

Figure 13.5. Electron spin resonance spectra of normal human cervix (a) and of carcinoma of the cervix (c). Powdered samples of whole tissue were examined at −180°C (data of Slater and Cook, 1969). The vertical arrow marks the position of the free electron spin at $g = 2{\cdot}0023$. [Trace (b) is the background signal obtained from the cavity in the absence of tissue samples.]

Moreover, the Russian workers agree with Brennan *et al.* (1966) in ascribing the triplet signal to an unshared electron on a nitrogen atom. The authors conclude: "some new rather stable paramagnetic complex absent from normal tissues seems to be involved in the metabolism of tumour cells". If similar observations can be made with human tumour samples then the matter requires urgent study.

Brand *et al.* (1967a, 1967b) have studied the sarcomatous transplantable tumours that can be produced in mice by the insertion of thin plastic film (vinyl chloride acetate, 15 mm x 22 mm) under the skin. It is well known that during the polymerisation process of plastic manufacture numerous free radicals are trapped in the growing lattice (the 'cage-effect'); the insertion of plastic film under the skin can thus be considered to be equivalent to the insertion of spatially constrained free-radical moieties. The fact that the earliest changes in cell morphology (described as 'pre-malignant cells') occur in the immediate vicinity of the plastic film insert 1–8 months before the appearance of macroscopically visible tumours suggests that the carcinogenic process is associated with the plastic material itself; some controversy exists as to whether the carcinogenic action of the film is purely physical in nature or whether chemical properties of the plastic have significant importance. The development of the tumours is strongly dependent on the size of the inserted film and this suggests that physical factors such as alterations in the blood and nutrient supply, and chronic irritation leading to increased mitotic rate, are of considerable importance. Oppenheimer *et al.* (1953, 1955) have observed that there is a higher incidence of tumours when the plastic film is irradiated before insertion under the skin and their findings are consistent with the trapped free radicals within the film being of some significance to the overall pathological process. The trapping of free radicals in polymer cages during plastic polymerisation processes is discussed by Ingram (1958).

In relation to such free radicals in polymer cages it is noteworthy that Schwartz *et al.* (1965) have detected long-lived free radicals after the irradiation of bone and blood samples. As the authors comment, the existence of persistent radiation-induced radicals at room temperature in aqueous solution may represent a significant amount of chemical energy that is capable of producing damage long after the actual irradiation process has ceased. These findings of Schwartz *et al.* (1965) may be relevant to reports that whole body irradiation (with mixed γ- and neutron-radiation) produces material in human plasma that caused chromosome breaks when added to leucocytes in culture (Goh and Sumner, 1968).

Several workers have concerned themselves with the question: do free radicals themselves produce tumours when administered to experimental animals? For example, Boyland and Sargent (1951) administered the stable free-radical diphenylpicrylhydrazyl (see Figure 3.2) intradermally to mice and found that it did not produce localised greying of the hair;

a positive effect on hair greying had previously been shown with other materials to be related to carcinogenic activity. A number of organic hydroperoxides and peroxides have been tested for carcinogenicity by Van Duuren *et al.* (1963), Dunn (1965) and Melzer (1967). Lauryl peroxide, benzoyl peroxide 2,4-cholestadiene peroxide and 7,12-dimethyl-7,12-peroxybenz(*a*)anthracene were all inactive. Although the transannular peroxide ascaridole (Figure 2.8) is weakly carcinogenic in mice it is evident that the stable free radicals, peroxides and hydroperoxides so far tested have not been efficient in inducing tumours.

It is easy to suggest explanations for the failure of such experiments to induce cancer. First, the free radicals tested (for example diphenyl picryl hydrazyl) were stable in nature; as a consequence they were relatively inert in terms of their reactivity with biologically important molecules. Secondly, even had they been chemically very reactive, the very shortness of their half-life in solution at 37°C would be unfavourable for penetration to the genetic material in the nucleus if produced originally as the result of some metabolic event elsewhere in the cell, in the endoplasmic reticulum for example. Even if they should reach the vicinity of the genome whilst still possessing a considerable degree of chemical reactivity, then, as will be seen in a later Section, it appears to be important that the free radical should be capable of producing a *particular* type of lesion in the genome. Clearly the stable free radicals tested for carcinogenicity did not satisfy these criteria.

Chemically induced tumours may involve damage to DNA. Other possibilities involving protein modifications or disturbances to *t*-RNA are discussed for example by Pitot (1969), Miller (1970) and Craddock (1970). The mechanisms involved in producing the damage, and the nature of the damage itself need not necessarily be the same for different classes of carcinogen. Most of the following discussion will be restricted to the aromatic polycyclic hydrocarbons as examples of powerful carcinogens; as a working hypothesis it will be assumed that these materials produce their oncogenic effect primarily through an interaction with DNA. In such a situation involving disturbance to DNA function several minimal conditions would seem to be required. Firstly, the damage to DNA should be subtle enough in character to be unrecognised by repair processes; recent work suggests that DNA-polymerase is involved in repair of damaged DNA helices (De Lucia and Cairns, 1969). Secondly, the damage to DNA should lead to a misreading during replication of the damaged DNA molecule so that an inherited alteration in the sequence of nucleotide bases is obtained. Thirdly, this altered sequence should be non-lethal. Fourthly, the altered base sequence in DNA should result in a faulty transfer of information via the transcription by RNA polymerase to *m*-RNA; these disturbances then producing errors in translation. Fifthly, the translation errors should lead to cellular observations including by necessity those components of the cell membrane concerned with self-self recognition. A failure to reproduce this surface property accurately (either by errors in synthesis of specific components or by faulty arrangement within the membrane) would remove the constraint on organ size (in this context the 'organ' is the growing colony of transformed cells). Under normal conditions, organ size appears to be regulated by such factors as contact inhibition and circulatory inhibitors of mitosis (see Bullough, 1965) that interact in a tissue in a specific manner with cell membrane receptors. Although the chemical nature of such specific membrane receptors is not known, Neville (1968) has isolated a tissue specific protein from rat liver cell membranes; this protein constituted approximately 10% of the total membrane protein and was not significantly present in cell membranes obtained from a poorly differentiated tumour [3].

13.3 Interaction of carcinogens with DNA

What evidence is there that carcinogens in general, or a carcinogen in particular, reacts under physiological conditions with nucleic acids? There has been, in fact, a rapid increase in reports dealing with this subject; most of them have been concerned with covalent binding to DNA. Agents studied in this respect that have been found to bind covalently to DNA have included polycyclic hydrocarbons, aflatoxin, azo-dyes and nitrosamines (Figure 13.6). It is not possible to deal comprehensively with this work, study by study [4]; instead the salient features will be described briefly below.

[3] Changes in membrane bound antigens in various types of tumour have been reported but these are not necessarily related to the hypothetical specific receptors controlling organ size. Arcos *et al.* (1969) have found transient changes in the properties of the mitochondrial membrane during the early stage of carcinogenesis induced by amino azo dyes.

In addition, changes in the membrane composition of other intracellular particles have often been found in tumour tissues (e.g. in endoplasmic reticulum and in lysosomes; see Wallach, 1969). All of these changes may reflect some fundamental aberration of membrane synthesis perhaps involving glycoprotein metabolism (for review see Spiro, 1969). Although such changes would presumably have manifold effects on cell metabolism (the changes in endoplasmic reticulum composition could alter *m*-RNA activity; see Pitot, 1969) it is the change in cell surface self-self recognition sites that presumably is responsible for uncontrolled growth. It is this change in feedback control of organ size that differentiates the tumour cell from other types of injured cell. Recent studies (e.g. with concavalin A; Shoham *et al.*, 1970) suggest that the recognition sites are glycoprotein in nature. Walborg *et al.* (1969) have described the isolation of a sialoglycopeptide from the cell membranes of Novikoff Ascites cells; their paper gives many background references to changes in cell surface properties in tumours.

[4] References include: Brookes and Lawley (1964); Warwick and Roberts (1967); Tada *et al.* (1967); Nagata *et al.* (1967); Harvey and Halonen (1968); Lesko *et al.* (1969); Gelboin (1969); Umans *et al.* (1969).

If the polycyclic hydrocarbons are considered as an illustrative example then it has been found that DNA does not interact with 3,4-benzpyrene unless conditions favourable for free-radical formation are present. Such conditions include: irradiation of the DNA–benzpyrene mixture either with X-radiation or with much longer wavelength radiation, provided that the incident energy can be absorbed by the polycyclic hydrocarbon; the addition of iodine, or of hydrogen peroxide and iron; the ascorbate model hydroxylation system; incubating the DNA–benzpyrene mixture in the presence of a liver microsomal system + NADPH + oxygen. Under such conditions a covalent binding of the benzpyrene to DNA has been observed. Unfortunately, such covalent binding in a series of related polycyclic hydrocarbons does not appear to correlate well with the carcinogenic activity of these compounds; nor for that matter does the formation of non-covalent *complexes* between the nucleic acid bases and polycyclic hydrocarbons (Harvey and Halonen, 1968). Clearly if covalent binding is significant for carcinogenicity it must be a particular *type* of binding that is relevant; moreover, from the results outlined above, the biologically important binding must be only a small *proportion* of the total covalent binding reactions that occur.

The situation with experiments on carcinogenesis that require chronic exposure times to the carcinogen is made considerably more difficult by the inductive effects of many such agents on the liver microsomal electron-transport chain. Most of the carcinogens are metabolised by the cytochrome P_{450} chain and this may increase or decrease their carcinogenic activity (see for example Wattenberg and Leong, 1970; Gelboin *et al.*, 1970). During chronic exposure to the agent (for example 3,4-benzpyrene) the activity of the P_{450} chain may increase considerably thereby complicating the overall situation. Further such effects often show an additional

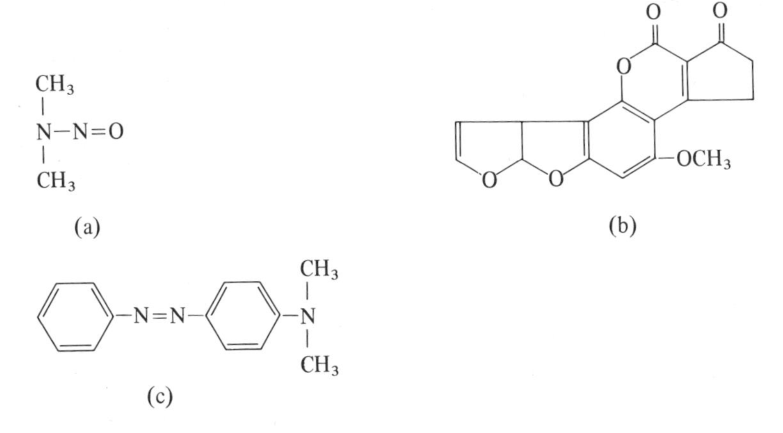

Figure 13.6. Structures of (a) dimethylnitrosamine, (b) aflatoxin B_1, and (c) butter yellow.

variation with the age and sex of the animal being studied[5]. Under such circumstances it is easy to appreciate that if a particular agent has to be metabolised by the cytochrome P_{450} chain to become carcinogenic then procedures that decrease the overall activity of the chain decrease the carcinogenic activity in parallel.

Returning now to the subject of covalent binding between polycyclic hydrocarbons and DNA, and its relationship if any to carcinogenic activity, some studies that have been made on the relationship between molecular structure and carcinogenicity can first be considered. A detailed discussion of this aspect is given by Hueper and Conway (1964). First, such studies have led to the concept already described concerning electronic distribution in the K- and L-regions. The physical size and conformation of the polycyclic molecule is also important with respect to carcinogenic action. Huggins *et al.* (1967) have stressed the need for **planarity** of the polycyclic structure in the induction of tumours[6]; high activity was associated with planar molecules in which the 'thickness' of the molecule did not exceed an apparently critical value of 4 Å. Planarity was also associated with a high activity in inducing the soluble liver enzyme menadione reductase. Similar findings with respect to planarity in relation to the induction of microsomal enzymes were reported by Arcos *et al.* (1961), who found in addition that the **surface area** of the molecule was a contributing factor to maximum activity.

The dependence of carcinogenic activity on planarity and on the size of the polycycle system (see Hueper and Conway, 1964) strongly suggests that the oncogenic response results from an intercalation of the polycycle into the DNA **double**-helix. The vertical separation between nucleotide base-pairs in the helix is approximately 3·4 Å; this can be doubled (with consequent distortion of the sugar phosphate backbone) by the intercalation of a planar molecule between successive base-pairs. This distance is compatible with the experimental findings of Huggins *et al.* (1967) that a thickness of less than 4 Å was a requisite for carcinogenic activity. Moreover, Haddow (1957) pointed out that the overall size of the purine–pyrimidine base-pairs was about that of carcinogenic polycyclic molecules; Boyland and Green (1962) showed clearly, with molecular models, how easily a polycyclic hydrocarbon could intercalate with little associated distortion of the DNA helix.

It is obvious that intercalation by itself is not a sufficient condition for the production of cancer: many planar molecules readily intercalate into DNA without being carcinogenic. It is easy to see why this is so. The binding forces involved in intercalation are weak even when the interactions

[5] Species differences in response to carcinogens are, of course, well known. A striking example is the extreme susceptibility of the rat to 7,12-dimethylbenzanthracene (Huggins *et al.*, 1965).

[6] For a review article on induction of mammary cancer by polycyclic hydrocarbons see Huggins and Yang (1962).

between the nitrogen bases and the polycyclic rings are considered. During the process of DNA replication, as the helix is progressively separated by a DNA polymerase the intercalation would be lost and thus have little or no effect on the DNA-replication process. Evidently, for a transmittable error in replication to occur some new covalent bonding must arise that affects the precision of DNA replication. Presumably, this bonding could affect either the nucleotide bases alone or involve the intercalator as well. It is important, however, that the bonding should lead to little distortion of the nucleotide helices so that **excision** by the repair enzyme does not occur prior to replication of DNA.

This leads to a consideration of the following question: what possible mechanisms involving polycyclic hydrocarbons are there to produce a change in the DNA helix, that are not easily recognised by the repair enzyme and which result in a genetically transmitted misreading of the code? One possible scheme involving free-radical intermediates is outlined below.

13.4 A speculative mechanism for the interaction between carcinogenic polycyclic hydrocarbons and deoxyribonucleic acid

That the nitrogen bases present in DNA can undergo free-radical reactions under suitable conditions has been discussed in Chapter 7. The rich variety of free radicals that may be formed is indicated by electron spin resonance examination of crystalline components of DNA undergoing exposure to ionising radiation. Free-radical centres may be formed on the bases or the sugar rings, with or without ring opening (see Zimmer, 1969). A general reaction is hydrogen addition or abstraction. In native DNA (and RNA) similar events probably occur. Certainly the nucleic acids are affected by exposure to ionising radiation and also to visible radiation in the presence of a photosensitiser. Singer and Fraenkel-Conrat (1966), for example, have shown that tobacco mosaic virus (an RNA-virus) loses its infectivity when illuminated in the presence of thiopyronin or, less efficiently, proflavin. The loss of infectivity was associated with a destruction of *guanine* residues.

Gräslund *et al.* (1969) have studied the effects of visible light irradiation on DNA in the presence of low concentrations of *acridine orange.* The results were consistent with the appearance of free radicals particularly in the *adenine–thymine*-rich regions of DNA.

Kodama and Nagata (1969) have extended such observations to the photosensitising action of polycyclic hydrocarbons, and the action of carcinogenic quinolines on DNA in an ethanolic solution. A selective photosensitised loss of guanine was described and intercalation was *not* required since the reaction occurred readily in single-strand DNA. During the photosensitising reaction polycyclic hydrocarbons were found to bind covalently to guanine but this binding under conditions *in vitro* did not correlate with carcinogenic activity of the polycyclic hydrocarbons. The solvent used may affect the base destroyed since Wacker *et al.* (1964)

found that photosensitisation of DNA by 3,4-benzpyrene in aqueous acetone caused a destruction of adenine in contrast to the destruction of guanine reported in the experiments of Kodama and Nagata.

In the above-mentioned experiments an external source of energy (e.g. visible light) was trapped by the photosensitising dye, which in effect conducted the energy to a susceptible region of the DNA molecule. Evidently, if base damage is required, the conduction is more efficient and far more probable if the dye is orientated close to the planar ring system of the acceptor base molecule. This suggests that the requirement for intercalation of the polycyclic hydrocarbons in respect of carcinogenesis must be to make more probable a relatively unlikely energy transfer. In other words, by intercalating in DNA *in vivo* the carcinogenic hydrocarbons conduct an energy input to the susceptible core of the DNA helix. This then raises two subsidiary questions: (a) what could serve as a source of the energy pulse *in vivo* corresponding to that of the radiation studies *in vitro*?; (b) what covalent change can be produced in DNA that would satisfy criteria previously stated with respect to the subsequent production of errors in replication?

It is possible that a polycyclic hydrocarbon could be photo excited directly by incident radiation (for example, in epidermis) or excited by photons emitted from a chemiluminescent reaction in the immediate vicinity of the DNA, or by energy transfer from a protein-free radical complex that would not itself be capable of intercalation. It seems intrinsically more probable, however, that excitation in tissues not directly exposed to light would proceed through a redox mechanism. It is well known that trace amounts of transitional elements occur in cells and that ascorbate and hydrogen peroxide are also present. The cell sap is also relatively rich in dissolved oxygen. The conditions are therefore suitable for a metal-catalysed excitation of the polycyclic hydrocarbon (PH) leading to structures such as $P^{\bullet}$, $PH^{\bullet -}$, $PH^{\bullet +}$ etc. Such an excited structure in the presence of oxygen could react to form a transannular peroxide. The formation of such compounds readily occurs under conditions *in vitro*; many decompose with the evolution of light. A transannular peroxide is unlikely to be formed, however, when the polycyclic hydrocarbon is intercalated; this follows from the limited space available between the successive pairs of nucleotide bases and the non-planar conformation of the transannular peroxide. If the excited polycyclic hydrocarbon can activate an oxygen molecule in the immediate vicinity (Figure 13.7) through the formation of a cyclic transition state then there is a possibility that a peroxide bridge forms between two opposing bases as shown in Figure 13.8.[7] The advantages of this reaction are: (a) the

[7] The N—O—O—N bonding arrangement shown in Figure 13.8 may be relatively stable whilst the nucleotide helices are associated, but may not be strong enough to maintain its character when the helices separate during replication. In such a case, the likely products are *N*-oxide or *N*-hydroxy derivatives of the respective bases.

oxygen bridge is now virtually coplanar with the bases; (b) there is minimum distortion of the base orientations, thus making it difficult to detect by the repair enzyme; (c) the covalent bond so formed, by preventing helix separation over a limited zone during transcription could lead to genetic error; (d) only those polycyclic compounds that can be excited easily by, for example, a redox reaction and that can donate electrons readily will enter the reaction; (e) the interaction will be greatly favoured by intercalation. All these points are consistent with the previous discussion of the essential properties of the carcinogenic hydrocarbons.

Of course, the presence of an intercalated hydrocarbon, a source of free radicals and DNA will lead to the formation of a variety of other covalent bond formations, for example polycycle-base linkage as found by Kodama and Nagata (1969) in the photosensitised system *in vitro*. Such covalent linkages would be readily apparent and would be excised by the repair

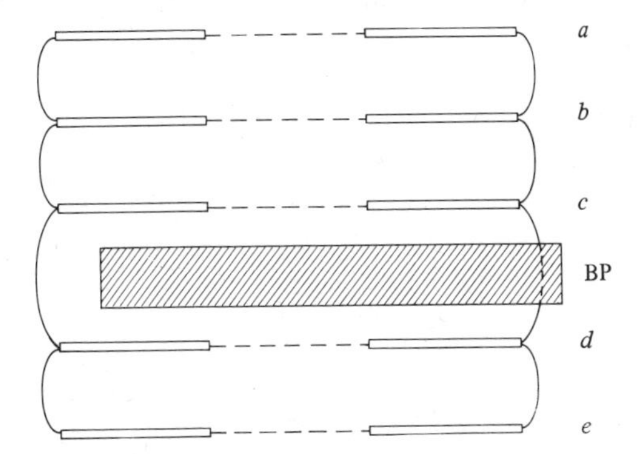

Figure 13.7. Representation of 3,4-benzpyrene (BP) intercalating between adjacent base-pairs of a DNA double helix. The bases are shown as horizontal rectangles lying in a plane at 90° to the paper with the sugar-phosphate backbone shown as loops. The normal separation distance of the base-pairs (approximately 3·4 Å) is shown between *a*–*b*, *b*–*c*, and *d*–*e* whereas 3,4-benzpyrene has produced an increased separation of the base pairs *c*–*d* by intercalating between them.

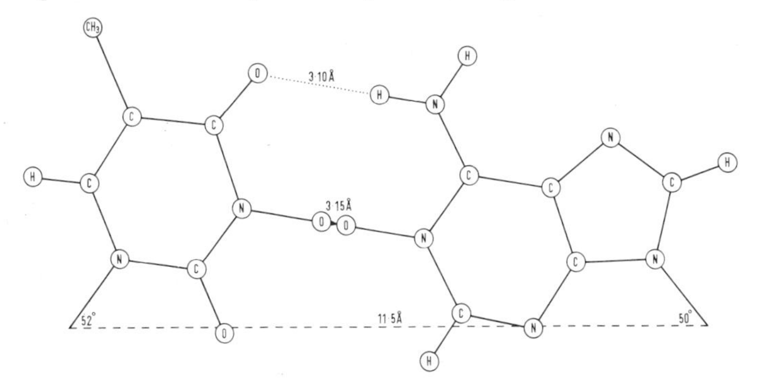

Figure 13.8. Structure of the hypothetical peroxide bridging across the corresponding bases of a DNA base-pair. For details see the text.

enzyme and would thus have little or no permanent genetic effect. It is also easy to see from the model outlined above why it is that although free radicals are postulated to be of importance in this form of chemical carcinogenesis, the concomitant feeding of antioxidants had no significant protective action (Epstein *et al.*, 1967). The free radicals are formed *inside* the DNA helix in a molecule that, by its intercalation, is relatively frozen and thus inaccessible to antioxidants in solution. Protection could be expected from materials that by extensive prior intercalation prevent the later positioning of polycyclic hydrocarbons in the helix. The data of Huggins *et al.* (1964) have in fact demonstrated this point clearly: a number of polycyclic hydrocarbons (e.g. chrysene, retene, 6-aminochrysene) were effective in reducing the incidence of mammary cancer in rats dosed with the potent carcinogen 7,12-dimethylbenzanthracene.

The hypothesis given above in regard to the carcinogenic action of the polycyclic hydrocarbons is interesting because it suggests a mechanism by which an alteration in the genome may occur and affect the first replication stage. Experimental tests of the mechanism would include searching for disturbances in specific cell membrane receptor materials (probably glycoprotein in character) following transformation by polycyclic hydrocarbons; and identification of oxygen substitution onto nucleotide-base nitrogens arising from degradation of the N—O—O—N arrangement. In the latter connection it is of interest that Brennan *et al.* (1966) and Emanuel *et al.* (1969) have reported unusual ESR signals in certain tumours that appear to result from a new radical species involving nitrogen.

It is an obvious step to consider extending the above argument concerning the polycyclic hydrocarbons to other carcinogens that may interact with DNA. The aflatoxins, for example, are derivatives of furocoumarin and the latter substance is known to react with nucleic acid bases upon irradiation as described in Section 14.4 to yield cyclobutane structures. It can be imagined therefore that aflatoxin may undergo excitation whilst intercalated so that it forms inter-strand linkages across the helices. Dall'Acqua *et al.* (1970) have recently reported that inter-strand cross-linkages are formed during the photoreaction of psoralen (a furocoumarin) with DNA. Dimethylnitrosamine, another powerful carcinogen[(8)], is known to be metabolised readily by liver microsomes and likely products include diazomethane and carbonium ion metabolites. These could conceivably form a methene bridge across base pairs (in addition to known alkylating effects on individual nucleotide bases) similar to the peroxide link suggested for the polycyclic hydrocarbons. Cross-linking of DNA strands has often been discussed in previous reports (e.g. see Peacock and Drysdale, 1965; Brookes, 1966) but the emphasis here has been on cross-linking in conjunction with minimal distortion of the helices as discussed in Section 13.2.

[(8)] For reviews on the toxicity of aflatoxins and dimethylnitrosamine see Goldblatt (1969) and Magee and Barnes (1967).

13.5 Cigarette smoking and free-radical studies

It has been known for some time that pyrolysis occurring during the smoking of cigarettes leads to the formation of a variety of polycyclic compounds including benzpyrene in small amounts (Wynter and Hoffmann, 1958). Similar materials are produced by the pyrolysis of a variety of organic matter other than tobacco as shown by Ingram (1961). The formation of these polycyclic hydrocarbons is accompanied by the production of unpaired electrons that give appreciable electron-spin-resonance signals; these unpaired electrons appear mainly associated with the condensed ring compounds (Lyons *et al.*, 1958; Lyons and Spence, 1960). An analysis of various pyrolytic deposits by Lyons and Spence gave the values for the numbers of free electrons per gram of deposit shown in Table 13.1. Whereas the soot radicals were very stable to light and not affected by solvent extraction, the cigarette-tar radicals were markedly light-sensitive and appreciable numbers of radicals were extractable with benzene or acetone. The authors conclude that the radicals in cigarette smoke would be more likely to interact with cell components (for example the lining of the respiratory tract) than 'soot radicals', which are tightly bound into the relatively large soot particles.

Table 13.1. Free radical content of various pyrolytic deposits. Data of Lyons and Spence, 1960.

	Number of radicals g^{-1}
Domestic chimney soot	5×10^{18}
Diesel soot	2×10^{19}
Cigarette side-stream smoke	5×10^{14}
Cigarette main-stream smoke	10^{15}

The identity of the free radicals in cigarette smoke has been investigated by Forbes *et al.* (1967), who also studied the effects of concentrated sulphuric acid on the electron-spin-resonance spectra of cigarette-smoke residues. They suggest that at least two radical types are present in sulphuric acid solution: a short-lived radical similar to the anthracene cation radical and a longer-lived radical similar to, but not identical with, the 3,4-benzpyrene cation radical. 3,4-Benzpyrene itself gave a much increased free-radical signal when heated to 290°C with subsequent cooling. The mechanism of this effect is discussed by Forbes and Robinson (1968), who suggest that the pyrolysis of 3,4-benzpyrene produces an azuleno[1,2,3-*cd*]phenalene radical (Figure 13.9).

The existence of free radicals in cigarette smoke is a topic currently attracting interest and further information about the identities and stabilities of the radical components in cigarette tars is required. It would also be interesting to study the effects of low concentrations of purified samples of the free-radical components that occur in cigarette smoke on cultured mammalian epithelial cells *in vitro*. Do these components have particularly marked inhibitory action on *ciliary action*, for example, or are they especially potent in producing malignant transformation *in vitro* as demonstrated already with 3,4-benzpyrene (see Sivak and Van Duuren, 1968; Di Paolo *et al.*, 1969)? Another possibility to consider for the action of such components *in vivo* is that they may induce the cytochrome P_{450} system normally present in lung tissue in very small amounts. As a consequence of such induction a 'lethal synthesis' of other extraneous components to carcinogenic forms may occur.

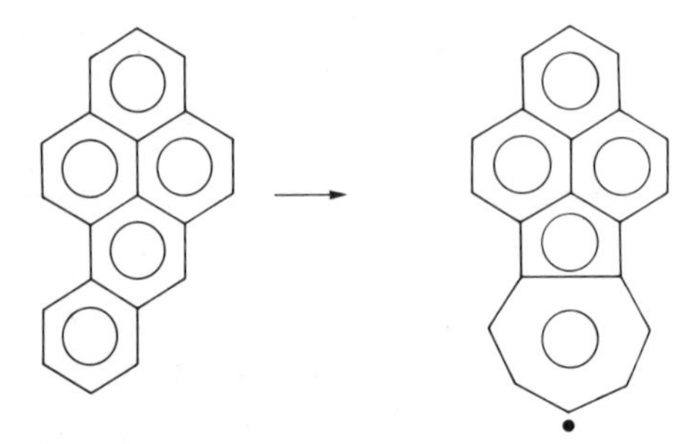

Figure 13.9 Formation of the azuleno(1,2,3-*cd*)phenalene radical from 3,4-benzpyrene (Forbes and Robinson, 1968).

Cigarette smoke contains **nitrogen oxides** in potentially toxic amounts as well as polycyclic materials. **Nitric oxide** is known to be less toxic than nitrogen dioxide and Norman and Keith (1967) showed that the major oxide in cigarette smoke was the lower oxide, which was present at about 700 p.p.m. Nitrogen dioxide was present in exceedingly low amounts if at all. This is an important finding since Thomas *et al.* (1968) have shown that rats exposed to 1 p.p.m. of **nitrogen dioxide** for 4 h showed evidence of lipid peroxidation in their unsaturated fatty acids extracted from lung tissue. The changes in lung lipids *in vivo* were partially prevented by a large dose of α-tocopherol. There are implications in this study of the onset of chronic inflammatory changes in the lungs (e.g. chronic bronchitis) of heavy smokers.

A further point relevant to the findings of Norman and Keith is that Neurath (1967) observed that a secondary amine would react with equimolar proportions of nitric oxide and nitrogen dioxide to produce nitrosamines under conditions that apply in cigarette smoking. Johnson *et al.* (1968), however, could find no evidence of nitrosamines in tobacco smoke by using a sensitive gas-chromatographic procedure.

References

Arcos, J. C., Conney, A. H., Buu-Hoi, N. P., 1961, *J. biol. Chem.*, **236**, 1291.

Arcos, J. C., Mathison, J. B., Tison, M. J., Mouledoux, A. M., 1969, *Cancer Res.*, **29**, 1288.

Boyland, E., Green, B., 1962, *Br. J. Cancer*, **16**, 507.

Boyland, E., Sargent, S., 1951, *Br. J. Cancer*, **5**, 433.

Brand, K. G., Buoen, L. C., Brand, I., 1967a, *J. natn. Cancer Inst.*, **39**, 663.

Brand, K. G., Buoen, L. C., Brand, I., 1967b, *Nature, Lond.*, **213**, 810.

Brennan, M. J., Cole, T., Singley, J. A., 1966, *Proc. Soc. exp. Biol. Med.*, **123**, 715.

Brookes, P., 1966, *Cancer Res.*, **26**, 1994.

Brookes, P., Lawley, P. D., 1964, *Nature, Lond.*, **202**, 781.

Bullough, W. S., 1965, *Cancer Res.*, **25**, 1683.

Buu-Hoi, N. P., 1964, *Cancer Res.*, **24**, 1511.

Carruthers, W., 1961, *Acta med. scand.*, supplement number 369, p.8.

Craddock, V. M., 1970, *Nature, Lond.*, **228**, 1264.

Dall'Acqua, F., Marciani, S., Rodighiero, G., 1970, *FEBS Letters*, **9**, 121.

De Lucia, P., Cairns, J., 1969, *Nature, Lond.*, **224**, 1164.

Di Paolo, J. A., Nelson, R. L., Donovan, P. J., 1969, *Science*, **165**, 917.

Dunn, J. A., 1965, *Br. J. Cancer*, **19**, 496.

Duuren, B. L. van, Nelson, N., Orris, L., Palmes, E. D., Schmitt, F. L., 1963, *J. natn. Cancer Inst.*, **31**, 41.

Emanuel, N. M., Saprin, A. N., Shebalkin, V. A., Kozlova, L. E., Krugljakova, K. E., 1969, *Nature, Lond.*, **222**, 165.

Epstein, S. S., Joshi, S., Andrea, J., Forsyth, J., Mankel, N., 1967, *Life Sci.*, **6**, 225.

Forbes, W. F., Robinson, J. C., 1968, *Nature, Lond.*, **217**, 550.

Forbes, W. F., Robinson, J. C., Wright, G. F., 1967, *Can. J. Biochem. Physiol.*, **45**, 1087.

Gelboin, H. V., 1969, *Cancer Res.*, **29**, 1272.

Gelboin, H. V., Wiebel, F., Diamond, L., 1970, *Science*, **170**, 169.

Goh, K., Sumner, H., 1968, *Radiat. Res.*, **35**, 171.

Goldblatt, L. A., 1969, *Aflatoxin* (Academic Press, New York).

Gordy, W., 1958, in *Information Theory in Biology*, Eds. H. P. Hockey, R. L. Platzman, H. Quastler (Pergamon Press, Oxford), p.353

Gräslund, A., Rigler, R., Ehrenberg, A., 1969, *FEBS Letters*, **4**, 227.

Haddow, A., 1957, in *Canadian Cancer Research Conference*, Ed. R. W. Begg (Academic Press, New York), p.361.

Harman, D., 1962, *Radiat. Res.*, **16**, 753.

Harvey, R. G., Halonen, M., 1968, *Cancer Res.*, **28**, 2183.

Heidelberger, C., 1970, *Cancer Res.*, **30**, 1549.

Hueper, W. C., Conway, W. D., 1964, *Chemical Carcinogenesis and Cancers* (Charles C. Thomas, Springfield, Illinois).

Huggins, C., Ford, E., Jensen, E. V., 1965, *Science*, **147**, 1153.

Huggins, C., Grand, L., Fukunishi, R., 1964, *Proc. natn. Acad. Sci. U.S.A.*, **51**, 737.

Huggins, C., Pataki, J., Harvey, R. G., 1967, *Proc. natn. Acad. Sci. U.S.A.*, **58**, 2253.

Huggins, C., Yang, N. C., 1962, *Science*, **137**, 257.

Ingram, D. J. E., 1958, *Free Radicals as Studied by Electron Spin Resonance* (Butterworths, London).

Ingram, D. J. E., 1961, *Acta med. scand.*, supplement number 369, p.43.

Johnson, D. E., Millar, J. D., Rhoades, J. W., 1968, *National Cancer Institute, Monograph 28*, p.181.

Kensler, C. J., Dexter, S. O., Rhoads, C. P., 1942, *Cancer Res.*, **2**, 1.

Kodama, M., Nagata, C., 1969, *Chem. biol. Interactions*, **1**, 99.

Lash, E. D., 1966, *Archs. Biochem. Biophys.*, **115**, 332.
Lesko, S. A., T'so, P. O. P., Umans, R. S., 1969, *Biochemistry, Easton,* **8**, 2291.
Lipkin, D., Paul, D. E., Townsend, J., Weissman, S. I., 1953, *Science,* **117**, 534.
Lyons, M. J., Gibson, J. F., Ingram, D. J. E., 1958, *Nature, Lond.*, **181**, 1003.
Lyons, M. J. Spence, J. B., 1960, *Br. J. Cancer,* **14**, 703.
Magee, P. N., Barnes, J. M., 1967, *Adv. Cancer Res.*, **10**, 164.
Melzer, M. S., 1967, *Biochim. biophys. Acta,* **142**, 538.
Miller, J. A., 1970, *Cancer Res.*, **30**, 559.
Nagata, C., Kodama, M., Tagashira, Y., 1967, *Gann,* **58**, 493.
Nebert, D. W., Mason, H. S., 1963, *Cancer Res.*, **23**, 833.
Neurath, G., 1967, *Experientia,* **23**, 400.
Neville, D. M., 1968, *Biochim. biophys. Acta,* **154**, 540.
Norman, V., Keith, C. H., 1967, *Nature, Lond.*, **205**, 915.
Oppenheimer, B. S., Oppenheimer, E. T., Danishefsky, I., Stout, A. P., Eirach, F. R., 1955, *Cancer Res.*, **15**, 333.
Oppenheimer, B. S., Oppenheimer, E. T., Stout, A. P., Danishefsky, I., Eirach, F. R., 1953, *Science,* **118**, 783.
Peacock, A. R., Drysdale, R. B., 1965, *The Molecular Basis of Heredity* (Butterworths, London), p.107.
Pitot, H. C., 1966, *A. Rev. Biochem.*, **35**, 335.
Pitot, H. C., 1969, *Archs Path.*, **87**, 212.
Pullman, A., Pullman, B., 1955, *Adv. Cancer Res.*, **3**, 117.
Reid, E., 1965, *Biochemical Approaches to Cancer* (Pergamon Press, Oxford).
Schuster, C. W., 1955, *Proc. Soc. exp. Biol. Med.*, **90**, 423.
Schwartz, H. M., Molenda, R. P., Lofberg, R. T., 1965, *Biochem. Biophys. Res. Commun.*, **21**, 61.
Shoham, J., Inbar, M., Sachs, L., 1970, *Nature, Lond.*, **227**, 1244.
Singer, B., Fraenkel-Conrat, H., 1966, *Biochemistry, Easton,* **5**, 2446.
Sivak, A., van Duuren, B. L., 1968, *Expl. Cell Res.*, **49**, 572.
Slater, T. F., Cook, J. W. R., 1969, in *Cytology Automation,* Ed. D. M. D. Evans (Livingstone, Edinburgh), p.108.
Spiro, R. G., 1969, *New Engl. J. Med.*, **281**, 991, 1043.
Tada, Mariko, Tada, Mitsuhiko, Takahashi, T., 1967, *Biochem. biophys. Res. Commun.*, **29**, 469.
Thomas, H. V., Mueller, P. K., Lyman, R. L., 1968, *Science,* **159**, 532.
Umans, R. S., Lesko, S. A., T'so, P. O. P., 1969, *Nature, Lond.*, **221**, 763.
Utsumi, K., Yamamato, G., Inaba, K., 1965, *Biochim. biophys. Res. Commun.*, **105**, 368.
Vithayathil, A. J., Ternberg, J. L., Commoner, B., 1965, *Nature, Lond.*, **207**, 1246.
Wacker, A., Dellweg, H., Träger, L., Kornhauser, A., Lodemann, E., Türck, G., Selzer, R., Chandra, P., Ishimoto, M., 1964, *Photochem. Photobiol.*, **3**, 369.
Walborg, E. F., Lantz, R. S., Wray, V. P., 1969, *Cancer Res.*, **29**, 2034.
Wallach, D. F. H., 1969, *New Engl. J. Med.*, **280**, 761.
Warwick, G. P., Roberts, J. J., 1967, *Nature, Lond.*, **213**, 1206.
Wattenberg, L. W., Leong, J. L., 1970, *Cancer Res.*, **30**, 1922.
Woolum, J. C., Commoner, B., 1970, *Biochem. biophys. Acta,* **201**, 131.
Wynter, E. L., Hoffmann, D., 1958, *Cancer N.Y.*, **11**, 1140.
Zimmer, K. G., 1969, *Phys. Med. Biol.*, **14**, 545.

Radiation-induced tissue injury

14.1 Introduction

Tissue injury can be produced by **electromagnetic radiation** covering a broad span of wavelengths[1]. The electromagnetic spectrum from γ-radiation to microwave sources is shown diagrammatically in Figure 14.1.

The penetration of radiation through skin tissue has been studied by various authors and is discussed by Blum (1964). With radiation of wavelength 7500 Å about 99% of the incident radiation is absorbed by 2·5 mm of tissue; with shorter wavelengths the absorption effect is increased so that 99% of incident radiation of wavelength 3000 Å is absorbed by 0·1 mm thickness of skin[2].

Ultraviolet radiation of wavelength less than 2900 Å probably does not penetrate deeply into the epidermis and the typical sunburn response in the dermal vasculature (erythema) to light at 2970 Å is probably due to a diffusion process from the epidermis (see Van der Leun, 1966). As the wavelength of the ultraviolet radiation is decreased still further there is a strong absorption of radiation in the region of 2800 Å, due mainly to the protein content of the tissue. In contrast radiation of very much shorter or very much longer wavelength passes through tissue more readily: for example, the use of X-radiation in diagnostic radiology and the use of visible radiation for transillumination examination of the maxillary sinuses.

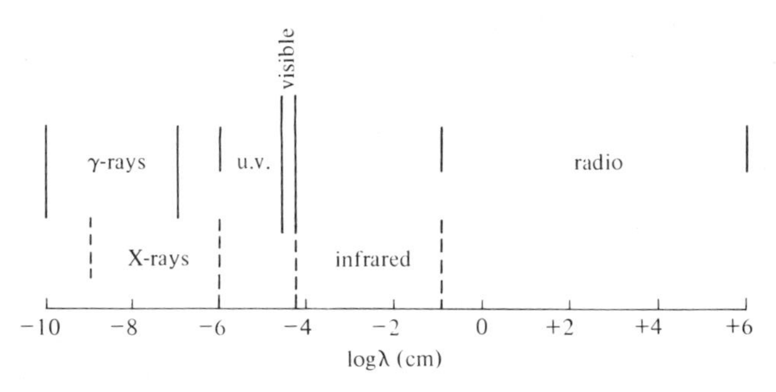

Figure 14.1. Diagrammatic representation of the electromagnetic radiation spectrum from γ-radiation to radio waves. The visible spectrum occupies the range 390–700 nm (1 nm = 10^{-7} cm).

[1] Radiation damage produced by particulate bombardment (for example by protons, neutrons or mesons) will not be considered here.

[2] Human epidermis is usually about eight cells deep; this corresponds to a thickness of about 0·1 mm.

14.2 Electromagnetic radiation of high energy

Radiation damage produced by X-rays, γ-radiation and short-wavelength ultraviolet radiation will be considered under this heading. This matter has been studied extensively in connection with protection against the harmful effects of atomic radiation, against accidental over-dosage with X-irradiation, and in view of the therapeutic potential of high-energy radiation for tumours. The energies associated with quanta of radiation of these types are enormous, and are easily capable of producing homolytic bond dissociation provided that the energy can be trapped by a molecular species during passage of the radiation through the irradiated tissue. Since the interaction between high-energy radiation and a molecular species often results in the ejection of an electron with the formation of a positively charged ion, such radiation is called *ionising radiation*. The production of free radicals when aqueous solutions are exposed to high-energy radiation is well known and a suitable introductory review of this is by Boag (1965). The numerous studies on protection against the deleterious effects of ionising radiation are reviewed by Bacq (1965).

The energies involved in the absorption of ionising radiation are usually so large that the effects on tissue can be quite unexpected and varied. It is not proposed to deal extensively with this question here and the interested reader is referred to Bacq and Alexander (1961). A few examples only will be described to illustrate the diversity of effects involved, particularly with respect to the effects of high-energy radiation on tissue components *in situ*. The effects of high-energy radiation on isolated *tissue components* or on pure materials (e.g. nucleic acid bases) have already been outlined in Chapter 6.

14.2.1 Mitochondrial changes following γ-irradiation

Noyes and Smith (1959) found that whole-body γ-radiation of rats at a rate of 35 rd min^{-1} up to a total of 1000 rd, led to a rapid decrease in

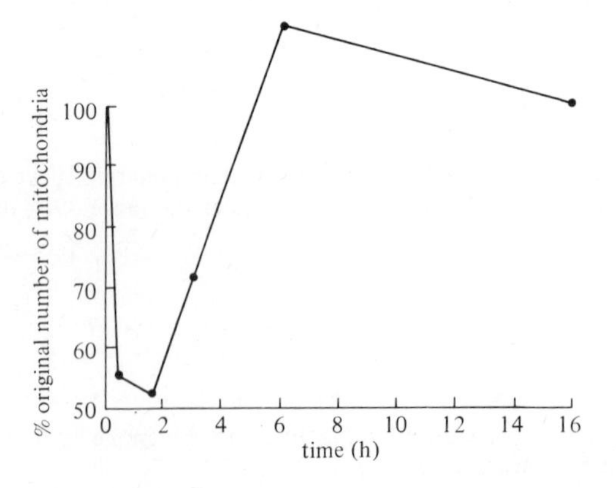

Figure 14.2. Effect of whole body γ-irradiation on the number of mitochondria in rat liver (data of Noyes and Smith, 1959).

mitochondrial numbers in the liver. The number fell by approximately 40% within 15 min of dosing and had recovered by 6 h (see Figure 14.2). These changes are very rapid and are of interest in view of the values previously quoted for rat liver mitochondrial turnover, based on protein labelling, that indicated a half-life of approximately 12 *days* (Fletcher and Sanadi, 1961).

14.2.2 Liver changes *in situ*

The irradiation of human liver with 3000–5000 rd over a 6-week period during therapy for hepatic tumours was found to produce hyperaemia and centrilobular necrosis. Later a fibrous obliteration of the small hepatic veins resembling *veno-occlusive disease* was observed (Reed and Cox, 1966).

14.2.3 Changes in blood components after γ-irradiation

Extra-corporeal irradiation of blood is used as a therapy for certain blood dyscrasias. This procedure ensures that the high-energy radiation required for treatment will not damage overlying tissue as occurs in the radiation of solid tissue or in whole body irradiation. However, Swartz *et al.* (1965) and Swartz (1965) have found that γ-radiation of blood (and bone) can lead to long-lived radical species. The possibility of endothelial injury from the return of these radical species in blood suspensions to the circulation cannot be overlooked. As mentioned in Chapter 12 Harman and Piette (1966) have suggested that chronic production of free radicals in the vasculature may lead to age changes characteristic of atheroma.

14.2.4 Proton irradiation of liver

Ghidoni and Thomas (1968) have studied the disturbances in liver during whole-body irradiation with 32 MeV protons equivalent to 6000 rd. The radiation penetrated approximately 1 cm into the animals, but membrane changes in the mitochondria, endoplasmic reticulum and peroxisomes in the liver were observed. Within a few hours of irradiation bleb formation in the sinusoidal spaces was seen that was possibly responsible for the observed changes in hepatic blood flow.

14.2.5 Lysosomal membrane changes

The effects of ultraviolet radiation *in vitro* on lysosomal membranes have been studied by Desai *et al.* (1964). A correlation was found between the concentration of free radicals produced by lipid peroxidation and enzyme release from the lysosomes. Similar changes in lysosomal enzyme release were found when free-radical production was achieved by γ-irradiation, or by the addition of hydrogen peroxide to the lysosomal suspensions. These findings are illustrated in Figure 14.3.

The participation of free-radical intermediates has been well established in the production of the effects described in some of the examples above. In others it can be assumed for the energies involved in the absorption of the quanta of radiation used are so large that homolysis would be a likely consequence. It is evident, however, that for some examples

(Section 14.2.3) the final event observed (here a form of veno-occlusive disease with fibrous obliteration of small hepatic veins) must evolve from a most complex sequence or network of effects of which little is known.

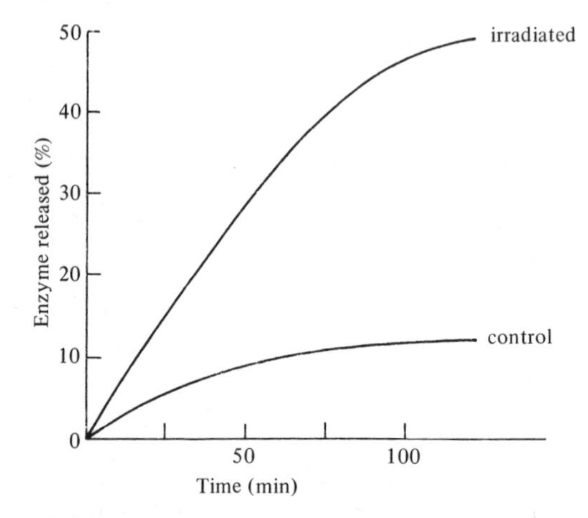

Figure 14.3. Effect of ultraviolet radiation (366 nm) on the release of arylsulphatase from rat liver lysosomes at 37°C (data of Desai *et al.*, 1964).

14.2.6 Nucleotide changes

The reduced forms of the nicotinamide–adenine dinucleotides are known to be particularly sensitive to high-energy radiation in solution with loss of coenzyme function. Susceptibility of reduced NADPH to attack by $CCl_3^\bullet$ has already been described in Chapter 10 (Section 10.8.3); the straightforward destruction of reduced nicotinamide–adenine dinucleotides by ionising radiation in solution has been studied by Barron *et al.* (1954), and with ultraviolet radiation by Myers (1960). In tissue suspensions, however, destruction of the nucleotides is accompanied by changes in the activity of catabolic enzymes that complicate the interpretation of the results. In irradiated thymocytes, for example, Scaife (1963) found that there was a decrease in total NAD that was associated with an **increase** in the activity of NAD-*glycohydrolase.*

Land and Swallow (1968) have studied the mechanism of destruction of the oxidised nucleotides by high-energy electrons in pulse-radiolysis experiments and have obtained evidence that the process involves the formation of a free-radical form of NAD with subsequent dimerisation (see also Burnett and Underwood, 1968).

$$\text{NAD} + e^- \longrightarrow \text{NAD}^\bullet \qquad (14.1)$$

$$2\text{NAD}^\bullet \longrightarrow (\text{NAD})_2 \qquad (14.2)$$

The NAD-dimer has no coenzyme activity and yet shows considerable absorption of light at 340 nm.

14.3 Visible radiation

With visible radiation the energies available for chemical change are far less than those dealt with in the previous Section. Further, many substances exhibit sharp absorption bands in the visible region (for example porphyrins that have a Soret band at approximately 405 nm); light-radiation (particularly monochromatic) with its major intensity located away from such a peak would not be significantly trapped and its energy would not be capable of promoting a chemical change. This is another way of stating the Grotthuss–Draper law described in Section 2.3. This can be circumvented to some extent by the addition to the system of a molecular species that will absorb the incident light and, as a result, become excited; the excited molecule may then pass on its excess of energy to the original reactants under study. A classical example of such a photosensitised reaction has been described in Section 2.3 for the dissociation of molecular hydrogen by mercury. Many such photosensitised reactions are dependent on the presence of oxygen; where such reactions involve degradative changes in living organisms or tissue it is usual to speak of the reaction as a **photodynamic effect** [(3)].

A large number of photodynamic effects have been described under laboratory conditions since the observations of Raab (1900) that *Paramecia* were rapidly killed by low concentrations of acridine in the presence of light that normally had no effect on the organisms. The lethal action of such systems on *micro-organisms* has been widely studied and is discussed in detail by Blum (1964). The kinetics of *virus* inactivation by photodynamic action are discussed by Hiatt (1967).

The chemical mechanisms underlying such photodynamic effects have attracted much attention. To summarise a complex situation: the photosensitiser molecule absorbs incident radiation, becoming excited to an excited singlet state (S*) or, by a so-called forbidden transition to a triplet state (T). This may occur with the photosensitiser either in free solution or when bound to the material that finally undergoes degradation. Two pathways are now open to the system: the photosensitised excited molecule may react with oxygen, or may transfer energy to the other molecular species that itself then reacts with oxygen. These changes are illustrated in Reactions (14.3)–(14.8), in which PS is the photosensitiser, and S_{red} is the molecule undergoing photodynamic change in its original reduced form:

$$\mathrm{PS} + h\nu \longrightarrow \mathrm{PS}^{*}(\text{excited}) \qquad (14.3)$$

$$\mathrm{PS}^{*} \longrightarrow \mathrm{PS}^{\bullet} \qquad (14.4)$$

$$\mathrm{PS}^{\bullet} + \mathrm{O_2} \longrightarrow \mathrm{PSO_2^{\bullet}} \qquad (14.5)$$

$$\mathrm{PSO_2^{\bullet}} + \mathrm{S_{red}} \longrightarrow \mathrm{PS} + \mathrm{SO_2^{\bullet}} \qquad (14.6)$$

$$(\mathrm{PS}^{*} + \mathrm{S_{red}}) + \mathrm{O_2} \longrightarrow \mathrm{PS} + \mathrm{SO_2^{\bullet}} \qquad (14.7)$$

$$\mathrm{PS}^{*} + \mathrm{S_{red}} \longrightarrow \mathrm{PS_{red}} + \mathrm{S_{ox}} \qquad (14.8)$$

[(3)] Such effects must be clearly distinguished from normal *physiological* processes that involve light-absorption, such as photosynthesis or retinal function.

In the absence of oxygen it can be seen that the photosensitiser may catalyse the oxidation of the substrate [Reaction (14.8)]. This is clearly shown in Figure 14.4, in which NADH is oxidised by protoporphyrin in the presence of light but absence of oxygen. When oxygen is admitted to the system the normal porphyrin colour reappears.

The homolytic nature of such photosensitised decompositions is now well known and in this Chapter a few examples of naturally occurring photodynamic effects observed in livestock and in man will be discussed. Of course, since visible radiation does not penetrate very well into the body (see Section 14.1) the most profound clinical effects are on the skin. It is possible that similar effects may arise by the exposure of other organs to light during special treatments (for example during operations or by bronchoscopy etc.).

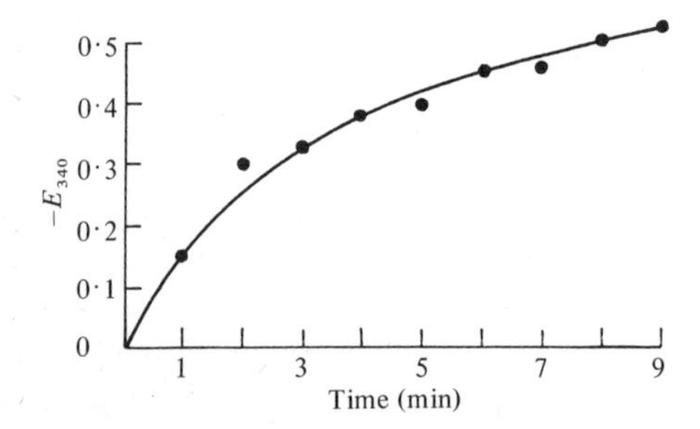

Figure 14.4. Photosensitised oxidation of NADH (0·5 mM) by chlorophyll (10 μM) in alcoholic solution with time of irradiation with red light (data of Krasnovsky, 1967).

14.3.1 Photosensitisation in man

It has been known since antiquity that certain diseases predispose to an abnormal sensitisation of the skin to light and the outstanding observations by Fischer on **porphyrin** photosensitisation about 50 years ago have led to recent proposals concerning the chemical mechanisms involved. In certain types of porphyria (for review see Rimington *et al.*, 1967) there is a marked and even extreme sensitivity to sunlight. The lesions may vary from relatively innocuous ones to severe disfigurement as described by Fischer. Magnus *et al.* (1959) showed that the most effective wavelength for producing skin reaction in these patients coincided with the Soret band (wavelength approximately 405 nm), indicating that it was an interaction of the porphyrin in the excited state with some cellular component that was responsible for the rapid disturbances in the epidermis. This wavelength region must be distinguished from the short wavelength ultraviolet region (about 290 nm) that is responsible for erythema in normal subjects. The question arises: what intracellular changes could be *initiated* by the production in the epidermis of the porphyrin in its excited state that would result in changes consistent with the clinical observations? In 1966

Slater and Riley suggested that a likely mechanism would be the interaction of the excited porphyrin or a derived free-radical product with unsaturated lipid components of epidermal cell lysosomes in the presence of oxygen. A consequence of such an interaction would be a lipid peroxidative change of the lysosomal membranes resulting in a liberation of intralysosomal hydrolases. The liberation of considerable amounts of the lysosomal acid hydrolases into the epidermal cell supernatant can be confidently expected to result in deleterious changes such as those observed clinically in photosensitisation. The free-radical nature of the initial porphyrin-dependent process was demonstrated by several experimental procedures. First, studies *in vivo* had shown that the photosensitisation response was dependent on oxygen (see Blum, 1964); secondly, with an *in vitro* system Slater and Riley (1966) showed that brief periods of illumination with visible light in the presence of very low concentrations (nM) of a porphyrin phylloerythrin (Figure 14.5) caused rupture of epidermal lysosomes. This lysosomal damage was prevented by excluding oxygen from the incubation medium, or by the addition of low concentrations of free-radical scavengers such as α-tocopherol, promethazine or vitamin C. Further, a copper derivative of phylloerythrin, which was nonfluorescent, was not effective in producing lysosomal damage in the system *in vitro*. Some of the results obtained are elucidated in Figure 14.6. Independently of the work of Slater and Riley outlined above was the study by Allison *et al.* (1964, 1966), who showed that dyes like eosin could be taken up by lysosomes and that irradiation of skin in the presence of a photosensitisor led to changes in *endothelium*. The endothelial changes were accompanied by gross permeability changes that may be responsible for the oedamatous response characteristic of photosensitisation disturbances. Thus the evidence for a free-radical attack on lysosomes in skin as a primary mechanism for photosensitisation is compelling. These findings

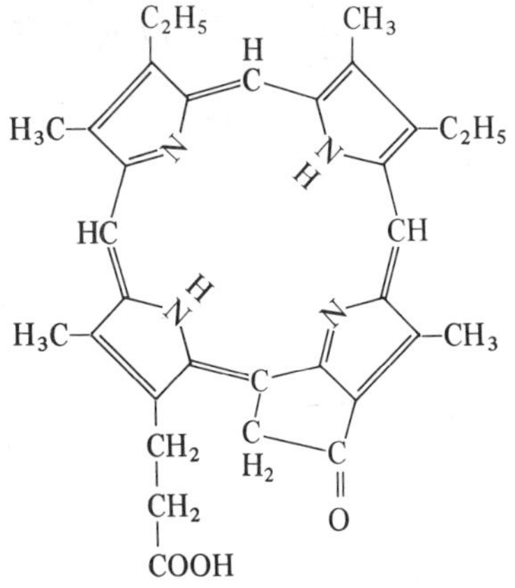

Figure 14.5. Structure of phylloerythrin.

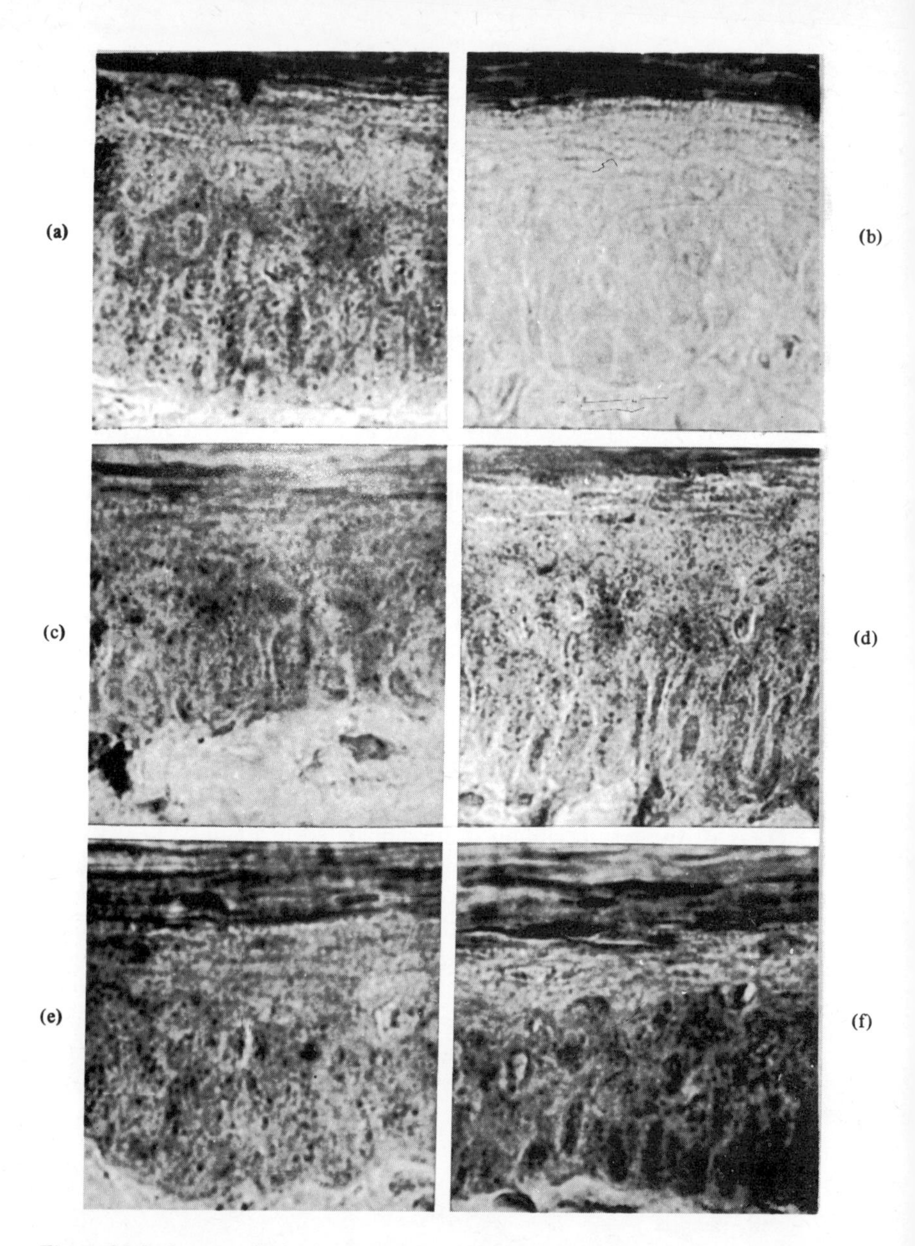

Figure 14.6. Sections (7μ) of rat tail skin stained for acid-phosphatase activity (60 min incubation at 37°C, Gomori technique) in the lysosomes of epidermal cells. The lysosomes show up as small dark granules by this technique. In (a) the skin section was incubated under standard conditions for 15 min with a low concentration (5 μM) of the photosensitiser phylloerythrin but without irradiation by visible light. In (b) the conditions were the same as in (a) but the section was irradiated for 10 s during incubation. In (c)–(f) the conditions were similar to (b) with the following exceptions: (c) phylloerythrin was replaced by a non-fluorescent copper-phylloerythrin derivative (10 μg ml^{-1}); (d) the irradiation in the presence of phylloerythrin (50 μM) was done after flushing out oxygen in the incubation medium with nitrogen; (e) ascorbate (1 mM) was included in the incubation medium; (f) promethazine (1 mM) was included in the incubation medium. Results obtained by T. F. Slater and P. A. Riley.

offer the hope that parenteral therapy of photosensitisation diseases may be possibly based on the use of free-radical scavengers as opposed to current practice where rather objectionable barrier creams are used to shade the underlying tissue from light. In general, quite thick layers of skin-cream preparations have to be applied to decrease *effectively* the intensity of light reaching the skin (Robertson, 1969). One free radical scavenger that has already been used parentally to successfully attenuate photosensitisation in humans is β-carotene. Matthews-Roth *et al.* (1970) have administered this substance to patients with clinical erythropoietic protoporphyria and have observed a marked clinical improvement.

Human skin contains the pigment melanin within intracellular granules called melanosomes (see Nicolaus, 1968). These are synthesised by specialised cells (the melanocytes) that later extrude the melanosomes; these particles are subsequently taken up by epidermal cells into structures that are probably lysosomal in nature. The presence of melanin is generally believed to be protective in that melanin exhibits a very broad absorption of ultraviolet and visible radiation and this acts as an effective screen to protect the underlying *dermis*. It is well known that *continued* exposure of Caucasians to sunlight increases the darkness of the skin, primarily through the increased number of melanosomes.

Melanin has been found to contain free-radical centres (Commoner *et al.*, 1954; Mason *et al.*, 1960; Longuet-Higgins, 1960; Blois, 1969) and exposure to light increases the intensity of the electron spin resonance signals. It has been suggested that a function of melanin is to protect the epidermis against radicals produced by radiation through a trapping process (Daniels, 1959). Since the melanin is normally confined within a membrane-limited particle, however, this speculation does not appear very likely. Another suggestion has been that the melanin itself would be potentially dangerous to cell components owing to its free-radical content unless it were so confined (Slater and Riley, 1966).

The effects of ultraviolet, visible and laser radiation on human skin samples has recently been studied by Pathak and Stratton (1968). Some of their data are shown in Figure 14.7. It can be seen that no significant electron spin resonance signal is apparent in unpigmented skin but that increasing doses of ultraviolet radiation result in an increasing signal. Laser radiation produced no signal in unpigmented skin, but a considerable augmentation of the normal melanin signal in pigmented skin was observed after exposure to a laser flash.

The production of free-radical signals in unpigmented skin (Figure 14.7) is of considerable interest in view of the known association of exposure of Caucasians to the tropical sun in Australia with the incidence of skin cancer. The solar radiation received at ground level in the tropics includes light of a considerably shorter wavelength than received in temperate zones at comparable times of the year. In the more temperate regions solar energy falls off rapidly below about 400 nm and virtually none is

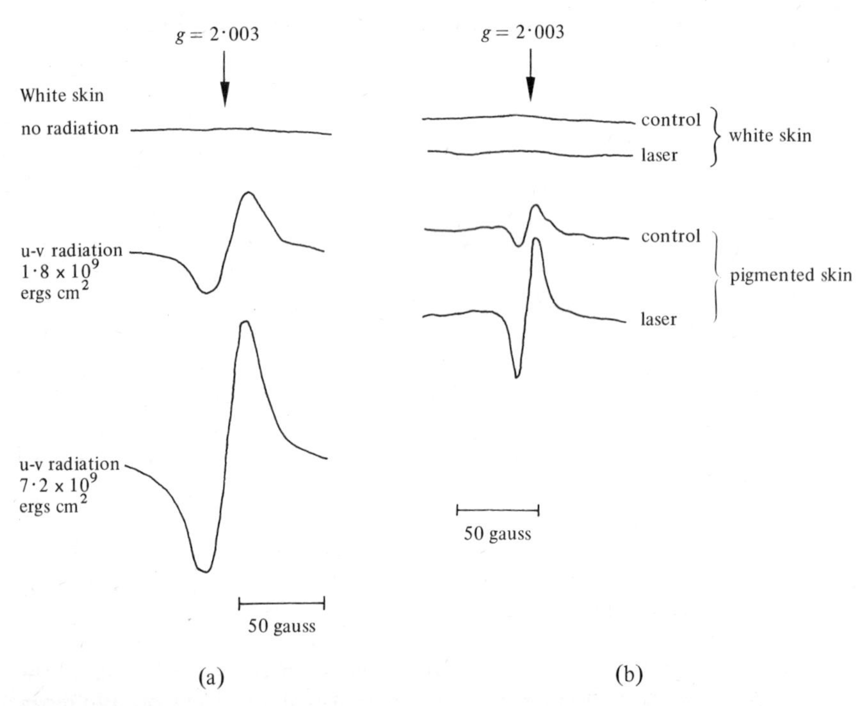

Figure 14.7. Electron spin resonance spectra of (a) 'white' skin with and without ultraviolet radiation and (b) 'white' and pigmented skin with and without laser irradiation at 694 nm. Measurements were made at −196°C. Data from Pathak and Stratton (1968).

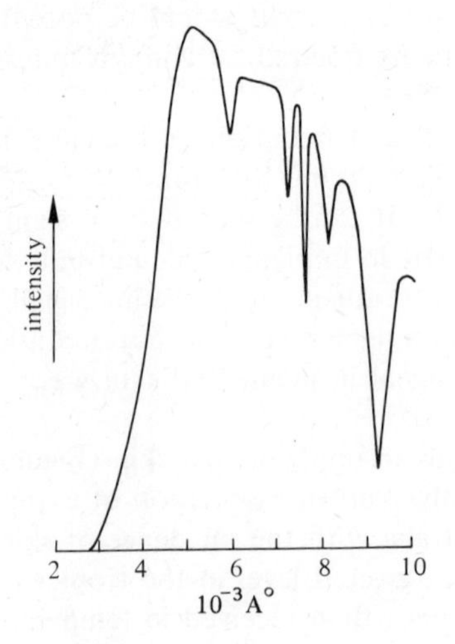

Figure 14.8. Spectrum of sunlight received at earth's surface with the sun at 60° from the zenith (from Blum, 1964).

present below 310 nm (see Figure 14.8). The tropical sunlight is greatly augmented in the ultraviolet region that is responsible for the erythematous response.

This last observation is of relevance to the disease **Xeroderma pigmentosum**, which is inherited via an autosomal recessive gene. Here the affected patient is conspicuously photosensitive from very early life and, without treatment, widespread skin tumours develop and there is a poor prognosis. Although an action spectrum for this affliction has not been obtained, the relationship of exacerbations with the season in temperate zones, and with exposure to tropical sunlight, suggests that it is light of about 300 nm that is responsible.

It is possible that in this context there is an inherited defect either in the stability of the lysosomal membrane or in the presence of a photosensitiser that absorbs in the region of 300 nm and which is responsible for the chronic photosensitised damage. Repeated exposure of skin epithelium to such radiation-induced trauma is apparently associated with a greatly increased incidence of skin carcinoma, probably arising from free-radical attack on genetic material by mechanisms similar to those discussed in Section 7.4. A review of Xeroderma pigmentosum has been published by El-Hefnawi and Rasheed (1966).

14.3.2 Photosensitisation in animals

As a second example, the mechanism of action of the fungal toxin **sporidesmin** will be discussed (Figure 14.9). In New Zealand, after a long dry summer the warm rainy season is accompanied by an extensive growth on the pasture of a fungus *Pithomyces chartarum*. Sheep eating the infected grass develop acute liver damage with an associated photosensitisation. The underlying basis of the disturbance is that sporidesmin produces a liver lesion resembling acute cholangiolitis; this causes an obstructive jaundice with the retention of components normally excreted in the bile. In herbivores the retention of bile produces a large increase in the plasma concentration of the chlorophyll degradation product **phylloerythrin** (Figure 14.5). Phylloerythrin subsequently accumulates in the epithelium of the skin and it is this porphyrin that is responsible for photosensitisation and which, since the first signs of oedematous swelling are on the

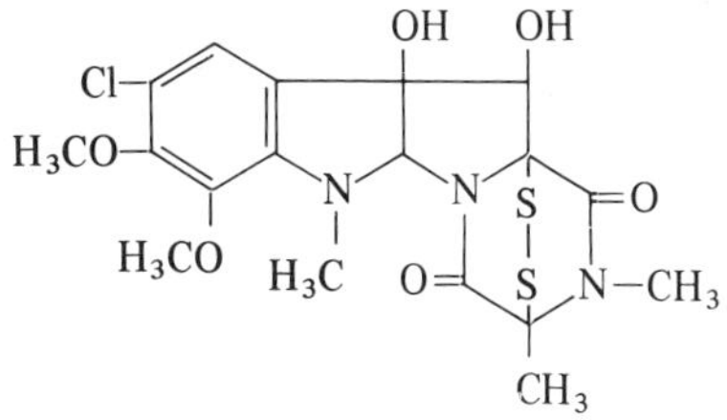

Figure 14.9. Structure of sporidesmin (see Hodges *et al.*, 1963).

unprotected parts of the face, has led to the term *facial eczema* for this disease. The rat is not as sensitive to sporidesmin as are the sheep and rabbit but extensive changes in the permeability of various capillary plexuses occur so that ascites and pleural effusions are a feature of sporidesmin intoxication in the rat (Rimington *et al.*, 1962; Slater *et al.*, 1964).

The underlying cause of the photosensitisation in sheep and rabbits is the prolonged inhibition of bile flow associated with severe liver damage. Although sporidesmin does not produce a comparable damage to the liver in the rat it does produce a transient decrease in bile flow. This effect can be studied with the isolated perfused liver preparation and is rapid in its onset (Figure 14.10). This liver disturbance is associated with gross changes in the structural organisation around the *biliary canaliculi*. Normally the canaliculus is lined with numerous microvilli (Figure 14.11) but 45 min after treatment with sporidesmin the microvilli have largely disappeared and the canaliculi have distorted and become fragmented (Figure 14.11). This evidence suggests that sporidesmin is having a profound effect on the membrane lining the canaliculus. Previous studies of the liver *in situ* with fluorescent microscopic techniques had shown that low doses of sporidesmin inhibited the secretion of an intravenously injected porphyrin from the liver cell into the bile; higher doses prevented uptake of porphyrin into the liver itself (Reese and Rimington, 1964). The finding that sporidesmin is a potent irritant and produces a long sustained increase in capillary permeability (Slater *et al.*, 1964), supports

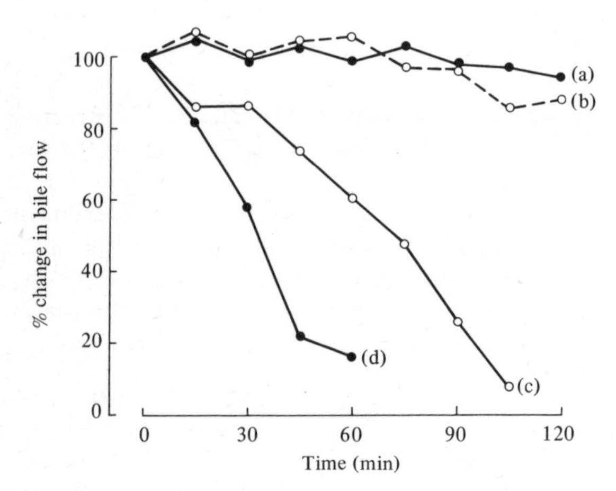

Figure 14.10. Changes in bile flow in isolated perfused rat liver (unpublished data of T. F. Slater, B. C. Sawyer and M. N. Eakins):
(a) control, no additions to perfusing medium; (b) 0·2 ml of ethanol in 9·8 ml of Krebs–Ringer bicarbonate buffer added at time 0; (c) 2 mg of sporidesmin added in the same volume of solution as for (b); (d) 2 mg of icterogenin added in the same volume of solution as for (b).

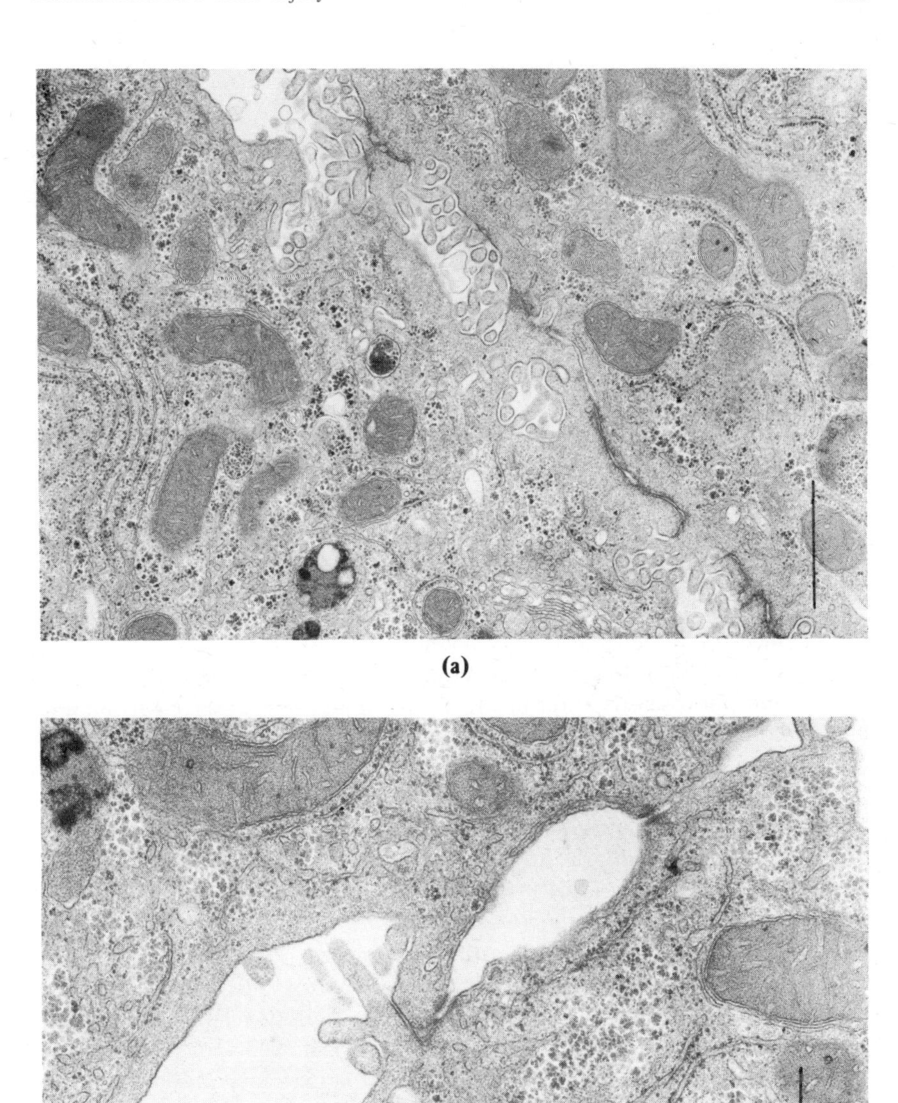

(a)

(b)

Figure 14.11. Electron micrographs of liver cells. In (a) the sample was taken from a control liver maintained in an isolated perfusion system for 3 h. The liver was then washed through with saline and then with cacodylate buffer containing glutaraldehyde. The canaliculus contains numerous microvilli. In (b) a liver was treated similarly to (a) except that 2 mg of sporidesmin was added to the perfusion medium after 90 min. The canaliculus is almost devoid of microvilli (unpublished data of T. F. Slater, B. C. Sawyer, M. N. Eakins and G. Bullock).

the view that sporidesmin mediates its diverse effects via actions, either direct or indirect, on membrane function. There is some evidence that the disulphide bridge in sporidesmin is essential for its activity; this indicates that sporidesmin possibly reacts with thiol groups in the membrane that are important for membrane structure [Reaction (14.9)].

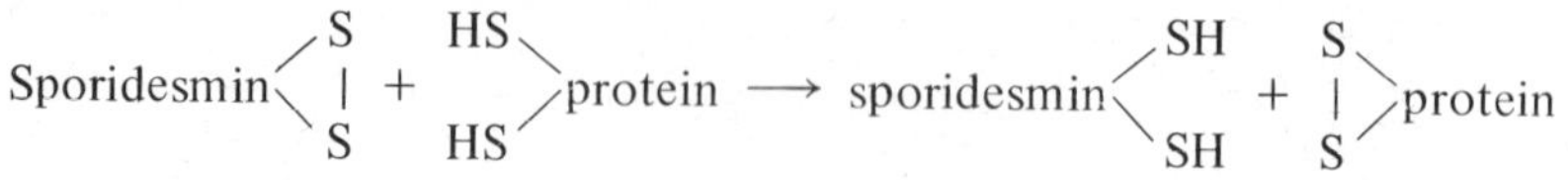

$$\text{Sporidesmin}\left\langle\begin{matrix}S\\|\\S\end{matrix}\right. + \left.\begin{matrix}HS\\ \\HS\end{matrix}\right\rangle\text{protein} \longrightarrow \text{sporidesmin}\left\langle\begin{matrix}SH\\ \\SH\end{matrix}\right. + \left.\begin{matrix}S\\|\\S\end{matrix}\right\rangle\text{protein} \qquad (14.9)$$

Brief mention can be made at this point to another naturally occurring agent that produces photosensitisation in livestock, **icterogenin** (Figure 14.12). This substance was isolated by Rimington and colleagues, who were studying the disease of cattle in South Africa known as *Geel-dik-kop* or yellow thickhead (an allusion to the jaundice and oedema of the face that is apparent). Geel-dik-kop is associated with ingestion of the plant *Tribulus terrestris*; a similar course of events follows the ingestion of the shrub *Lippia rehmanni*. From the latter source Rimington *et al.* (1937) isolated icterogenin. Although icterogenin produces jaundice and photosensitisation in sheep liver injury was observed to be minimal by conventional microscopical examination. Like sporidesmin, icterogenin produces a very rapid inhibition of bile flow in the isolated perfused liver (Figure 14.10) and characteristic changes in the appearance of the biliary canaliculus (Figure 14.13). The similarity of its structure to the natural choleretics such as taurocholate (Figure 14.14) suggests that icterogenin may be competing with taurocholate (or other bile salts) for a site on the canalicular membrane concerned with biliary secretion. If so, then it is likely that icterogenin has first to undergo reduction of the carbonyl oxygen atom at C-3 to produce a C-3 hydroxyl derivative. This is made more probable by the finding that the closely related epimers of a substituted oleonolic acid (Figure 14.15) showed very different biological activities. The 3β-hydroxyl compound is extremely potent as a cholestatic agent whereas the 3α-epimer is inactive (Brown *et al.*, 1963). Further studies on the mechanisms of action of these substances should yield valuable information concerning the mechanisms underlying biliary secretion.

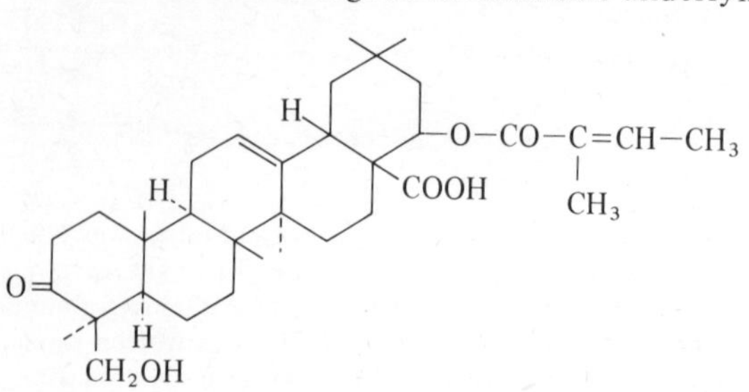

H
H
$-O-CO-C=CH-CH_3$
COOH
CH_3
O=
H
CH_2OH

Figure 14.12. Structure of icterogenin.

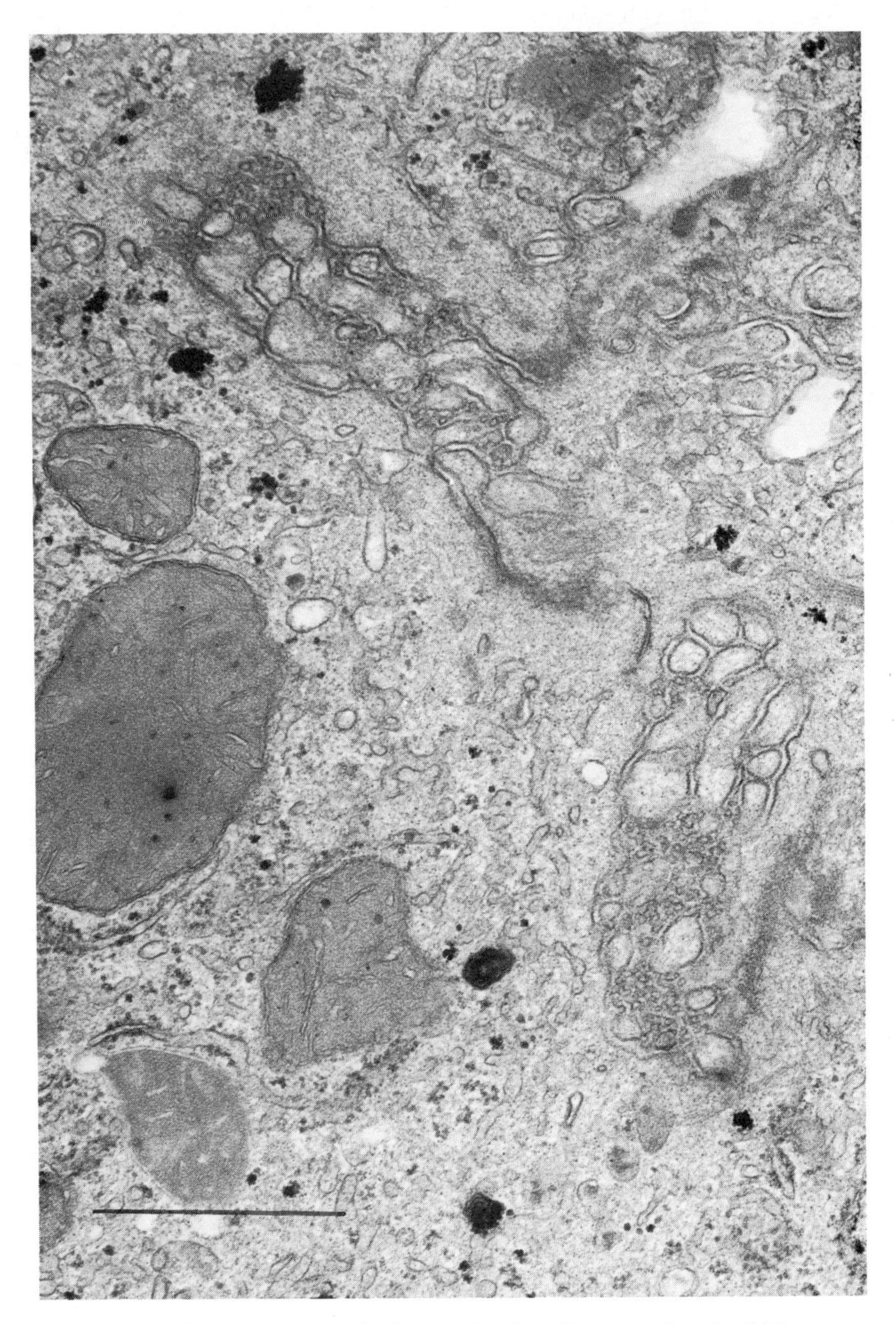

Figure 14.13. Electron micrograph of a sample of rat liver treated as for (a) in Figure 14.11 except that icterogenin (2 mg) was added to the perfusion medium after 90 min. The canaliculi are almost filled with swollen microvilli and extraneous particulate matter of unknown composition (unpublished data of T. F. Slater, B. C. Sawyer, M. N. Eakins and G. Bullock).

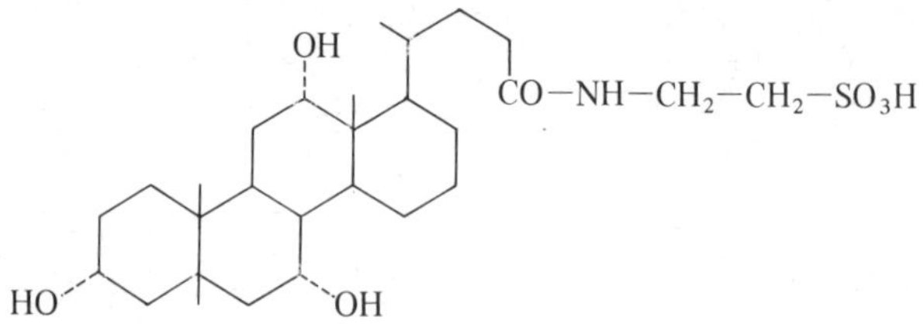

Figure 14.14. Structure of taurocholic acid.

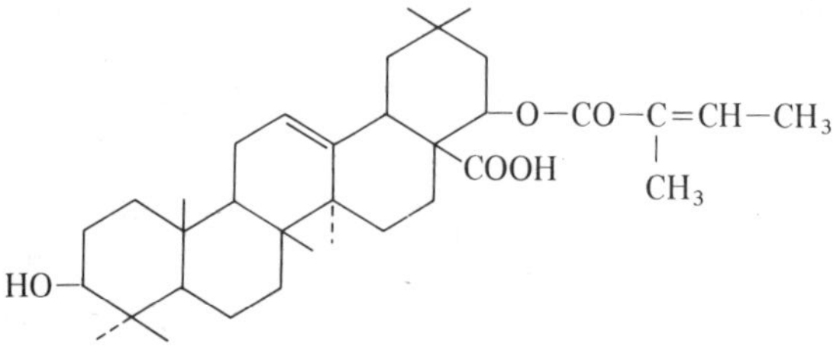

Figure 14.15. Structure of 22β-angeloyloxy-3β-hydroxyolean-12-en-28-oic acid (22β-angeloyloxyoleanolic acid).

14.3.3 Effects on *Tetrahymena pyriformis*

An interesting application of photodynamic action has been made by Epstein *et al.* (1965). They found that the protozoan *Tetrahymena pyriformis* was killed by irradiation in the presence of the photosensitiser benzpyrene (Figure 13.2). The organism was protected against this photodynamic effect by antioxidants. Epstein and colleagues have further developed this biological reaction as a screening procedure to detect antioxidant activity in a wide variety of materials.

14.3.4 Retinal damage

Santamaria (1967) has studied the action of high-intensity visible radiation on adult cattle **retinae**. An oxygen-dependent inhibition of mitochondrial respiratory enzymes was found; similar treatment of mitochondria from liver and other tissues was without effect. This suggests that the retinal mitochondria are specially susceptible to a photodynamic inhibition. In this connection it is known that retinal mitochondria behave differently from liver mitochondria in their response to swelling agents (Wang *et al.*, 1963), especially to vitamin A.

The neonatal retina is known to be susceptible to high partial pressures of oxygen. In the past, treatment of premature infants with increased oxygen concentrations was found to be associated with a high incidence of severe retinal damage known as **retrolental fibroplasia** (for review see Duke-Elder and Dobree, 1967). The damage commences with an oxygen-dependent destruction of the small retinal blood vessels infiltrating the retina. When the increased oxygen atmosphere is replaced by air, the retina is invaded by an unorganised network of capillaries associated with a fibrosis.

The mechanisms underlying the sequence of events is not clear but a speculative scheme has been put forward by Riley and Slater (1969) and Slater and Riley (1970). In their scheme it is envisaged that the high oxygen concentration stimulates a process of lipid peroxidation in the lipoprotein membranes of the developing retina. This peroxidation may involve the lysosomal fraction since it is known from the studies of Sledge and Dingle (1965) and of Allison (1965) that *hyperoxic* conditions increase lysosomal membrane permeability. The initiation of the peroxidative mechanism was suggested to proceed by a photochemical step but, on general grounds, could equally well be through a metal-catalysed redox reaction. The point was made (Riley and Slater, 1969) that *adult* retinal tissue contains a localised high concentration of **ubiquinone** (Pearse, 1965); the main physiological function of this substance in the retinal zone may be that of a tissue antioxidant. The retina contains numerous arrays of lipoprotein membranes, it is vascularised so that oxygen is present in the tissue and it is, of course, exposed to local quite high intensities of light. Many oxidative systems in other tissues (for example liver) develop only after birth so that it is possible that antioxidant protective mechanisms (i.e. a localised high concentration of ubiquinone) are likewise deficient in the *premature* retina. In such a situation exposure to raised oxygen partial pressure and light may be sufficient to start the vaso-obliterative phase which can be regarded as the primary lesion.

The studies of Locke and Reese (1952) provide some indication that light in fact aggravates retinal damage in premature infants exposed to light. The incidences of retrolental fibroplasia in infants with or without eye patches to exclude light are plotted in Figure 14.16, where it can be

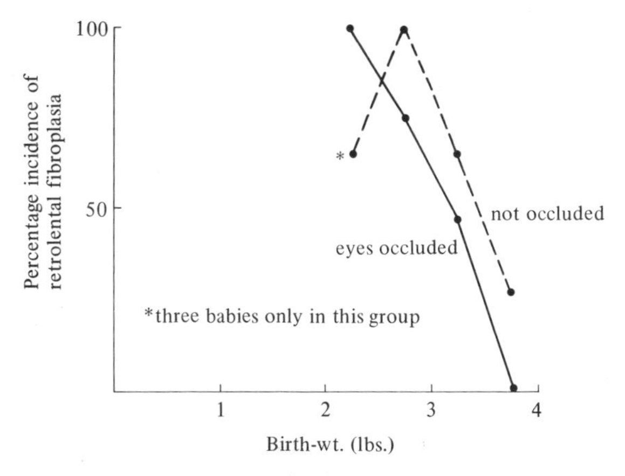

Figure 14.16. Effect of decreasing exposure to light (eyes occluded within 24 h of birth) on the incidence of retrolental fibroplasia in neonates (data from Locke and Reese, 1952). The result marked * is for only three babies.

seen that the incidence was lower in babies with their eyes covered. These studies were performed before the importance of high partial pressure of oxygen to the onset of the disease was realised and, as a consequence, the concentrations of oxygen used were probably very high. It would be of interest to repeat the experiment of Locke and Reese (1952) with oxygen concentration controlled at 30%, the generally accepted upper limit for safety in terms of retinal damage. Moreover, Owens and Owens (1949) have reported that pre-dosing infants with α-tocopherol had a beneficial action on the incidence of retrolental fibroplasia; these experiments were also performed before discovery of the role of oxygen and so need careful re-evaluation. Riley and Slater (1969) suggested that the aggravating effect of light on the course of the disease had not been disproved and that in the circumstances it would be wise to protect infants' eyes from exposure to light (particularly short wavelength light-radiation) as far as possible. Observations that are relevant to the above discussion have been reported by Sisson *et al.* (1970). These workers studied the effects of blue light on the retinae of new-born piglets. Retinal damage was observed in unshielded piglet eyes but not in shielded control eyes. Phototherapy of hyper-bilirubinaemia in the new-born infants entails using light with at least some contribution from the blue end of the visible spectrum, and the authors conclude that infants so treated should have their eyes covered during exposure to radiation. Since these experiments were performed in the absence of high oxygen pressures it can be expected that even more marked retinal disturbance would result where infants on high oxygen therapy received similar amounts of incident radiation on their eyes. A most interesting development concerning the damaging effect of visible radiation on the retina has recently been reported by Noell and Albrecht (1971), who found that when the retina is continuously illuminated for more than 40 hours severe damage to the visual cells and pigment epithelium is produced. The suggestion is made that the visual cell is a metastable, highly differentiated cell and that the diurnal cycle of light and darkness may be important for the maintenance of its proper function.

Newborn infants have also been *exposed* to high-intensity light in attempts to decrease the concentration of plasma bilirubin. The catabolism of foetal haemoglobin to bilirubin increases rapidly at birth; the neonatal liver is unable to handle the bilirubin load owing to the low activity of the enzyme glucuronyl transferase. This enzyme normally converts bilirubin into the glucuronide derivative that is excreted in the bile. The enzyme increases in activity over the first few weeks of extra-uterine life. Shortly after birth therefore there is a rise in plasma unconjugated bilirubin and this may be large enough to cause a bilirubin encephalopathy with permanent mental retardation. Bilirubin is photo-sensitive and is thereby converted into water-soluble products that can be excreted in the urine. Consequently, one method of decreasing plasma bilirubin concentrations in such infants has been to subject them to whole body light-exposure. Other methods based on induction of liver glucuronyl transferase have been described in Section 11.4.

14.4 Photosensitisation changes not dependent upon oxygen

The sensitising action of **psoralen** (Figure 14.17) and related derivatives has been known for a long time. A recent example has come from the work of Caporale *et al.* (1967), who showed that 8-methoxypsoralen sensitised the guinea-pig eye to small doses of ultraviolet radiation that normally were harmless. These furocoumarins are unusual in that the sensitisation reaction is not dependent on oxygen (see Musajo, 1967). The interesting discovery was made by Musajo's group that the furocoumarins would combine with thymine or cytosine upon irradiation to give cyclobutane structures (Figure 14.18). The chemistry of this type of reaction has been closely examined by Farid and Krauch (1967) in terms of the general reaction between dicarbonyls and olefinic bonds (Figure 14.19). The resemblances in structure between the furocoumarins

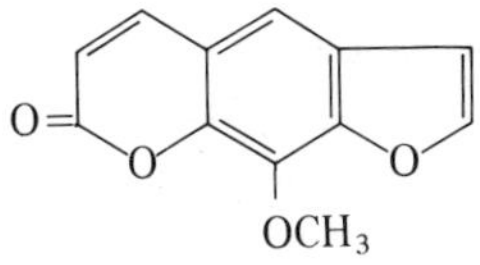

Figure 14.17. Structure of 8-methoxypsoralen ('Psoralen').

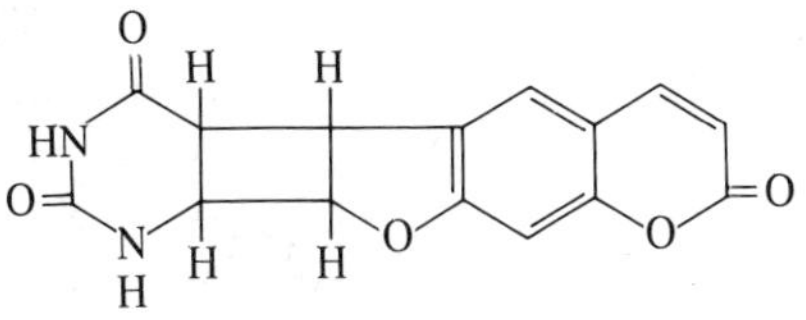

Figure 14.18. Cyclobutane addition product formed by photo-irradiation of cytosine and psoralen (see Musajo, 1967).

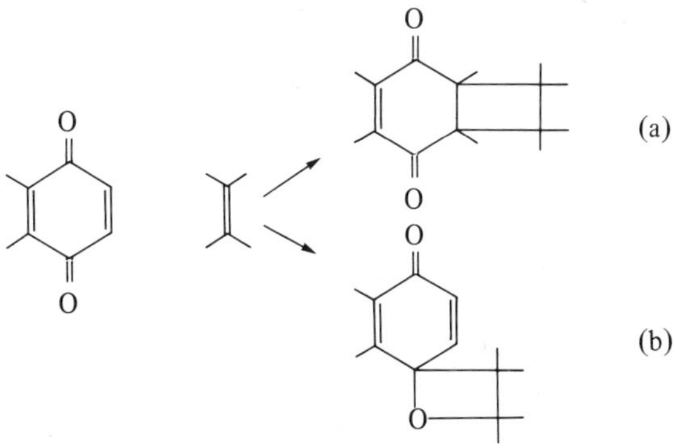

Figure 14.19. Reactions between 1,4-naphthaquinone-like materials and olefins to yield cyclo-addition products of the cyclobutane (a) or spiro-oxetane (b) types (see Farid and Krauch, 1967).

studied for photosensitisation and the carcinogen aflatoxin B_1 (Figure 13.6) are noteworthy. Aflatoxin is fluorescent and is known to bind to DNA *in vivo* (see Rees, 1966).

14.5 Laser radiation

So far we have spoken about the effects of polychromatic light on the organism as occurs, for example, with natural sunlight. With the development of the **laser**, however, the possibility of applying very large amounts of monochromatic pulsed energy occurred (see Berns and Rounds, 1970). This may lead to severe heating effects when directed on tissue and the directional aim of the beam can be made very precise. Such properties have led to the use of laser radiation in attempts to *attach* detached retina. In a different application Fine *et al.* (1968) showed that laser light of wavelength 6943 Å could penetrate abdominal muscle relatively well, after the skin was reflected, and could cause severe localised injury (centrilobular necrosis and other gross damage) to the underlying *liver*. Illustrative of the precision with which such laser beams may be used are the experiments of Bessis and colleagues (see Moreno *et al.*, 1969), and Amy and Storb (1965) in which selective damage to *intracellular* organelles of cells in culture was obtained. In such situations although the quantum of radiation energy may be insufficient to cause homolytic fission (wavelength 6943 Å is equivalent to 41 kcal mol^{-1}) the local heating effects may result in free-radical formation by thermal dissociation.

The effects of radiation from a ruby laser on the inactivation of enzymes in solution have been studied by Igelman *et al.* (1965), Klein *et al.* (1965) and Rounds (1965). The radiation was not effective in decreasing enzyme activity unless a photosensitiser such as methylene blue was added to trap the radiation (at 694·3 nm or 347·1 nm). Dashman and Stellar (1969) have shown that oxygen has only a partial role in the photosensitised destruction of chymotrypsin under such conditions.

14.6 Chemiluminescent reactions

It is not proposed to review this subject here[4] but simply to draw attention to a few points of interest relevant to the preceding discussion. Many of the reactions described in this Chapter depend on a utilisation of incident radiant energy to promote homolysis. The radiation energy may, however, be produced internally by a **chemiluminescent** reaction. For example, Steele and colleagues (Williams and Steele, 1965; Vorhaben and Steele, 1967) have shown that chemiluminescence occurs with a mixture of riboflavin, hydrogen peroxide, ascorbate and copper ions, all of which occur under normal physiological conditions. Another example concerns the aromatic endoperoxides. Many polycyclic aromatic hydrocarbons

[4] For review see McCapra (1966).

absorb oxygen upon irradiation; the process is usually reversible and some compounds emit light as the oxygen is evolved.

A recent discovery has shown that photochemical reactions may proceed without the existence of light at all! The process is aptly described as 'photochemistry without light' (see White *et al.*, 1969). The thermal decomposition of the four-membered ring peroxide (Figure 14.20), an oxaoxetane derivative, results in the formation of one of the products in an excited state (see McCapra, 1968). This may decay with evolution of light or, as shown by White *et al.* (1969), may immediately react with an acceptor molecule to give products equivalent to those found by radiation of the acceptor molecule. For example, decomposition of the oxaoxetane (Figure 14.20) in the presence of anthracene led to the appearance of typical anthracene fluorescence. The above reactions are of considerable biological interest, in view of the suggested formation of *five*-membered ring peroxides during lipid peroxidation (see Figure 3.6).

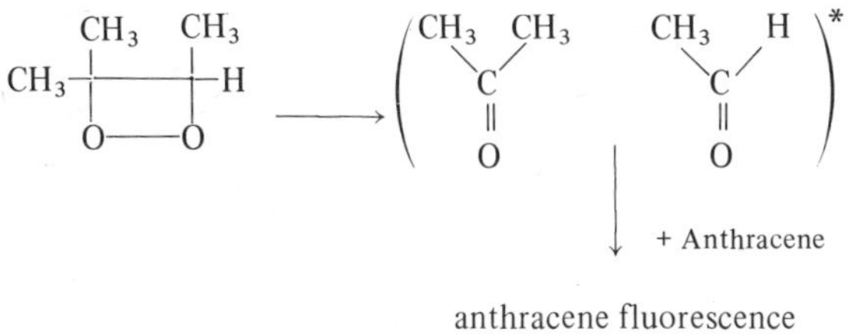

CH_3 CH_3 | CH_3 CH_3 | CH_3 H *
CH_3—C—C—H → C=O , C=O
O—O

+ Anthracene

anthracene fluorescence

Figure 14.20. Thermal decomposition of an oxaoxetane ring to produce an excited ketone product that interacts with anthracene to produce a typical anthracene fluorescence (see White *et al.*, 1969).

References

Allison, A. C., 1965, *Nature, Lond.*, **205**, 141.
Allison, A. C., Magnus, I. A., Young, M. R., 1966, *Nature, Lond.*, **209**, 874.
Allison, A. C., Young, M. R., 1964, *Life Sci.*, **3**, 1407.
Amy, R. L., Storb, R., 1965, *Science*, **150**, 756.
Bacq, Z. M., 1965, *Chemical Protection Against Ionising Radiation* (Charles C. Thomas, Springfield, Illinois).
Bacq, Z. M., Alexander, P., 1961, *Fundamentals of Radiobiology*, second edition (Pergamon Press, Oxford).
Barron, E. S. G., Johnson, P., Cobure, A., 1954, *Radiat. Res.*, **1**, 410.
Berns, M. W., Rounds, D. E., 1970, *Scient. Am.*, **222**, 98.
Blois, M. S., 1969, in *Solid State Biophysics*, Ed. S. J. Wyard (McGraw-Hill, London), p.245.
Blum, H. F., 1964, *Photodynamic Action and Diseases Caused by Light* (Hafner, New York).
Boag, J. W., 1965, *Physics Med. Biol.*, **10**, 457.
Brown, J. M. M., Rimington, C., Sawyer, B. C., 1963, *Proc. R. Soc. B.*, **157**, 473.
Burnett, R. W., Underwood, A. L., 1968, *Biochemistry, Easton*, **7**, 3328.
Caporale, G., Musajo, L., Rodighieto, G., Baccichetti, S., 1967, *Experientia*, **23**, 1.
Commoner, B., Townsend, J., Pake, G. E., 1954, *Nature, Lond.*, **174**, 689.
Daniels, J. F., 1959, *J. invest. Derm.*, **32**, 147.

Dashman, T., Stellar, S., 1969, *Life Sci.*, **8**, 495.
Desai, I. D., Sawant, P. L., Tappel, A. L., 1964, *Biochim. biophys. Acta,* **86**, 277.
Duke-Elder, S., Dobree, J. H. (Eds.), 1967, *System of Ophthalmology, volume 10,* in *Diseases of the Retina* (Henry Kimpton, London).
El-Hefnawi, H., Rasheed, A., 1966, *Gaz. Egypt Soc. Dermatol. and Venereol.,* **1**, 189.
Epstein, S. S., Saporoschetz, I. B., Small, M., Park, W., Mantel, N., 1965, *Nature, Lond.,* **208**, 655.
Farid, S., Krauch, C. H., 1967, "Radiation research", *Proceedings of the Third International Congress on Radiation Research,* Ed. G. Silini (North-Holland, Amsterdam), p.869.
Fine, S., Edlow, J., MacKeen, D., Feigen, L., Ostrea, E., Klein, E., 1968, *Am. J. Path.,* **52**, 155.
Fletcher, M. J., Sanadi, D. R., 1961, *Biochim. biophys. Acta,* **51**, 356.
Ghidoni, J. J., Thomas, H., 1968, *Radiat. Res.,* **36**, 327.
Harman, D., Piette, L. H., 1966, *J. Geront.,* **21**, 560.
Hiatt, C. W., 1967, "Radiation research", *Proceedings of the Third International Congress on Radiation Research,* Ed. G. Silini (North-Holland, Amsterdam), p.857.
Hodges, R., Ronaldson, J. W., Taylor, A., White, E. P., 1963, *Chemy Ind.,* 42.
Igelman, J. M., Rotte, T. C., Schechter, E., Blaney, D. J., 1965, *Ann. N. Y. Acad. Sci.,* **122**, 790.
Klein, E., Fine, S., Ambrus, J., Cohen, E., Netal, E., Ambrus, C., Bardos, T., Lyman, R., 1965, *Fedn Proc. Fedn Am. Socs exp. Biol.,* **24**, supplement, p.104.
Krasnovsky, A. A., 1967, "Radiation research", *Proceedings of the Third International Congress on Radiation Research,* Ed. G. Silini (North-Holland, Amsterdam), p.813.
Land, E. J., Swallow, A. J., 1968, *Biochim. biophys. Acta,* **162**, 327.
Leun, J. C. van der, 1966, *Ultraviolet Erythema,* Doctorate Thesis, University of Utrecht.
Locke, J. C., Reese, A. B., 1952, *Archs Ophthal.,* **48**, 44.
Longuet-Higgins, H. C., 1960, *Archs. Biochem. Biophys.,* **86**, 231.
Magnus, I. A., Porter, A. D., Rimington, C., 1959, *Lancet,* i, 912.
Mason, H. S., Ingram, D. J. E., Allen, B., 1960, *Archs. Biochem. Biophys.,* **86**, 225.
Matthews-Roth, M. M., Pathak, M. A., Fitzpatrick, T. B., Harber, L. C., Kass, E. H., 1970, *Trans. Ass. Am. Physns.,* **83**, 176.
McCapra, F., 1966, *Q. Rev. chem. Soc.,* **20**, 485.
McCapra, F., 1968, *Chem. Commun.,* 155.
Moreno, G., Lutz, M., Bessis, M., 1969, *Int. Rev. Exp. Pathol.,* **7**, 99.
Musajo, L., 1967, "Radiation research", *Proceedings of the Third International Congress on Radiation Research,* Ed. G. Silini (North-Holland, Amsterdam), p.803.
Myers, D. K., 1960, *Can. J. Biochem. Physiol.,* **38**, 1255.
Nicolaus, R. A., 1968, *Melanins* (Hermann, Paris).
Noell, W. K., Albrecht, R., 1971, *Science,* **172**, 76.
Noyes, P. P., Smith, R. E., 1959, *Expl. Cell Res.,* **16**, 15.
Owens, W. C., Owens, E. U., 1949, *Am. J. Ophthal.,* **32**, 1631.
Pathak, M. A., Stratton, K., 1968, *Archs. Biochem. Biophys.,* **123**, 468.
Pearse, A. G. E., 1965, in *Biochemistry of the Retina*, Ed. C. N. Graymore (Academic Press, London), p.110.
Raab, O., 1900, *Z. Biol.,* **39**, 524.
Reed, G. B., Cox, A. J., 1966, *Am. J. Path.,* **48**, 597.
Rees, K. R., 1966, *Proc. R. Soc. Med.,* **59**, 755.
Reese, A. J. M., Rimington, C., 1964, *Br. J. exp. Pathol.,* **45**, 30.
Riley, P. A., Slater, T. F., 1969, *Lancet,* ii, 265.
Rimington, C., Magnus, I. A., Ryan, E. A., Cripps, D. J., 1967, *Q. Jl. Med.,* **36**, 29.
Rimington, C., Quin, J. I., Roets, G. C. S., 1937, *Onderstepoort J. vet. Sci. Anim. Ind.,* **9**, 225.

Rimington, C., Slater, T. F., Spector, W. G., Strauli, U. D., Willoughby, D. A., 1962, *Nature, Lond.*, **194**, 1152.
Robertson, D. F., 1969, *World Medicine*, **4**, 13.
Rounds, D. E., 1965, *Fedn Proc. Fedn Am. Socs exp. Biol.*, **24**, supplement, p.116.
Santamaria, L., 1967, "Radiation research", *Proceedings of the Third International Congress on Radiation Research*, Ed. G. Silini (North-Holland, Amsterdam), p.839.
Scaife, J. F., 1963, *Can. J. Biochem. Physiol.*, **41**, 1469.
Sisson, T. R. C., Glauser, S. C., Glauser, E. M., Tasman, W., Kuwabara, T., 1970, *J. Pediat.*, **77**, 221.
Slater, T. F., Riley, P. A., 1966, *Nature, Lond.*, **209**, 151.
Slater, T. F., Riley, P. A., 1970, *Lancet*, ii, 467.
Slater, T. F., Strauli, U. D., Sawyer, B. C., 1964, *Res. vet. Sci.*, **5**, 450.
Sledge, C. B., Dingle, J. T., 1965, *Nature, Lond.*, **205**, 140.
Swartz, H. M., 1965, *Radiat. Res.*, **24**, 579.
Swartz, H. M., Molenda, R. P., Lofberg, R. T., 1965, *Biochem. biophys. Res. Commun.*, **21**, 61.
Vorhaben, J. E., Steele, R. H., 1967, *Biochemistry, N.Y.*, **6**, 1404.
Wang, D. Y., Slater, T. F., Dartnell, H. J. A., 1963, *Vision Res.*, **3**, 171.
White, E. H., Wiecko, J., Roswell, D. F., 1969, *J. Am. chem. Soc.*, **91**, 5194.
Williams, J. R., Steele, R. H., 1965, *Biochemistry, Easton*, **4**, 814.

15

Redox radical reactions in tissue damage

15.1 Introduction

In this Chapter the cellular disturbances produced by exposure to *excessive* concentrations of **iron** and **copper** will be considered in relation to the homolytic reactions known to be catalysed by these transitional metals (see Section 2.4). It has already been mentioned (Section 2.4) that inorganic iron catalyses homolytic reactions in tissue suspensions *in vitro* as evidenced by a stimulation of lipid peroxidation: for example, the increased lipid peroxidation in mitochondrial suspensions resulting from the addition of iron (Figure 12.2). Iron also stimulates lipid peroxidation of microsomal suspensions both with NADPH (Figure 4.3) or ascorbate as electron-donors. In many instances copper ions will behave similarly to iron salts in initiating homolytic reactions, as is well known with the copper-catalysed autoxidation of ascorbate in solution. The presence of free[(1)] copper or iron ions inside the cell may be expected therefore to initiate free-radical reactions that are coupled with such electron-donors as NADPH and ascorbate. The concentrations of metal ions required for such catalytic activities may be extremely small. Latarjet *et al.* (1963), studying the variable action of disuccinoyl peroxide on the transforming activity of DNA, traced the cause of variability to contaminating metal ions at 10^{-11} M concentration in the reagents. A further illustration concerns the catalytic action of copper on the oxidation of orcinol to produce the dye **orcein**: the catalysis was found to be more or less specific for copper and could be used to determine copper in minute concentrations (Slater,1961).

Although in this Chapter the effects of excessive accumulations of iron and copper within the body will be considered in terms of the activities of the metals in stimulating homolytic reactions, it is important to remember how *essential* these elements are as trace nutrients. This is widely known for iron, of course, since iron deficiency presents clinically as a commonly encountered anaemia: iron is an essential component of iron–porphyrin structures that constitute haem and the cytochromes[(2)].

(1) 'Free' as opposed to chelated strongly with a ligand such as EDTA. In such 'bound' form many of the catalytic properties of the metal ion disappear (for example, the inhibitory action of EDTA on the iron-catalysed lipid peroxidation in microsomal suspensions. In a few cases, however, the chelated form may be more active [for example, the stimulatory action of EDTA on the copper-catalysed destruction of ascorbate (Younathan and Frieden, 1961)].

(2) One treatment for iron deficiency involves oral ingestion of a mixture of a ferrous salt with vitamin C, which, as outlined above, and discussed more fully in Chapter 2, is an active hydroxylating reagent (Fenton's reagent). It may be expected to attack unsaturated lipid that it encounters on its journey down the alimentary tract. Presumably the mucosaccharide coating of the tract prevents direct attack on the cell walls. The rationale behind this treatment is the belief that ferrous iron is better absorbed than is ferric iron by the intestinal tract.

The requirement for copper in man is less widely appreciated although its role in several important oxidative enzymes (for example cytochrome oxidase, catalase, tyrosinase) is well established. Various disorders in animals are known to be associated with copper deficiency, however, of which the best known example is that of swayback in lambs, in which there is a progressive demyelination and degeneration of the motor and sensory tracts[(3)].

It is believed that the therapeutic use of iron extends back for at least 3000 years and its current uses have been extensively reviewed, for example by Wallerstein and Mettier (1958) and by Hallberg *et al.* (1970). As a consequence of the widespread usage of iron preparations by women, with occasional careless disposal or storage of the tablets, accidental poisoning by iron is one of the commonest causes of intoxication in children. In recent years efficient chelating agents for removing iron from the body have been developed; for example, **desferrioxamine-B**, which has a very high binding constant (10^{31}) for ferrous iron (Figure 15.1). Iron (and copper) salts, in common with various other transitional metal ions, cause a number of enzyme inhibitions (see Dixon and Webb, 1964) utilising diverse mechanisms of action; the oxidation of essential thiol groups is a well-known example. An early example of this inhibitory action was described by Altmann and Crook (1953), who showed that inhibition of respiratory chain preparations by heavy metal ions could be overcome by the addition of EDTA to the incubation medium. Obviously relatively unspecific and gross metabolic disturbances would be expected to follow the overwhelming accumulations of iron or copper that may occur in acute

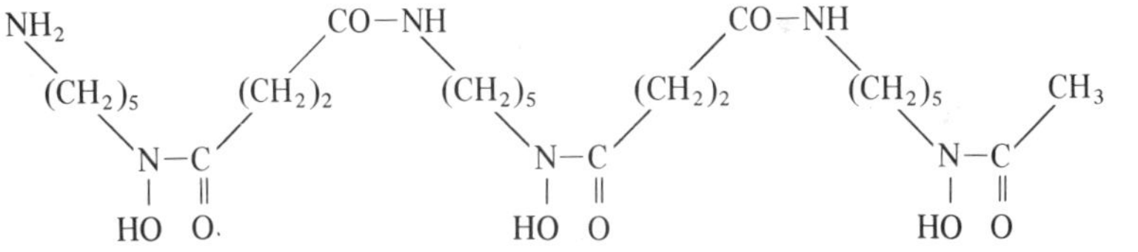

$NH_2-(CH_2)_5-N(OH)-C(=O)-(CH_2)_2-CO-NH-(CH_2)_5-N(OH)-C(=O)-(CH_2)_2-CO-NH-(CH_2)_5-N(OH)-C(=O)-CH_3$ (a)

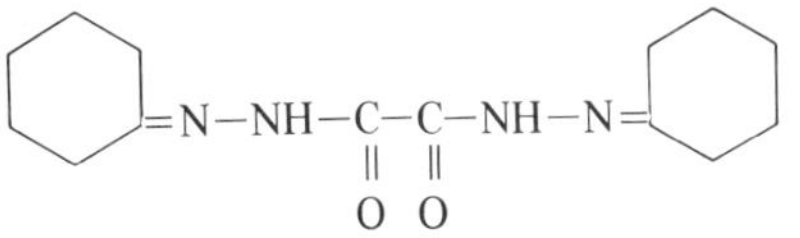

$C_6H_{10}=N-NH-C(=O)-C(=O)-NH-N=C_6H_{10}$ (b)

Figure 15.1. Structures of (a) desferrioxamine-B and (b) cuprizone.

(3) Some work relative to copper deficiency is that of Suzuki (1969), who administered the specific copper-chelator **cuprizone** to rats. Cuprizone (bis-cyclohexanone oxalyldihydrazine, Figure 15.1) caused a dramatic enlargement of mitochondria in the liver so that their diameters approached or even exceeded that of the nucleus; some brain damage was also observed to occur. Similar experiments have been reported by Rifkin and Gahagan-Chase (1970) using α,α'-dipyridyl. Intraperitoneal injection of this strong iron-chelating agent produced enlarged bizarre shaped mitochondria with dilated cristae.

poisoning. For example, Witzleben (1966) injected *intravenously* into rabbits 50–80 mg of ferrous sulphate/kg and found that this treatment caused convulsions, hyperventilation and an increasing accumulation of iron in the periportal parenchymal cells in the animals that survived for more than a few hours. Electron-microscopic examination of the liver showed early mitochondrial damage, that was followed later by periportal necrosis.

Our concern in the remainder of this Chapter, however, will be with less acute manifestations of iron and copper disturbances in which mechanisms of cellular disturbance that are dependent on homolytic reactions may be operative.

15.2 Iron toxicity

It is known that iron salts administered by mouth are absorbed in the gut and are stored in the form of histochemically stainable iron, mainly in the reticuloendothelial system; in liver the iron-staining reaction conspicuously marks out the *Kupffer cells*. However, under certain conditions that are clinically observed as well as experimentally obtainable, iron can be forced into the *parenchymal* cells themselves. These conditions must be clearly distinguished from the acute experiments of Witzleben, previously mentioned, where an *intravenous* injection of large quantities of iron was given.

In **haemochromatosis**, for example, there is an excessive deposition of iron in the liver parenchyma (and other tissues) and this is associated with *fibrosis*. The aetiology of haemochromatosis has recently been subjected to a searching analysis by McDonald and colleagues in Boston (McDonald, 1963; McDonald *et al.*, 1963; McDonald and Pechet, 1965; McDonald *et al*;. 1965). McDonald was led to the conclusion that, on the one hand, there was an association between haemochromatosis and the daily intake of iron and, on the other hand, there was a relationship between chronic alcoholism and haemochromatosis. An examination of the iron content of alcoholic beverages from a variety of different sources showed that some were extremely rich in iron. When combined with a high daily consumption of the alcoholic beverage, iron intakes reach astonishing proportions. For example, in Rennes (France) the local population were observed to drink large quantities of cider and wine in amounts that averaged 5 *litres* per day in warm weather. This led to intakes of iron at concentrations of approximately 80 mg l^{-1}, largely in the ferrous form. It may be calculated that over a period of 20 years, with a percentage iron absorption of about 12% of that ingested, the total iron absorbed in such cases may easily reach 70 g, an amount quite compatible with that observed in several cases of haemochromatosis. A similar large intake of iron occurred in members of the Bantu tribe of South Africa, who drank large quantities of home-brewed beer fermented in cast-iron pots.

Attempts to produce large parenchymal deposits of ferrous iron in rats were unsuccessful even when large amounts of ethanol were included with the diet. The excess of iron accumulated in the reticuloendothelial system and gave rise to *haemosiderosis* rather than haemochromatosis. McDonald *et al.* (1965) found that if rats were placed on a diet deficient in lipotropic factors and with a high endogenous iron content they developed liver damage that was analogous to that seen in human haemochromatosis with cirrhosis; these changes in rat liver were retarded by the inclusion of choline or folic acid. Thus the indications were that excessive iron could enter the liver parenchyma provided that the liver cell membrane was damaged by exposure to a suboptimum diet.

The role of parenchymal cell injury in facilitating the entry of iron has also been studied by Kent *et al.* (1963a), who found that **ethionine** or ***N*-2-fluorenylacetamide** increased iron deposition in the liver. The iron deposition was localised histochemically in the peribiliary region of the parenchymal cells and appeared to be in lysosomal particles. The lysosomal localisation was confirmed (Kent *et al.*, 1963b) by electron microscopy; it was observed that as the iron loading of the lysosomes was increased complex particle structures were produced that eventually encroached on surrounding intracellular structures. An interesting development of this work that localised the increased parenchymal cell iron in the lysosomes was performed by Kent *et al.* (1965); they followed the movements of iron-labelled lysosomes during mitotis that is prevalent in liver after partial hepatectomy. Characteristic movements of the parenchymal cell lysosomes during mitosis were observed, the initial mainly peribiliary location changing to a juxtanuclear position surrounding the centrioles during early mitosis. In late mitosis the lysosomes surrounded the daughter nuclei and eventually regained a peribiliary location in interphase. The intracellular localisation of the excess of parenchymal iron in clinical haemochromatosis has been reported by Essner and Novikoff (1960) to be lysosomal in distribution. Such a localisation is to be expected not only from the studies *in vivo* of Kent *et al.* described above but from the work *in vitro* of Koenig (1969), in particular, who has shown that lysosomes have a general avidity for *cations*. The lysosomal fraction has an important role in the physiological turnover of cell components. In the normal course of events lipid-rich membrane material is digested inside a lysosome by lipases, phospholipases etc. (for review of lysosomal enzymes see Tappel, 1969), often in conjunction with autophagy in which a region of the cytoplasm is surrounded by a limiting single membrane. The *autophagosome* (or 'autophagic cytosegresome') so formed then acquires acid-hydrolase activity. An electron micrograph illustrating the appearance of autophagic cytosegresomes is shown as Figure 15.2. A comprehensive review on autophagy is by Ericsson (1969).

If the amount of lipid material ingested in this way exceeds the rate of metabolism then the accumulating lipid-rich deposit becomes susceptible to secondary reactions such as peroxidation of the unsaturated fatty acid residues in the lipid. This process of lipid peroxidation of material *within*

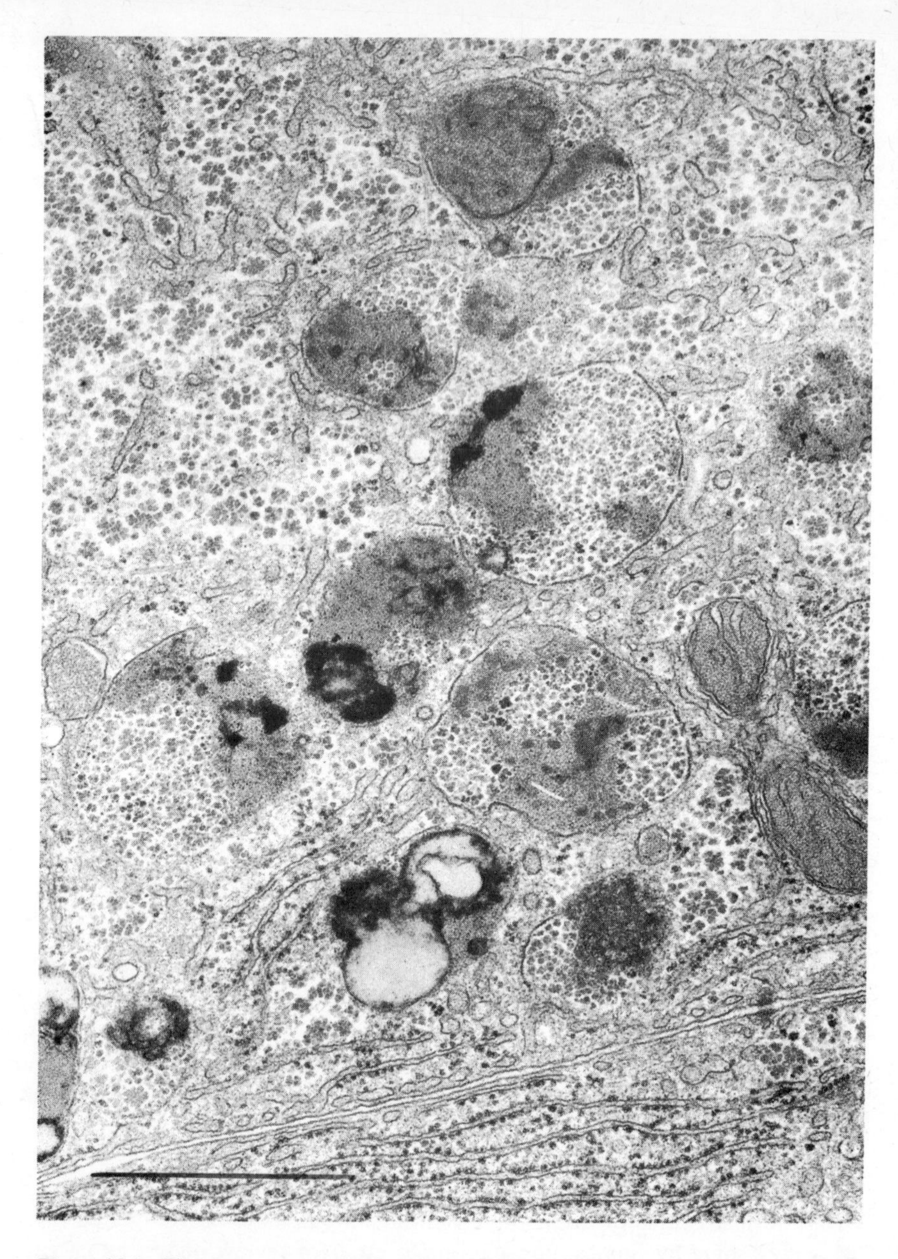

Figure 15.2. Electron micrograph of a section of rat liver showing numerous membrane-bounded structures (cytosegresomes). A liver obtained from a female rat was perfused with a mixture (130 ml) of Krebs–Ringer bicarbonate buffer, amino acids, lactate, glucose, albumin and 25% (*v*/*v*) fresh difibrinated whole rat blood at 37°C (see Slater *et al.*, 1968). Sporidesmin (2 mg) was added after 90 min perfusion as a solution in 10 ml of Krebs–Ringer bicarbonate containing 5% (*w*/*v*) albumin. After 45 min when bile flow had been severely depressed the liver was removed and perfused first with cold 0·9% (*w*/*v*) NaCl solution, then fixed with a 4% (*v*/*v*) glutaraldehyde–cacodylate (0·1 M, pH 7·2) solution. After post-fixation with osmium, sections (silver-grey) were cut and examined using an A.E.I. electron microscope model EM.6B. The horizontal line indicates a length of 1 μm (unpublished results of T. F. Slater, M. N. Eakins, B. Sawyer and G. Bullock).

the lysosome results in oxidised fatty material with characteristic colouration. This is the mechanism of formation of the lysosomal lipid pigment long known as lipofuschin[4].

The extent of autophagy in liver parenchyma is increased by various procedures that damage the cell; for example, hypoxia, starvation, high doses of **glucagon** and, in the isolated perfused liver, hyperoxia (Abraham *et al.*, 1968). The experimental treatments described earlier that result in an increased parenchymal accumulation of iron, as well as clinically manifest haemochromatosis, also involve liver disturbances and may therefore be expected to be associated with an increased turnover of lipid-rich material through autophagy. Concomitantly there is an increased accumulation of iron within the lysosomal fraction. The net result is that there is an association in time and space of material rich in unsaturated fatty acids, iron and oxygen; these conditions are favourable for an increased lipid peroxidation within the autophagic particle. This sequence of events probably accounts for the increased lipid pigment observed in haemochromatosis [and, incidentally, in Wilson's disease (see Section 15.3), where there is a similar situation but with copper accumulation in place of iron].

Of course the experimental situation where iron is forced rapidly into animals in conjunction with severe liver injury must be distinguished from the human situation where iron accumulation and low-grade hepatic disturbances are spread over many years. In the latter situation it can be imagined that superimposed on the increased lipofuschin formation there would be an increased lysosomal injury due to the increasing accumulation of iron in the liver, for it is known that iron and copper salts *in vitro* have a labilising action on the lysosomal membrane (see Koenig, 1969). One consequence of this would be the slow leakage of lysosomal acid hydrolases into the cytoplasm and into the bile. Bile normally contains lysosomal hydrolases in small amounts[5] but the activities may be greatly increased by exposure to volatile anaesthetics such as ether (Slater, 1966). Since chronic iron overload may result in the leakage into bile of acid hydrolases (for reasons outlined above) this could lead to the presence in bile of partially degraded protein and carbohydrate-containing material in amounts greatly in excess of those normally present. Perhaps, then, in the slow but sustained elimination of partially degraded and antigen-rich material from the parenchymal cells into the bile (or lymph), with a danger of stimulating reaction in the collecting ducts of the portal tracts, an essentially *intraparenchymal* reaction may lead to a spatially displaced reaction involving the whole organ, as is manifest, for example, in cirrhosis. Repeated exposure of the liver to foreign protein has been

[4] The general accumulation of such pigment in normal subjects with increasing age has led to the use of the term 'age-pigment' for lipofuschin (see Section 12.6).

[5] Bile is also relatively rich in the breakdown products of liver cell metabolism: for example, nucleotides.

shown to lead to cirrhosis (see Blackwell, 1965). The cirrhogenic response to repeated small doses of carbon tetrachloride may share a similar mechanism: the lysosomes become overloaded with peroxidised lipid owing to the continued cytoplasmic injury and iron uptake is increased because of the lytic action of carbon tetrachloride on the erythrocytes.

Certain other aspects of iron toxicity are worth mentioning here. For some time it was fashionable to treat iron-deficient patients by the intramuscular or subcutaneous injection of iron in colloidal suspensions. However, it was found that a similar procedure in experimental animals[6] could give rise to sarcomas (see *British Medical Journal*, 12th March 1960, p.788). The relevance of the findings with animals to the clinical situation has, however, been attended by some disagreement, as has been the value of subcutaneous screening tests for carcinogenic activity (Shimkin, 1940; Grasso and Golberg, 1966). In view of the effectiveness of iron salts to initiate homolytic reactions (Chapter 2) it seems possible that the iron–colloid-induced sarcomas arise from free-radical motivated changes in the fibroblast genome and where the normal cellular protective mechanisms are overcome by the localised very high concentrations of iron.

A further example of disturbances possibly involving homolytic reactions catalysed by iron arises from the treatment of iron deficiency in **piglets**: piglets are generally the only species of livestock in which iron-deficiency anaemia is an economic problem; the anaemia has been treated successfully by intramuscular injection of iron–colloid. The toxicity of iron is greatly decreased when it is bound to some material of high molecular weight, owing to the slow rate of release of iron into the tissue. However, even the use of iron-complexes is not completely without risk of fatalities; the aetiology of the toxic action of iron–colloid in these cases has been investigated particularly by Tollerz (1965). It was found that iron injections were particularly toxic to piglets born from a sow that was deficient in vitamin E, and some evidence was obtained that antioxidants were at least partially protective in decreasing the iron toxicity. The clinical symptoms are muscular weakness and prostration, which, unless the piglets are forcibly fed, results in death by starvation. Patterson *et al.* (1969, 1971) have studied the biochemistry of the muscular degeneration. They found that in vitamin E-deficient piglets an increasing severity of illness was accompanied by an increasing concentration of iron in the muscle and by an increasing content of malonaldehyde (Table 15.1). Further correlations were found between the concentrations of plasma potassium, that was assumed to have leaked from damaged muscle, serum acid hydrolases and

[6] But with repeated and greatly increased doses of iron complex. For example, Roe and Carter (1967) found that two injections equivalent to 375 mg of iron/kg body wt. gave local tumour formation. In the work of Roe *et al.* (1964) rats were given 24 weekly injections equivalent to a total dose of 600 mg of iron/rat.

muscle malonaldehyde. The authors suggest that **lipid peroxidation** of intracellular membranes including the lysosomes is involved in the muscle degeneration. Since vitamin E deficiency accentuates iron toxicity it seems probable that intracellular antioxidant protective mechanisms are disturbed in the susceptible piglets. In this case the muscle degeneration appears directly due to the stimulation of homolytic processes by injected iron.

Table 15.1. Effects of injecting an iron-dextrose mixture intraperitoneally into 2-day-old susceptible piglets (data of D. S. P. Patterson and colleagues, personal communication). Malonaldehyde (MAA) was determined by the thiobarbituric acid method; 'true peroxides' by iodometry with an amperiometric end-point in muscle samples frozen immediately in liquid nitrogen after death. The dose of iron-dextrose was 47 mg of Fe/kg body wt.

Treatment	Muscle iron (μg/g)	Muscle peroxides		Plasma K^+ (μequiv./l)
		MAA (nmol/g)	'True peroxide' (nmol/g)	
Control (dextrose)	11 ± 1·6	2·4 ± 0·4	23·2 ± 3·3	4·9 ± 0·3
Iron-dextrose (clinically affected animals)	22 ± 2·9[b]	6·9 ± 1·9[a]	93·7 ± 38·6	6·7 ± 1·9
Iron-dextrose (unaffected animals)	23 ± 3·4[b]	2·3 ± 0·4	25·3	4·7 ± 0·3

[a] $P < 0{\cdot}05$ for difference between control and treated groups.
[b] $P < 0{\cdot}01$ for difference between control and treated groups.

15.3 Copper toxicity

Changes seen with acute and chronic copper toxicity to some extent parallel those discussed above for iron. It has been mentioned in Chapter 2 that copper will also catalyse homolytic reactions, although the specificity involved is often very different from that seen with iron. The general toxicology of copper and a description of copper-deficiency diseases is given by Underwood (1962).

Copper is a potent inhibitor of many important enzyme systems *in vitro*; *oral* administration of large amounts of copper sulphate leads to abdominal irritation, prostration and eventual death from circulatory failure. Peters and Walshe (1966) have studied the toxicity of very small amounts of copper (10–25 μg) *in vivo* in pigeons, given by *subarachnoid injection*. Such administrations produced convulsions and death. Particularly strong inhibition of microsomal adenosine triphosphatase activity was found by these workers, who were concerned with explaining the number of lesions found in patients with hepatolenticular degeneration. In this condition (also known as **Wilson's disease**) there is an inherited disturbance of copper metabolism and large amounts of this metal are found in the liver, brain and other tissue. The liver changes progress to portal cirrhosis. The morphological changes in the liver in human cases of Wilson's disease have been described by Schaffner *et al.* (1962), Goldfischer (1963, 1965) and by Goldfischer and Moskal (1966). All of these investigations showed that the excess of liver copper was concentrated in peribiliary lysosomes. Normally bile contains small quantities of copper and it has been suggested that a failure in biliary excretion is one factor that leads to an increased accumulation of copper in the peribiliary bodies in Wilson's disease (Underwood, 1962). The overall situation is more complex, however, as illustrated by the findings of Vogel and Kemper (1963, 1966). These authors found that copper in low concentrations uncoupled oxidative phosphorylation in brain, myocardium and skeletal muscle mitochondria, but not in liver, spleen, kidney or lung. Extraction of spleen mitochondria led to the isolation of a specific polypeptide with strong chelating activity for copper that is suggested to monitor copper toxicity in this and the other tissues. The mechanism by which patients with Wilson's disease accumulate their high concentrations of copper (they are in positive copper balance) is unknown, but there seems little doubt that the accumulation of copper in the liver lysosomes could be responsible for the increased autoxidation of lipid that occurs and which leads to an excess of lipofuschin pigment in these patients. Since copper salts also labilise the lysosomal membrane *in vitro*, as mentioned above for iron, it may be that haemochromatosis and Wilson's disease share a common lysosomal dysfunction in relation to the onset of portal cirrhosis.

Copper is potentially very toxic and yet accumulates in the perinatal period in the liver (colostrum contains much higher concentrations of iron and copper than later milk), so that it seems reasonable to assume that some protective mechanism must be operative in the neonatal period. Indeed, Porter *et al.* (1962) have found a liver mitochondrial protein in neonates that contains a very high content of copper, and which they have named **mitochondrocuprein**. The concentration of this protein decreased with increasing age after birth. Since bile does not flow until shortly before birth it is possible that mitochondrocuprein is a protective store for copper before the opening up of the biliary excretion route.

In this Chapter the toxic effects of iron and copper have been considered mainly in terms of their actions on the liver and on liver lysosomes. The increased pigment characteristic of haemochromatosis and Wilson's disease is quite probably a reflection of increased lipid peroxidation initiated by the high localised concentrations of the transitional metals present in these diseases.

If the speculations (Section 15.2) concerning the role of chronic lysosomal damage in initiating portal tract reaction have any basis in fact then it should be possible to retard the development of fibrosis in susceptible individuals by inhibiting the cause of the lysosomal instability (for example by decreasing the high concentrations of iron and copper).

Iron and copper excess may now be treated by direct removal, through chelation with desferral or D-penicillamine respectively. It is of obvious importance to know whether by reducing the liver concentrations of these metals the extent of liver fibrosis can be held in check.

References

Abraham, R., Dawson, W., Grasso, P., Goldberg, L., 1968, *Expl. molec. Path.*, **8**, 370.
Altmann, S. M., Crook, E. M., 1953, *Nature, Lond.*, **171**, 76.
Blackwell, J. B., 1965, *J. Path Bact.*, **90**, 245.
Dixon, M., Webb, E. C., 1964, *Enzymes* (Longmans, Green, London).
Ericsson, J. E. E., 1969, in *Lysosomes in Biology and Pathology*, volume 2, Eds. J. T. Dingle, H. B. Fell (North-Holland, Amsterdam), p.346.
Essner, E., Novikoff, A. B., 1960, *J. Ultrastruct. Res.*, **3**, 374.
Goldfischer, S., 1963, *Am. J. Path.*, **43**, 511.
Goldfischer, S., 1965, *Am. J. Path.*, **46**, 977.
Goldfischer, S., Moskal, J., 1966, *Am. J. Path.*, **48**, 305
Grasso, P., Goldberg, L., 1966, *Fd. cosmetic Toxic.*, **4**, 297.
Hallberg, L., Harwerth, H. G., Vannotti, A., (Eds.), 1970, *Clinical Symposium on Iron Deficiency* (Academic Press, New York).
Kent, G., Minick, O. T., Orfei, E., Volini, F. I., Madera-Orsini, F., 1965, *Am. J. Path.*, **46**, 803.
Kent, G., Minick, O. T., Volini, F. I., Orfei, E., de la Huerga, J., 1963a, *Lab. Invest.*, **12**, 1094.
Kent, G., Volini, F. I., Orfei, E., Minick, O. T., de la Huerga, J., 1963b, *Lab. Invest.*, **12**, 1102.
Koenig, H., 1969, in *Lysosomes in Biology and Pathology*, volume 2, Eds. J. T. Dingle, H. B. Fell (North-Holland, Amsterdam), p.111
Latarjet, R., Ekert, B., Demerseman, P., 1963, *Rad. Res. Supplements*, number 3, 253.
McDonald, R. A., 1963, *Archs. intern. Med.*, **112**, 184.
McDonald, R. A., Becker, J. P., Pechet, G. S., 1963, *Archs. intern. Med.*, **111**, 315.
McDonald, R. A., Jones, R. S., Pechet, G. S., 1965, *Archs. Path.*, **80**, 153.
McDonald, R. A., Pechet, G. S., 1965, *Am. J. Path.*, **46**, 85.
Patterson, D. S. P., Allen, W. M., Berrett, S., Sweasey, D., Done, J. T., 1971, *Zenbtl. VetMed. A.*, in the press.
Patterson, D. S. P., Allen, W. M., Berrett, S., Sweasey, D., Thurley, D. C., Done, J. T., 1969, *Zenbtl. VetMed. A.*, **16**, 199.
Peters, R., Walshe, J. M., 1966, *Proc. Roy. Soc. B.*, **166**, 273.
Porter, H., Johnston, J., Porter, E. M., 1962, *Biochim. biophys. Acta*, **65**, 66.
Rifkin, R. J., Gahagan-Chase, P. A., 1970, *Lab. Invest.*, **23**, 480.
Roe, F. J. C., Carter, R. L., 1967, *Int. J. Cancer*, **2**, 370.
Roe, F. J. C., Haddow, A., Dukes, C. E., Mitchley, B. C. V., 1964, *Br. J. Cancer*, **18**, 801.
Schaffner, F., Sternlieb, I., Barka, T., Popper, H., 1962, *Am. J. Path.*, **41**, 315.
Shimkin, M. B., 1940, *J. natn. Cancer Inst.*, **1**, 211.
Slater, T. F., 1961, *Nature, Lond.*, **192**, 420.
Slater, T. F., 1966, "Experimental study of the effects of drugs on the liver", *Proceedings of the European Society for the Study of Drug Toxicity, Rome*, volume 7, p.30.
Slater, T. F., Sawyer, B. C., Delaney, V. B., Bullock, G., 1968, *Biochem. J.*, **110**, 15.
Suzuki, K., 1969, *Science*, **163**, 81.

Tappel, A. L., 1969, in *Lysosomes in Biology and Pathology,* volume 2, Eds. J. T. Dingle, H. B. Fell (North-Holland, Amsterdam), p.207.
Tollerz, G., 1965, *Studies on the Tolerance to Iron in Piglets and Mice* (Department of Medicine, Royal Veterinary College, Stockholm).
Underwood, E. J., 1962, *Trace Elements in Human and Animal Nutrition,* second edition (Academic Press, New York).
Vogel, F. S., Kemper, L., 1963, *Lab. Invest.,* **12**, 171.
Vogel, F. S., Kemper, L., 1966, *Am. J. Path.,* **48**, 713.
Wallerstein, R. O., Mettier, S. R., 1958, *Iron in Clinical Medicine* (University of California Press, Berkeley).
Witzleben, C. L., 1966, *Am. J. Path.,* **49**, 1053.
Younathan, E. S., Frieden, E., 1961, *Biochim. biophys. Acta,* **46**, 51.

Index